SEDIMENTARY STRUCTURES

TITLES OF RELATED INTEREST

Aeolian geomorphology
W. G. Nickling (ed.)

Biogenic textures in sedimentary rocks
R. Bromley

Cathodoluminescence of geological materials
D. J. Marshall

Chemical fundamentals of geology
R. Gill

Deep marine environments
K. Pickering *et al.*

A dynamic stratigraphy of the British Isles
R. Anderton *et al.*

Experiments in physical sedimentology
J. R. L. Allen

Geomorphology and soils
K. Richards *et al.* (eds)

Geomorphology in arid regions
D. O. Doehring (ed.)

Geomorphological field manual
V. Gardiner & R. Dackombe

Geomorphological techniques
A. S. Goudie (ed.)

Glacial geomorphology
D. R. Coates (ed.)

Hillslope processes
A. D. Abrahams (eds)

The history of geomorphology
K. J. Tinkler (ed.)

Image interpretation in geology
S. Drury

Introduction to theoretical geomorphology
C. Thorn

Karst geomorphology and hydrology
D. C. Ford & P. W. Williams

Marine geochemistry
R. Chester

Mathematics in geology
J. Ferguson

Pedology
P. Duchaufour (translated by T. R. Paton)

Perspectives on a dynamic Earth
T. R. Paton

A practical approach to sedimentology
R. C. Lindholm (eds)

Principles of physical sedimentology
J. R. L. Allen

Rock glaciers
J. Giardino *et al.* (eds)

Rocks and landforms
J. Gerrard

Sedimentology: process and product
M. R. Leeder

Soils of the past
G. Retallack

Tectonic geomorphology
M. Morisawa & J. T. Hack (eds)

Volcanic successions
R. A. F. Cas & J. V. Wright

SEDIMENTARY STRUCTURES

SECOND EDITION

J.D. COLLINSON & D.B. THOMPSON

University of Bergen

Senior Lecturer in the Department of Education at the University of Keele

London
UNWIN HYMAN
Boston Sydney Wellington

Published by the Academic Division of
Unwin Hyman Ltd
15/17 Broadwick Street, London W1V 3FP, UK

Unwin Hyman Inc.,
8 Winchester Place, Winchester, Mass. 01890, USA

Allen & Unwin (Australia) Ltd,
8 Napier Street, North Sydney, NSW 2060, Australia

Allen & Unwin (New Zealand) Ltd in association with the
Port Nicholson Press Ltd,
60 Cambridge Terrace, Wellington, New Zealand

First published in 1982
Second edition 1989

British Library Cataloguing in Publication Data

Collinson, J. D. (John David)
Sedimentary structures. — 2nd ed.
1. Sedimentary rocks
I. Title II. Thompson, D. B. (David B.)
552'.5
ISBN 0–04–445171–7
ISBN 0–04–445172–5 (pbk.)

Library of Congress Cataloging in Publication Data

Collinson, J. D. (John David)
Sedimentary structures.
Bibliography: p.
Includes index.
1. Sedimentary structures. I. Thompson, D. B.
(David B.) II. Title.
QE472.C64 1988 552.5 88–19337
ISBN 0–04–445171–7
ISBN 0–04–445172–5 (pbk.)

Set in 9 on 11 point Times by Fotographics (Bedford) Ltd
and printed in Great Britain
by The University Printing House, Oxford

Preface to the second edition

Following the success of the first edition, published in 1982, we have been encouraged by the comments of colleagues, reviewers, students and, in particular, by Roger Jones of Allen & Unwin to improve and update our approach through the preparation of a second edition.

In the past 6 years the need for a series of first- and second-year undergraduate textbooks or study guides has grown as the backgrounds and needs of students become ever more diverse, particularly in view of the very different education systems of individual countries. If anything, the needs of beginners for a relatively simple, continuously flowing text with a large number of illustrations in different, but complementary styles, for guidance in what to do first in the field and then in the laboratory, has increased. New editions and advanced texts by authors such as Allen, Leeder, Friedman & Sanders and Reading, make ever greater demands upon beginning students and we hope, in this book, to bridge the gap between studies at school and those of the final years at university or college.

In the past 6 years understanding within many branches of sedimentology and allied disciplines has advanced rapidly, not least with regard to our understanding of sedimentary structures. As a result, considerable sections of Chapters 3 (properties of fluids and flows), 6 (sands and sandstones), 7 (gravels and conglomerates), 8 (structures of chemical and biological origin) and 9 (deformation structures) have been rewritten. These revisions take account of advances in experimental work and of a deeper appreciation of aeolian, shallow-marine, volcaniclastic, carbonate and mass flow processes, which has often resulted from studies of the ancient rock record. As in the first edition, however, we seek to provide a consensus, and limitations on length preclude detailed discussions of conflicting views on the origin of controversial structures.

We make no apology for trying to interest students and teachers in the methodology and processes whereby scientific understanding is achieved. The 'hands-on' experimental teaching strategies which we encourage by experiments and other practical investigations have been treated in greater detail in recent books by John Allen. However, we like to think that our book also goes some way towards fulfilling the need for a field manual devoted exclusively to sedimentary skills, techniques and problems. The problem-recognition, problem-solving approach of our data–response exercises is retained and augmented in this edition. However, we have to confess disappointment that so few people appear to have accepted our challenge and to have used those set out in the first edition.

The number of audio-visual aids which are available and which deal with sedimentary processes and environments has increased considerably in the last few years. We believe that while there is no substitute for first-hand experience of sedimentary processes and environments, a second-hand viewing of an experiment or a field investigation recorded on film or videotape is extremely useful. The current offerings of the Open University (UK) are particularly commended in this respect.

Many readers welcomed our previous attempts to provide a guide to the literature at several levels: for advanced level students of the GCE in UK, through essential reading for first- and second-year undergraduates, to further reading for final honours and postgraduate students. We have made a considerable effort to sift, shorten and update these lists, as well as to add items of lighter reading which we have come across in the interim, or have excavated from the more deeply buried strata of our memories.

While we hope that the present edition provides some improvement over the first, we are under no illusions that it is perfect. Therefore we welcome letters from our readers and critics, because only through receiving their views can we assess how far we still fall short of their hopes and our aspirations.

In preparing this edition we have benefited particularly from help and guidance provided by the following people. C. Boulter, Charlie Bristow, Ole Martinsen, Wojtec Nemec, John Pollard, Signe-Line Røe, Terry Scoffin, Ron Steel, Roger Suthren and Mike Talbot have provided valuable critical comments on either the first edition or on drafts for this edition. Lars Clemmensen, Tom Harland, Jon Ineson and Maurice Tucker have provided new photographs. Ellen Irgens and Jane Ellingsen have helped with drafting and reprographics and Joan Collinson has helped to prepare the manuscript.

J. D. COLLINSON
D. B. THOMPSON
November 1987

Acknowledgements

We gratefully acknowledge the following individuals and organisations who have supplied or given permission for the use of illustrative material. Numbers in parentheses refer to text figures unless otherwise stated. The sources of photographs which have not been taken by the authors are indicated in the figure captions.

Figure 2.1 adapted from a paper by J. C. Griffiths in *J. Geol.* **69** by permission of The University of Chicago Press, © 1961 by The University of Chicago; F. J. Pettijohn and Springer-Verlag (2.1); H. E. Reineck and Springer-Verlag (2.2, 3 & 6); Figure 2.4 adapted from a paper by F. M. Broadhurst and I. M. Simpson in *Geol. Mag.* **104**, by permission of Cambridge University Press (2.4); E. B. Wolfenden (2.4); Figure 2.8 adapted from a paper by J. V. Wright, A. L. Smith and S. Self in *J. Volcan. Geoth. Res.* **8**, (2.8); Dunbar, Dunbar, Roman & Anderson, P. A. (2.10); W. Nemec and Canadian Society of Petroleum Geologists (3.18, 7.4, 9 & 16); J. R. L. Allen and Elsevier (3.2, 3.5 & 6, 4.5–7, 4.9, 6.9, 6.16, 6.25, 10.8); G. V. Middleton and SEPM (3.16, 3.17, 7.14); Figure 3.14 adapted from a paper by G. V. Middleton in *J. Geol.* **84**, by permission of The University of Chicago Press © 1976 by The University of Chicago; Å. Sundborg and the Editor, *Geograf. Ann.* (3.15); F. Ricci-Lucchi and Zanichelli (4.1); S. Sengupta and SEPM (4.3); Grønlands Geologiske Undersøgelse (4.17b, 6.22, 6.24b, 8.6, 8.10, 8.11, 8.15, 9.14, 9.24); SEPM (6.21, 7.14, 9.36, 9.39 & 40); J. R. Boersma and the Editor, *Sedimentology* (6.12); J. C. Harms and SEPM (6.15, 6.30); H. E. Reineck and Elsevier (6.14); Figure 6.19 reprinted from *Proc. 8th Conf. Coastal Engng*. pp. 137–50 (D. L. Inman & A. J. Bowen, eds, 1963) by permission of the American Society of Civil Engineers and P. D. Komar (6.19); A. Kaneko and the Editor, *Repts. in Applied Mech., University of Kyoto* (6.20); Figure 6.23 reprinted from a paper by R. P. Sharp in *J. Geol.* **71** by permission of the University of Chicago Press © 1963 by The University of Chicago (6.23); A. V. Jopling and SEPM (6.38); the Editor *Geograf. Ann.* (6.39); P. J. McCabe and the Editor *Sedimentology*, (6.41); R. U. Cooke and Batsford (6.45); Elsevier (6.46); K. W. Glennie and Elsevier (6.47a & b); E. D. McKee and US Geol Survey (6.47c & d, 6.57); Figure 6.49 is based on a paper by R. J. Wasson and R. Hyde in *Nature* **304**, by permission © 1983 by Macmillan Magazines Ltd. (6.49); US Geol. Survey (6.50–52); R. A. Bagnold (6.53b); R. E. Hunter and the Editor *Sedimentology* (6.55a & b); L. B. Clemmensen and the Editor *Sedimentology* (6.58); S. L. Stokes and SEPM (6.59); R. G. Walker and Geological Association of Canada (6.63); BGS photographs reproduced by permission of the Director, British Geological Survey, NERC copyright reserved (7.13b); D. Piper (7.1a); R. G. Walker and SEPM (7.2, 7.3, 7.5, 7.15a, b & c, 7.18); E. Derbyshire and Butterworths (7.6, 7.–10 & 11); J. Lajoie and Geological Association of Canada (7.17); R. S. Arthurton and the Yorkshire Geological Society (8.1); R. S. Arthurton and the Editor *Sedimentology* (8.2); Figure 8.3 is based on a compilation by B. C. Screiber for *Sedimentary environments and facies* (H. G. Reading, ed.), by permission © 1986 by Blackwells Scientific Publications (8.3); N. P. James and Geological Association of Canada (8.5, 10.8d); W. V. Preiss and Elsevier (8.9); R. Gruhn and the Arctic Institute of North America (9.12); M. B. Edwards (9.22); G. E. Farrow and Springer-Verlag (9.30); A. Hallam and Springer-Verlag (9.33); A. Martinsson and the Editor, *Geol. J.* (9.34); J. D. Howard and SEPM (9.35); R. G. Bromley and Springer-Verlag (9.36); R. Goldring and SEPM (9.37); S. Simpson and Springer-Verlag (9.38), Table 9.1; C. K. Chamberlain and D. C. Rhoads (9.40); T. P. Crimes and Springer-Verlag (9.39–41); F. Surlyk and Grønlands Geologiske Undersøgelse (10.5, 10.6c); S. A. Greer and Senckenberg Institut fur Meeresgeologie und Meeresbiologie (10.6a); J. M. Coleman (10.6b); G. Kelling and Geologists' Association, South Wales Group (10.8b).

Contents

List of tables

1 Introduction to the study of sedimentary structures

1.1 The nature of this book

To give you an idea of what this book is about, you are invited to describe and interpret, using your present experience, the series of geological structures and relationships depicted in Figure 1.1. You might also think of what relevance their solutions could have for geologists engaged in exploring for and exploiting economic resources.

Whatever experience is brought to bear on this exercise, we suggest that you will have followed many of the steps that a fully trained sedimentologist would have taken in tackling the same problem. Many approaches will have been adopted, some of which we intend to sharpen and augment in the following pages.

(a) Several features will have been recognised and described on the basis of your everyday experience. Here you have an *information base* which is considerable but clearly inadequate to enable you to complete the task. This will often be the case whatever your later experience.

(b) You will have observed, measured, recorded, identified, compared, classified, predicted and inferred certain features and relationships: we would hope to enlarge this range of *intellectual skills and techniques* and involve you in developing specific practical and experimental skills.

(c) You will have tried to explain some of these features using your understanding of physical, chemical and biological processes which you see operating today;

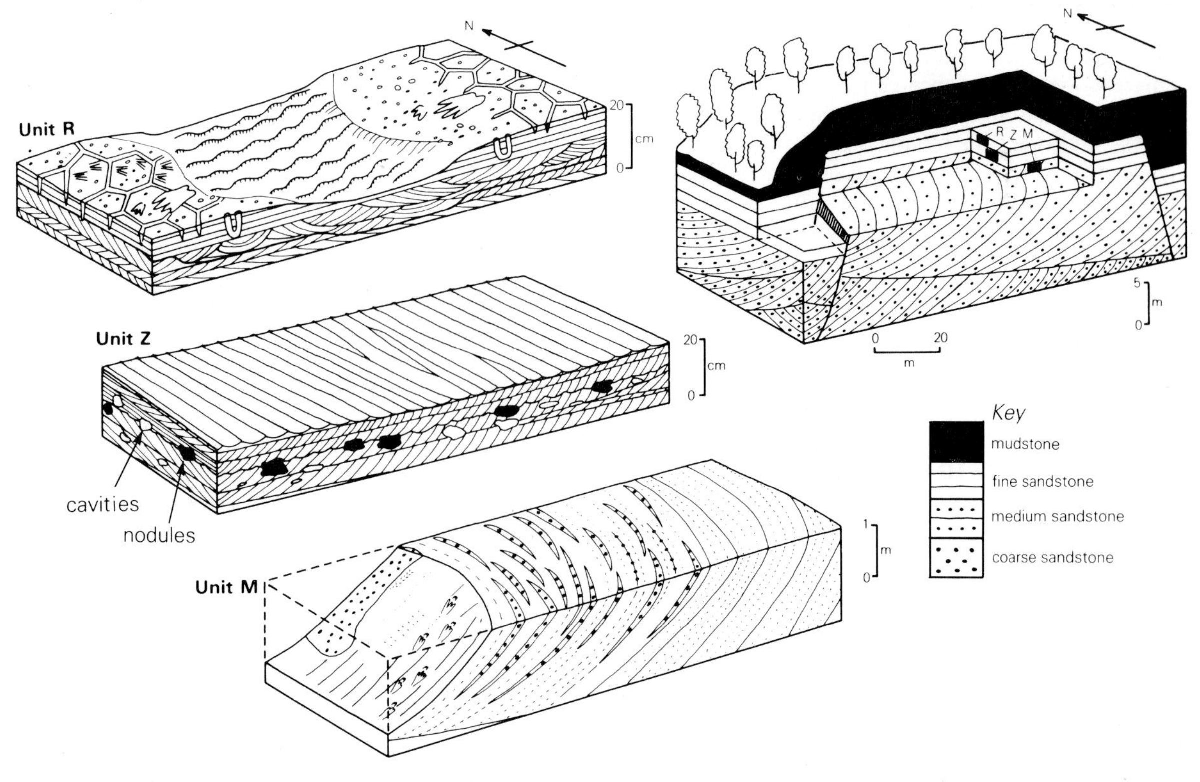

Figure 1.1 Sedimentary structures exposed in three blocks representative of units M, R and Z in a hypothetical quarry. Note the scale and orientation of the quarry walls.

you will have worked within a *set of current beliefs* about nature which suggests that it is orderly, has uniformity and that the present is the key to explaining the past.

(d) You might ask yourself whether you first took in a great deal of information at a glance, produced a guess or two, and then tested these 'half-baked' ideas by critical reference to the evidence that you thought you could recognise, or whether you first described each bit of the jigsaw and then came to a general idea of its meaning. In either case, working deductively (proving some things false on certain evidence) or inductively (going from the particular to the general), you were involved with the *process and method of scientific enquiry*.

(e) You will have attempted to sort a great many features into time–space relationships: a process of *historical ordering of events* at a particular place which is at the heart of the geological sciences and which helps to distinguish them from the other sciences.

Before we develop these points (particularly the first four) in greater detail, we advise beginners that this chapter deals with some methodological and philosophical questions the importance of which may not be fully realised until you have tackled the whole book, enlarged your experience considerably, and applied skills derived from other courses. Such students could best skim through the rest of this chapter and return to it later when they have mastered sufficient background to benefit. The skills, techniques, attitudes, values and sets of beliefs inherent in a discipline grow over a long period. The best students will beware of being hoodwinked by the consensus which we present, and will strive to develop better approaches of their own.

1.2 The relationships of sedimentary structures to sedimentology and geology

Sedimentology is the study of the nature and origin of both present-day and ancient sedimentary deposits. It includes sedimentary petrography (the description of composition and fabric) and is closely related to stratigraphy (and particularly palaeogeography). It differs from the latter in that it is not so concerned with the mechanics of correlation, or with the broader long-term time-space relationships of strata. Sedimentology draws upon and contributes to geological subdisciplines such as geochemistry, geophysics, mineralogy, palaeontology and tectonics, and to sciences such as biology, chemistry, civil engineering, climatology, fluid dynamics, geomorphology, glaciology, oceanography and soil science.

Sedimentary structures relate to all these subdisciplines and are generated from materials of a varied composition and are observably products of physical, chemical and biological processes. Certain processes are common to many present-day environments, but combinations of processes, often with particular directional properties, may be diagnostic of specific environments. Combinations of processes vary in kind and intensity laterally from sub-environment to sub-environment. Comparable changes of processes can be deduced to have taken place in the past if we learn to read the sedimentary structures that help to characterise different units in the rock record. The characterisation of rock units is based, however, on more than just the sedimentary structures and will commonly also involve palaeontological, compositional and textural features, some of which may require confirmation in the laboratory. Such characterisation allows similar rock units to be grouped together as *facies* with the implication that different facies will each be interpretable in terms of a set of processes relating to a particular environment.

It is important to realise that the definition of facies and the assigning of rock units to facies are not determined by absolute criteria but will be determined by the aims and circumstances of any particular investigation.

1.3 Sedimentary structures in relation to the nature of science and scientific methodology

From our example at the beginning of the chapter you will see that the subdisciplines of geology differ from those of the basic theoretical sciences in that they are not necessarily concerned with generating and testing universal laws. Established laws are commonly taken for granted and used to find and test particular problems relating to what happened successively at particular times and places. In that sense the science can be regarded as 'practical' rather than 'theoretical', but is none the less satisfying for that.

Sedimentologists work within a set of principles, held by all scientists, many of which are probably implicit in the way in which you tackled the initial exercise:

(a) *The principle of determinism:* that nature is regular and orderly.

(b) *The principle of the uniformity of nature:* that nature is constant with respect to its laws; that scientific laws are constant, i.e. invariable with respect to time, place and circumstances.
(c) *The principle of continuity:* that nature is continuous through space and time.
(d) *The principle of parsimony* (Ockham's principle): that the simplest hypothesis or theory is the best explanation of the facts.

In addition, sedimentologists, as part of their intellectual methodology, use procedures perhaps adopted intuitively by you in attempting the exercise. They attempt to develop worthwhile:

Conjectures or speculations: rapidly conceived, intuitive ideas about relationships of observed phenomena; hunches against which they can test the evidence at hand.

Hypotheses: untested explanations of observed phenomena, logically developed and tentatively adopted so that it may be possible to deduce what critical ideas, on present evidence, are false and which can be accepted for the time being as worthy of further testing. Good hypotheses predict a great many consequences, each of which can be tested in a variety of ways, perhaps by experiments. Distinguish *initial hypotheses* as the most tentative ones and recognise, in most investigations, the need for *multiple hypotheses*, i.e. many possible ideas about the solution of a problem, in case you become 'blinkered' by a single explanation. Remember that such ideas are all *working hypotheses* generated to explain certain relationships. The standing of any hypothesis changes as the evidence increases. Scientific 'truths', in the form of well developed, long-standing hypotheses, are still subject to constant scrutiny and revision and are not absolute.

Theories: coordinated sets of self-consistent hypotheses, each of which has been tested many times and remains a useful explanation of relationships. No, or extremely few, exceptions to any theory should exist. Beware, however, of the fact that many theories are known to last longer for social reasons than is justifiable with hindsight (e.g. Darwinian gradualism or the fixity of continents). Theories encompass and supersede one another.

Models: idealised simplifications set up to aid understanding of complex relationships between phenomena and processes, often to illustrate working hypotheses. We may draw up *actual models* based upon a modern environment (e.g. of a desert environment) using data from a particular basin in the Sahara, or an *inductive model* (e.g. a sand-sea and sand-dune dominated desert) based on the many basins in the Sahara. We may make *scaled, experimental models* to detect, under controlled conditions in a wind tunnel, the processes and variables responsible for particular structures, e.g. aeolian sand ripples and flat beds. *Mathematical models* attempt to simulate complex geological processes. In our desert example, the effects of changed wind direction and strength, increased sand flow rate and change of grain size on the shape of the sand sea, the sand dunes and the smaller structures upon them, could be predicted. *Visual models*, either diagrammatic or realistic, help us to see relationships and to picture processes, products and environments. Models may be *static*, descriptive of a particular time in the past, yet still be predictive of many relationships, as in a palaeogeographic map, or *dynamic*, attempting to show a changing pattern or dynamic equilibrium of processes and environment over a period of time or a *steady state* equilibrium over the same period.

One type of model, particularly important in arriving at an interpretation of a sequence of sediments is the *facies model*. This is a generalisation and simplification of the observed vertical and lateral relationships of the facies recognized in a sequence. It is an attempt to reduce the natural 'noise' of the relationships and reveal an underlying pattern which can then be compared with predictive actual models derived from studies of present-day environments. These ideas are elaborated in Chapter 10.

The values held by a group of scientists working in a particular field, their practices (methodology) and sets of beliefs (philosophy) constitute a *paradigm* and help to contribute to the scientific consensus. The general geological paradigm within which the proper interpretation of sedimentary structures could take place was generated roughly between 1785 and 1860. The basic principles of a specifically sedimentological paradigm can be traced to the work of H. C. Sorby and J. Walther between 1850 and 1900, but it has been fully developed only since the 1960s.

Two major events are associated with the names of Sorby and Walther. Henry Clifton Sorby (1826–1908) may truly be regarded as the father of sedimentology, for between 1850 and 1908 he pioneered most of the approaches which we develop in this book. He recognised the problems of understanding past deposits encountered in the rocks; that critical questions had first to be identified by making acute field observations and careful records in the light of a thorough understanding of processes. As an aid to observation, he made thin sections and used the polarising microscope.

As a better guide to understanding processes and products, he performed experiments, for example by generating ripples and cross lamination by current action. He was the first to measure the orientation of structures such as cross bedding in the field and, above all, he used his understanding to make environmental reconstructions and put them in a palaeogeographic context. Walther on the other hand, in his *Introduction to geology as a historical science* (1890–3), drew together many scattered observations on modern sediments and processes and demonstrated implicitly the power of the actualistic method as a basis for the study of sedimentary rocks. In addition, he established a further and most powerful stratigraphic principle: *Walther's principle of the succession of facies.* This principle states that unless the evidence indicates otherwise, we should expect the processes and environments that occur laterally adjacent to each other to be represented by facies that succeed each other gradationally in a vertical geological column (see Section 10.3.3). Alas! Sorby was ahead of his time, and Walther wrote in a language not readily accessible to the Anglo-Saxon world.

Reference and study list

References marked with an asterisk * are suitable for 16–19 year old advanced level students in schools and colleges in the UK as well as for undergraduates.

Standard texts on sedimentary rocks in which sedimentary structures are mentioned

Allen, J. R. L. 1970. *Physical processes of sedimentation.* London: Allen & Unwin. [Chapters 3–7 of the book contain analyses of processes in particular environmental settings.]

Blatt, H., G. V. Middleton and R. C. Murray 1980. *Origins of sedimentary rocks*, 2nd edn. Englewood Cliffs, NJ: Prentice-Hall.

Bouma, A. H. 1969. *Methods for the study of sedimentary structures.* New York: Wiley.

Conybeare, C. E. B. and K. A. W. Crook 1968. *Manual of sedimentary structures.* Canberra: Australian Dept. Nat. Dev. Bur. Mines Resources, Geol. & Geophys. Bull. 102.

Dunbar, C. O. and J. Rodgers 1957. *Principles of stratigraphy.* New York: Wiley. [A book that was ahead of its time in trying to introduce a methodology and approach which is commonplace today.]

Fairbridge, R. W. and J. Bourgeois (eds) 1978. *The encyclopaedia of sedimentology.* Stroudsburg, Pa: Dowden, Hutchinson & Ross. [A general reference with entries for many of the topics treated in this chapter.]

Friedman, G. M. and J. E. Sanders 1978. *Principles of sedimentology.* New York: Wiley. [The first chapter includes something of the history of the subdiscipline and its methodologies. Later, the authors dwell upon the applicability of the subject, often in relation to its links with other subjects.]

Leeder, M. R. 1982. *Sedimentology: Process and product.* London: Allen & Unwin.

Lindholm, R. 1987. *A practical approach to sedimentology.* London: Allen & Unwin.

Pettijohn, F. J. 1975. *Sedimentary rocks*, 3rd edn. New York: Harper & Row.

Pettijohn, F. J. and P. E. Potter 1964. *Atlas and glossary of sedimentary structures.* New York: Springer-Verlag.

Reading, H. G. (ed.) 1986. *Sedimentary environments and facies.* Oxford: Blackwell Scientific. [While the emphasis is on environments, structures are dealt with as a means of understanding the processes active in the environments.]

Reineck, H. E. and I. B. Singh 1980. *Depositional sedimentary environments*, 2nd edn. Berlin: Springer-Verlag. [A superbly illustrated textbook – highly recommended.]

Ricci-Lucchi, F. 1974. *Sedimentografia.* Bologna: Zanichelli. [Magnificent photographs of sedimentary structures of all kinds. Text in Italian, but not too difficult to understand if your experience of geological terms is reasonable.]

*Selley, R. C. 1976. *An introduction to sedimentology.* London: Academic Press. [A basic general text written in a distinctive, lively and humorous style, which is born of an author entertaining both himself and audiences as diverse as sceptical undergraduates and seasoned oil company geologists.]

Scoffin, T. P. 1987. *An introduction to carbonate sediments and rocks.* Glasgow: Blackie. [Succinct and well illustrated.]

Tucker, M. E. 1981. *Sedimentary petrology: an introduction.* Oxford: Blackwell Scientific.

Walker, R. G. (ed.) 1984. *Facies models*, 2nd edn. Geological Association of Canada. [A series of concise summaries of the major sedimentary environments, including several discussions of structures in their environmental context.]

Wilson, J. L. 1975. *Carbonate facies in geologic history.* Berlin: Springer-Verlag.

Books in which the economic applications of the study of sedimentary rocks are particularly apparent

Brenchley, P. J. and B. J. P. Williams (eds) 1985. *Sedimentology: Recent developments and applied aspects.* Spec. Publ. Geol. Soc. London 18. Blackwell Scientific. [The papers by Clemmey, Johnson & Stewart and Burchette & Britton are of direct economic significance.]

Illing, L. V. and G. D. Hobson (eds) 1981. *Petroleum geology of the continental shelf of north-west Europe.* London: Heyden.

James, C. H. (ed.) 1959. *Sedimentary ores ancient and modern* (revised). Leicester University Press.

Miall, A. D. (ed.) 1978. *Fluvial sedimentology.* Calgary: Mem. Can. Soc. Petrolm Geol. 5. [Selected papers should be read, particularly those by Minter & Turner.]

Sangster, D. F. 1972. *Precambrian volcanogenic massive sulphide deposits in Canada: a review*. Paper 72–22. Geol. Surv. Canada.

Selley, R. C. 1985. *Elements of petroleum geology*. London: Freeman.

Light reading

Ager, D. V. 1973. *The nature of the stratigraphic record*. New York: Wiley.

Darwin, C. H. 1891. *Journal of researches into the natural history and geology of the countries visited during the voyage round the world of HMS Beagle*. London: Routledge.

Geikie, A. 1962. *The founders of geology*. New York: Dover.

Hallam, A. (ed.) 1977. *Planet earth*. Oxford: Elsevier/Phaidon. [See particularly the chapter by Porter on the history of geology.]

Higham, N. 1963. *A very scientific gentleman. The major achievements of Henry Clifton Sorby*. Oxford: Pergamon Press.

Linklater, E. 1972. *The voyage of the Challenger*. London: John Murray (publishers). (London: Sphere Books, 1974.)

Porter, R. 1977. The making of geology. In *Planet earth*, A. Hallam (ed.), 288–309. Oxford: Elsevier/Phaidon.

Shepard, F. P. 1948. *The earth beneath the sea*. Oxford: Oxford University Press.

Further reading

(A) HISTORY OF SEDIMENTOLOGY AND SEDIMENTARY STRUCTURES

Cornish, V. 1914. *Waves of sand and snow*. London: T. Fisher Unwin. [Careful observations of natural bedforms with interesting early attempts to explain their occurrence.]

Darwin, C. H. 1842. *Structure and distribution of coral reefs*. London: Smith Elder. (Reprinted 1962 by the University of California Press.) [A classic study of the gross structures and origin of large carbonate structures.]

Daubrée, A. 1879. *Etudes synthetiques de géologie experimental*. Paris: Dunod. [The first experiments and models of physical processes of sedimentation.]

Gilbert, G. K. 1914. *Transportation of debris by running water*. Prof. Pap. 86. US Geol. Surv. [An early paper on experimental studies.]

Gressly, A. 1838. *Observations géologiques sur le Jura Soleurois*. Nouveaux Mem. Soc. Helv. Sci. Natur. 2. [The introduction of the facies concept.]

Hitchcock, E. 1858. *Ichnology of New England: a report on the sandstone of the Connecticut Valley, especially its foot prints*. Boston: W. White. [The first book to tackle the interpretation of suites of trace fossils.]

Kuenen, P. H. and C. I. Migliorini 1950. Turbidity currents as a cause of graded bedding. *J. Geol.* **58**, 91–127. [Its appearance ushered in the modern approach to sedimentology: a classic case wherein the study of the deposits of the past and some experiments in a small laboratory tank contributed enormously to the growth of a theory and a field of study.]

Lyell, C. 1830–1833. *Principles of geology, being an attempt to explain the former changes of the earth's surface by reference to causes now in operation*. London: Murray. [The introduction of the steady-state actualistic paradigm.]

Middleton, G. V. 1978. Sedimentology – history. In *The encyclopaedia of sedimentology*, R. W. Fairbridge and J. Bourgeois (eds), 707–12. Stroudsburg, Pa: Dowden, Hutchinson & Ross.

Rupke, N. A. 1983. *The great chain of history*. Oxford: Oxford University Press. [The development of basic ideas of stratigraphy in the context of prevailing philosophical and religious belief in the early 19th century.]

Shrock, R. R. 1948. *Sequence in layered rocks*. New York: McGraw-Hill. [A comprehensively illustrated manual of sedimentary structures; the rare and the unusual structures being given as much attention as the common and the pervasive structures.]

Summerson, C. H. (ed.) 1976. *Sorby on sedimentology, a collection of papers from 1851 to 1908 by Henry Clifton Sorby*. Miami: Univ. Miami Comparative Sedimentology Lab.

Walther, J. 1894. *Einleitung in die Geologie als Historische Wissenschaft*. Bd. 3: *Lithogenesis der Gegenwart*. (Vol. 3: *Modern lithogenesis*). Jena: Fisher Verlag, pp. 535–1055. [The first statement of the Principle of the Succession and Correlation of Facies and the first compendium on modern sediments, processes and environments to demonstrate the power of the actualistic method.]

(B) METHODOLOGY AND PHILOSOPHY

Albritton, C. C. Jr (ed.) 1975. *Philosophy of geohistory*: 1785–1970. Stroudsburg, Pa: Dowden, Hutchinson & Ross. [Papers by Hutton, Playfair, Lyell, Hubbert, Simpson and Hooykas outline the growth of thought relating to actualism and uniformitarianism. Students might begin by reading the conclusion at the end of Simpson's paper. Papers by Chamberlain, Gilbert, Davis, Johnson and Machin all deal with methodology in scientific investigations.]

Allen, J. R. L. 1970. *Physical processes of sedimentation*. London: Allen & Unwin. [Read the introduction for the philosophical–methodological approach of this author.]

Kitts, D. B. 1977. *The structure of geology*. Dallas: Southern Methodist University Press. [This book attempts to chart the distinctiveness of geology as an historical science in comparison with other sciences.]

Reading, H. G. Facies. In *Sedimentary environments and facies*. H. G. Reading (ed.), Ch. 2, 4–19. Oxford: Blackwell Scientific. [The author's views on the methodology of the sedimentologist.]

2 Bedding

2.1 The nature of bedding

2.1.1 Where to start: the recognition of sets of beds

When you approach a series of exposures of rock, we suggest that you might start by asking the following questions:

(1) Can I detect anything in the rocks that will reveal that they are bedded?
(2) Can I identify features that are characteristic of particular processes or environments of deposition?
(3) Can I detect any pattern of vertical and lateral change in the rocks that might suggest changing processes and thereby an environment of deposition?
(4) How far do the products of these processes and environments of deposition extend laterally?

2.1.2 The basis of this approach: the origins of bedding at the present day

In framing and trying to answer these questions you will rely to some extent upon your experience of laboratory experiments and of processes acting in modern depositional environments. Many bodies of sediment possess more or less planar bottom and top surfaces and are very extensive laterally in relation to their thickness. These units commonly reflect rather uniform physical conditions and sediment supply, and they are distinguished by composition, texture and internal structure from the bodies of sediment above and below (Fig. 2.1). Sedimentation units greater than 1 cm thick are known as **beds**, and their upper or lower surfaces are known as **bedding** or **bounding planes**, the lower bounding surface often being referred to as the **sole** and the upper as the **upper bedding surface** (Fig. 2.2). Beds may be further subdivided. Constituent units thicker than 1 cm are referred to informally as **layers** or **strata**. Below

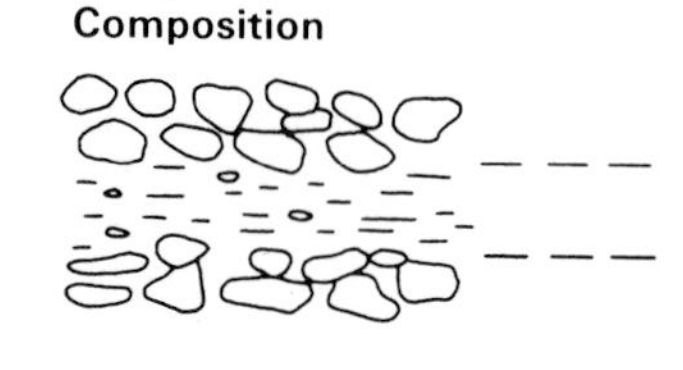

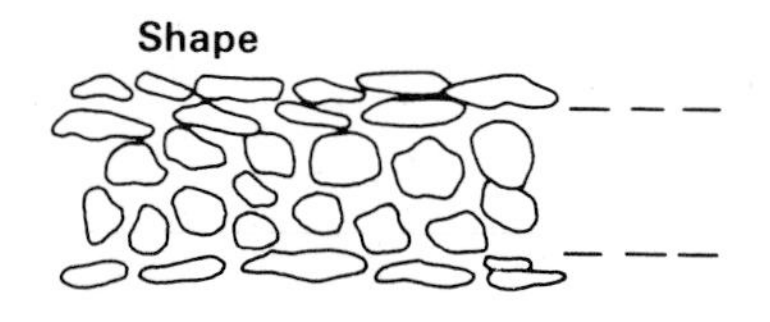

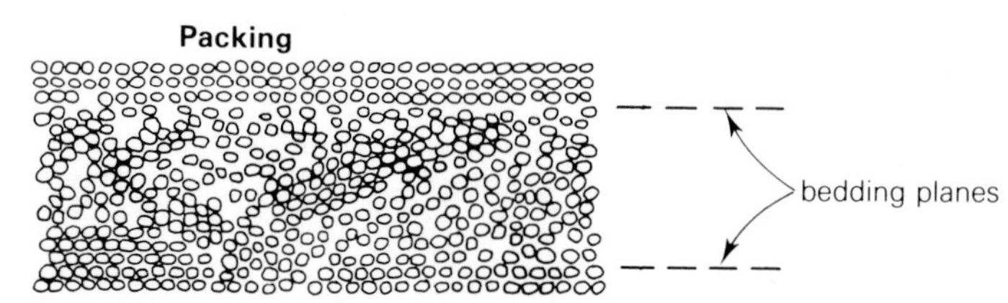

Figure 2.1 Bedding as the product of different combinations of grain composition, size, shape, orientation and packing (modified after Pettijohn, Potter & Siever 1972, and Griffiths 1961).

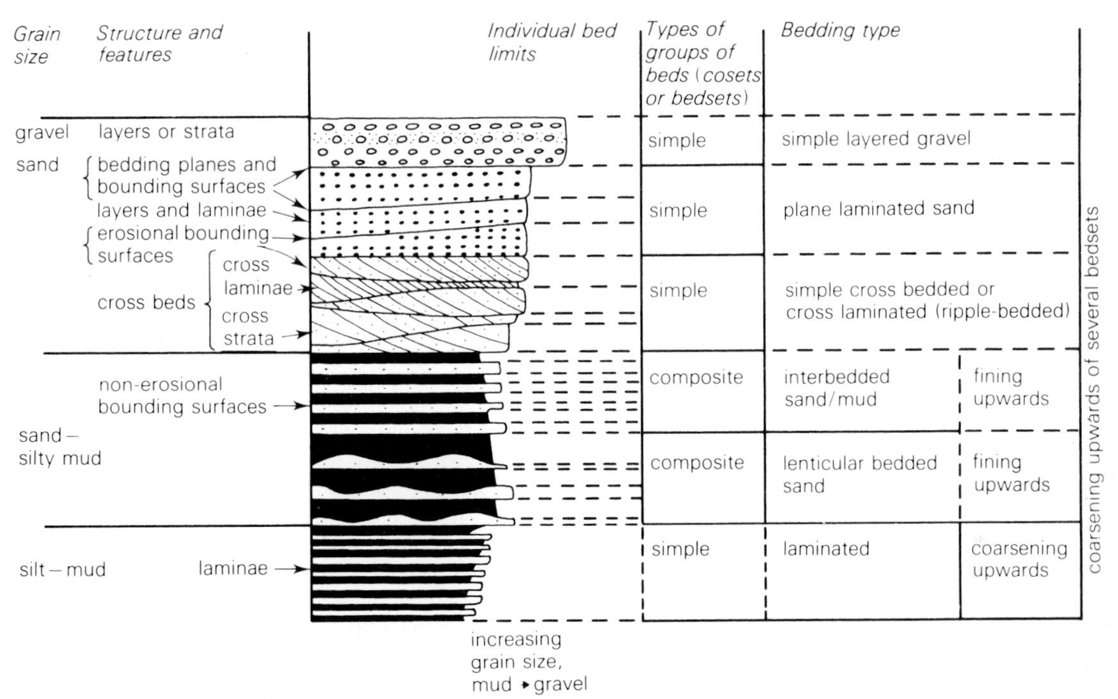

Figure 2.2 A scheme illustrating the terminology used in this book to describe sedimentation units (modified after McKee & Weir 1953, Campbell 1967, and Reineck & Singh 1973).

1 cm, sedimentation units are termed **laminae**: the smallest units visible in a sequence. Layers and laminae which help to make up beds and which are deposited at an angle to the main depositional surface are called **cross beds** (which include **cross laminae** or **cross strata**). The general phenomenon of inclined layers is termed **cross lamination** or **cross bedding,** depending on scale (see Ch. 6). Groups of beds may form **cosets** or **bedsets** which may be **simple** or **composite** (Fig. 2.2). The term bedding or bedding plane should not be used in ancient sediments for surfaces along which splitting takes place within a true bed on weathering; the term **splitting** or **parting plane** should be used (Fig. 2.3). Many splitting planes do, however, correspond to true bedding or bounding surfaces. Many beds and bedsets forming at the present day maintain their shape and thickness for considerable lateral distances, although all eventually thin out or change their nature either gradationally or suddenly when traced far enough. Their continuity, however, is often pervasive and striking. Furthermore, if areas of natural, present-day deposition are observed for long enough or if the variables of a laboratory experiment are changed, beds with different properties, produced by different physical conditions, will be deposited. Vertical sections of river floodplains, estuarine flats or beaches, which are exposed in the erosive banks of migrating channels, often show successions of such recently formed beds, the oldest at the base, the most recent at the top; each set of beds reflects changing physical conditions. Any sedimentary structure that cross-cuts a bedding feature – for example a channel downcutting into the layers – must postdate that feature. In this situation, fragments from an older bed could fall into and be incorporated within a younger bed. Close observation of a present-day bed may reveal further structures (for example symmetrical, wave-produced ripples or desiccation cracks) and organic traces (plant roots, shells orientated in life position, footprints, traces of burrowing and crawling). Such features may enable the bottom of a bed in the rock record to be distinguished from its top and hence its way-up recognised, an important observation in tectonically disturbed strata.

2.1.3 *Basic stratigraphic principles derived from present-day phenomena*

Observations of present-day processes and products allow us to erect several stratigraphic principles which aid recognition of both the presence and the attitude of

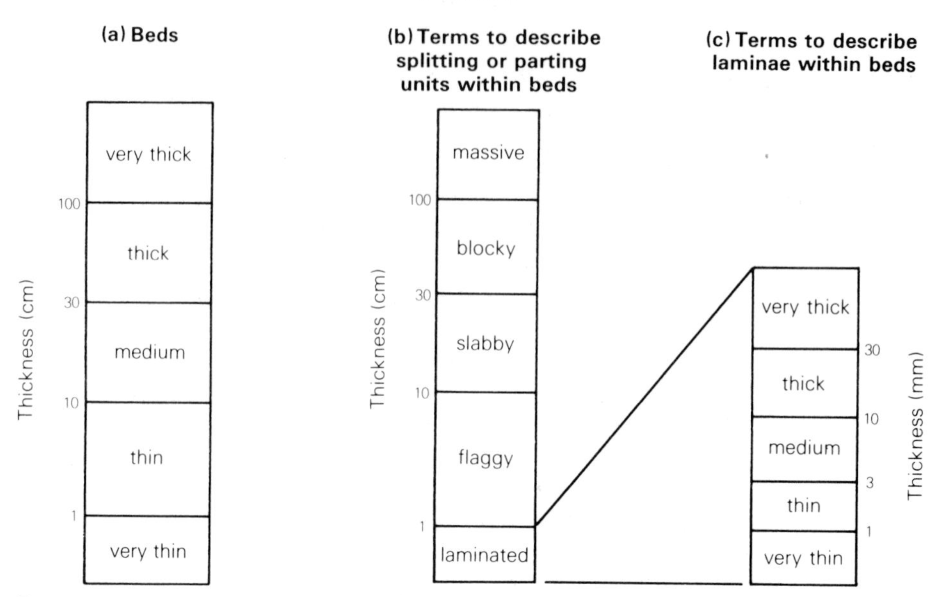

Figure 2.3 Terminology for thickness of beds and the description of units within beds due to splitting or parting, often after weathering (modified after Ingram 1954, Campbell 1967 and Reineck & Singh 1973).

beds in ancient sequences which have been subject to tectonic disturbance:

The principle of original horizontality of beds: that most beds were once laid down either parallel to the earth's horizontal or at very low angles. Exceptional beds deposited at angles up to 40° may be recognisable by their sedimentary structures.

The principle of original continuity of beds: that groups of beds are commonly laterally extensive and maintain their thickness and continuity for great distances. Individual beds and layers are commonly lenticular over short distances.

The principle of superposition of beds: that the younger beds are deposited on top of the older beds in a sequence.

The principle that the way-up of beds can be recognised by sedimentary structures that have characteristic attitudes with respect to the top or bottom of a bed.

The principle of included fragments: that fragments of an older bed can be included in a younger bed, but not *vice versa*.

The principle of cross-cutting relationships: that a feature which cuts across a bed must be younger than it.

The principle of strata identified and correlated by their included fossils: that strata may be dated and correlated by the sequence and uniqueness of their flora and fauna.

2.1.4. The application of basic stratigraphic principles to ancient rock sequences

The principles of original horizontality, original continuity, superposition and cross-cutting relationships will be sufficient to guide any initial exploration of rocks that have been tilted relatively little from the original plane of deposition. Where rocks are tilted at greater angles, the use of way-up features and sequences of fossils may permit the detection of overturned beds. In such cases,the term **younging** indicates the direction of the top of any sequence. A sequence could therefore be reported as 'younging to the east' for example. Application of the principles of original continuity, cross-cutting relationships and included fragments may also enable faults and channelling to be identified.

2.1.5 Preliminary observation and recording of bedding

Several levels of observation and investigation of bedding are possible, from the large-scale distant view to small-scale scrutiny with a hand lens. At first it is useful to scan exposures from a distance in order to decide the general attitude and orientation of the beds. It is also helpful to work out, at an early stage, the way-up of the succession and where, in general, the older and younger strata are to be found. This may be achieved at a distance by recognising the bases of cross-cutting features such as large-scale channels, but it is most likely to be confirmed by closer observation. It is possible from a distance to ask and give preliminary answers to the following questions:

(1) Can way-up be determined? Are the bases of any beds very irregular on a large scale? Are there major channels cutting into the underlying beds?
(2) Do the beds appear to form a comformable sequence throughout the exposures? Are there two groups of strata inclined at different angles or does one group have its lateral continuity terminated by a second group? Hence is there any likelihood of a major time gap, i.e. an unconformity?
(3) What is the approximate spacing of the more

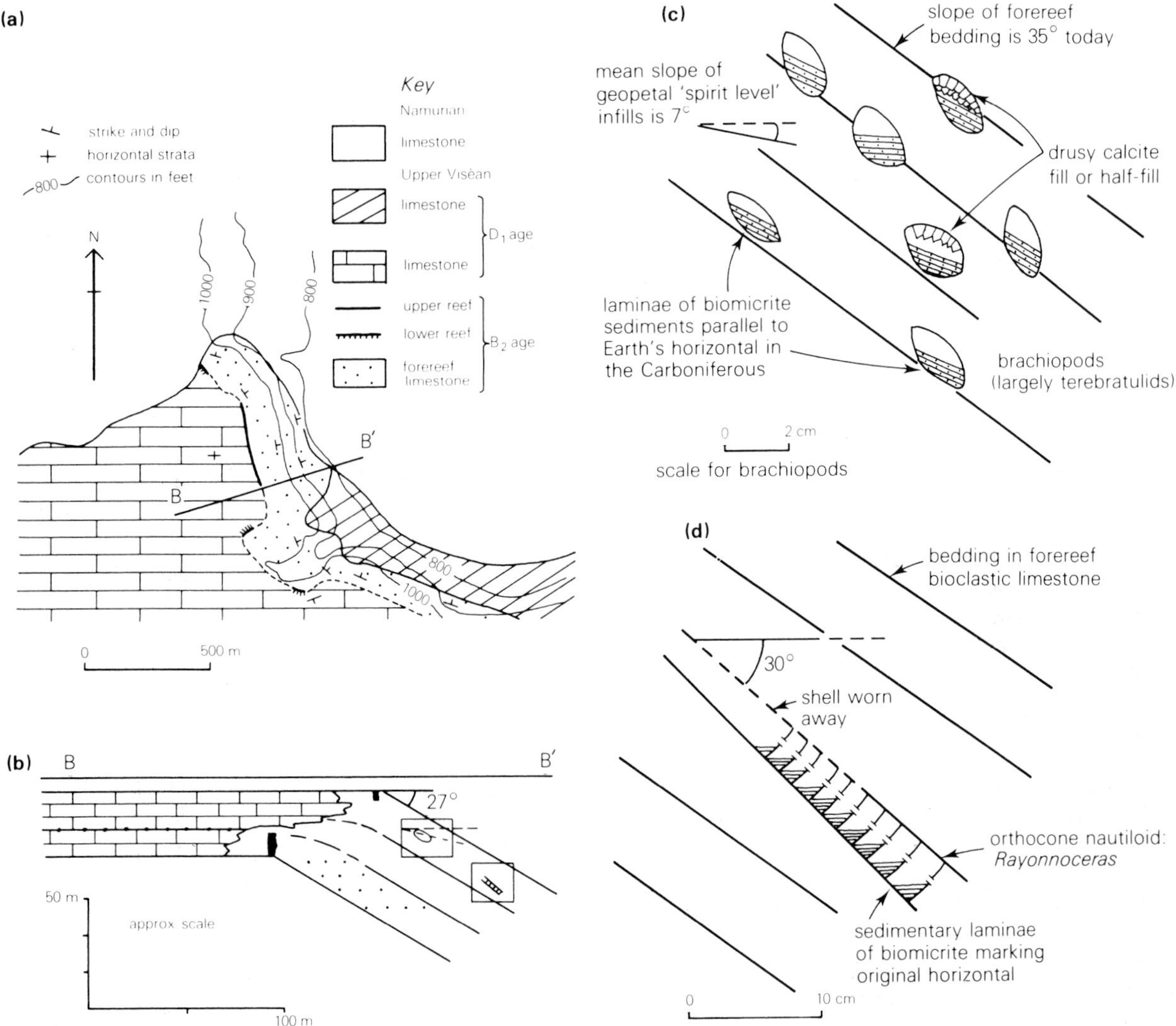

Figure 2.4 Way-up and geopetal (spirit-level) structures in Visean Carboniferous Limestones at Treak Cliff, Derbyshire, England: (a) geological map and (b) idealized section of the eastern part of the Castleton reef complex (modified after Wolfenden 1958); (c) geopetal infills of brachiopods on the fore-reef slope; (d) geopetal infills of a nautiloid on the fore-reef slope (after Broadhurst & Simpson 1967). The sedimentary laminae infilling the body cavity of the fossils were sedimented more or less parallel with the former horizontal of the Earth's surface and hence can be regarded as 'spirit level' measures of that horizontal. In the case of brachiopods, both the bedding planes of the fore-reef and the spirit level structures have been tilted by tectonic forces (to 35° and 7° respectively) since they originated in the Carboniferous. Hence the inclination of the fore-reef slope, as originally sedimented in the Lower Carboniferous, was 28° in the case of the brachiopods. In the part of the fore reef where the nautiloid was found the original depositional slope was about 30°. (See also Fig. 8.6.)

prominent bedding planes? Are the beds of uniform thickness? Do successive beds thicken or thin upwards? Are there patterns of repetition of vertical bed-thickness changes?

(4) Are there any signs of variation of grain size in vertical sequences? Are there beds, or a series of beds, that consistently 'fine upwards' or 'coarsen upwards'?

(5) Are vertical variations of overall composition suspected, e.g. limestone – shale – sandstone conglomerate? Are there patterns to this variation?

(6) Can the sequence be divided up into packages or units of contrasting aspect?
Are there any combined regular variations in the vertical sequence of bed thickness, grain size and composition?

(7) Do individual beds change thickness laterally and, if so, how? Are there any lateral changes of grain size and lithology?

Develop the habit of recording these preliminary observations and practise estimating the dimensions of the large beds and sequences of beds. From this kind of initial analysis, ideas will emerge concerning where to go to start detailed work, where to sample, and where to find key features and successions.

2.1.6 *Detailed observation and recording of bedding: methodology*

Attitude of beds. Detailed work on the outcrop should begin with the measurement of the attitude of beds (direction of strike and angle of dip) at many places, and these facts will be recorded both on a map *and* in a notebook. These data are necessary so that the measurements of inclination and alignment of sedimentary structures recorded in the field, which will include the effect of any tectonic rotation, can be restored (on a stereogram) to their original depositional attitude and orientation (see Appendix A). Usually corrections are necessary only for dip, but sometimes plunge needs to be considered. Look out particularly for structures that will show way-up, and test in detail the applicability of the principle of original horizontality. Some way-up structures are known as **geopetal** (spirit level) structures and, although quite rare, they can reveal the attitude of the original horizontal with some confidence, e.g. a brachiopod shell half-filled by sediment after death (e.g. Fig. 8.6). Such a surface will commonly coincide in attitude with bedding but in some cases it may diverge considerably. Beds formed on a reef front may, for example, have an original or initial dip of 30° or more (Fig. 2.4).

Successions. The basic aim of work on the units of a succession should be to observe and describe them accurately and concisely, and to divide them into beds and bedsets. Where possible, measure and record several laterally equivalent vertical successions in the same beds, selected to illustrate any lateral variability suspected from preliminary observation. This enables local variability to be distinguished from any suspected regional variability and ensures that regional trends are seen in proper perspective. Be sure that measurements of thickness are made normal to the bedding.

The division of the outcrop into measurable units may be rather arbitrary. The level of detail of the work will vary with the nature and scale of the questions being asked about the rocks and with the experience of the geologist, but an effort should always be made to apply consistent criteria and to sample evenly (Fig. 2.5). Develop systematic working procedures based on the stratigraphic principles already set down. Look laterally along any bed or layer to see what happens to continuity and thickness; look first beneath, then within, and then on top of the unit; investigate vertically beginning with the oldest beds. Methodical habits of measuring and recording generally enhance powers of observation and make pattern recognition easier; the search for patterns within field data then becomes second nature. Many of these points will be expanded and illustrated later.

In practice, recognition of bedding and the definition of upper and lower bounding planes may be difficult. Bedding has to be distinguished from cleavage, from joints and faults, and from colour banding due to diagenesis and weathering. Changes in composition or grain size are the best guide to identifying bedding and these may often be more apparent on weathered surfaces. Differential weathering and erosion may accentuate differences that are virtually invisible in fresh rocks. In other cases, deep weathering may obscure depositional structures. A few prominent subparallel splitting planes, or the suggestion of bedding planes in the shapes of an outcrop, may furnish the first clue to the orientation of bedding. Changes in colour, mineral composition, texture (grain size, grain-size variation, grain shape, porosity, packing, degree of cementation, hardness), internal structure (lamination, bedding) and orientation may serve to confirm or deny such an initial conjecture (Figs 2.1, 2, 5 & 6).

The practices recommended in Figure 2.5 relate to a method of working that is essentially one-dimensional: the logging of a vertical succession. It is often helpful, however, to photograph a group of beds in two dimensions from a series of positions equidistant from and perpendicular to the outcrop. In this way a mosaic of overlapping photographs can be assembled and the

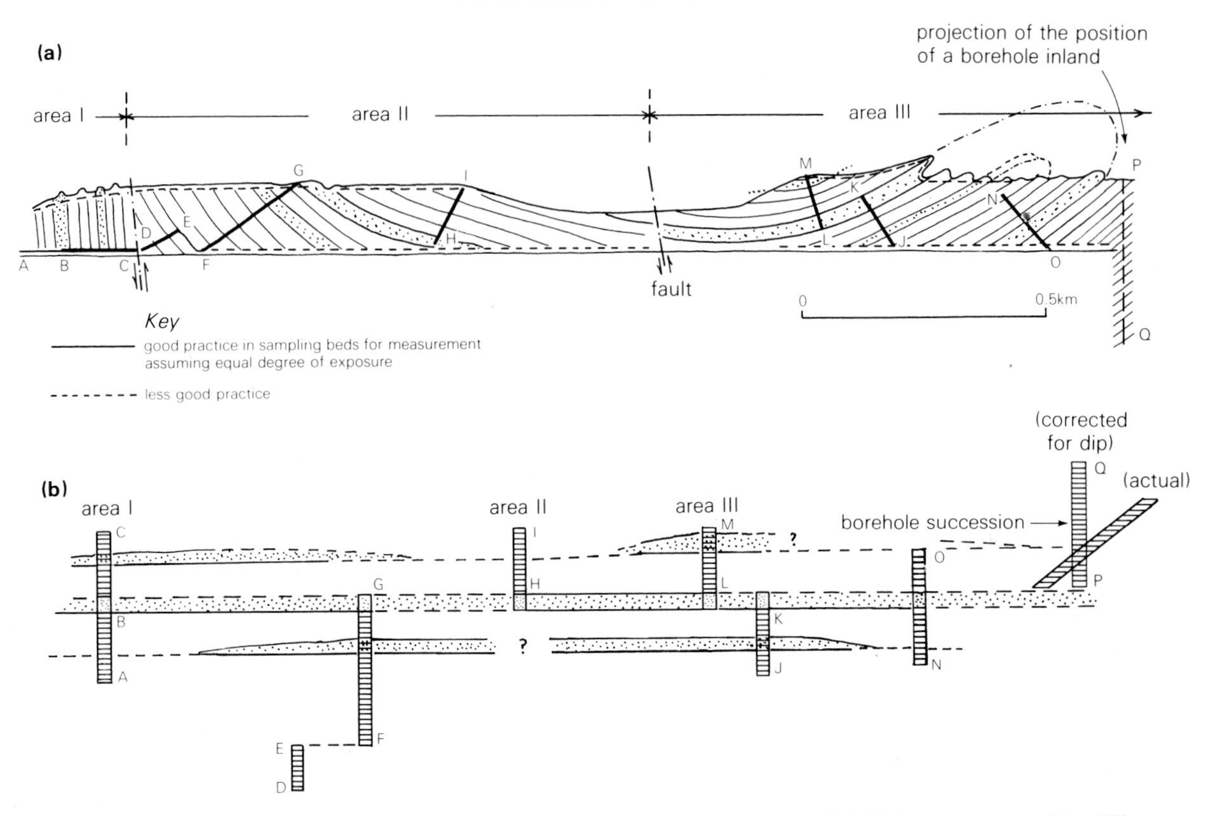

Figure 2.5 (a) Illustration of the way in which one might tackle measurement of folded strata exposed in cliffs or cuttings. (b) Restoration of the folded succession to pre-folding disposition. Notice how the absence of certain units in particular areas must be accounted for by lateral facies changes.

more important bedding structures may be drawn from the panorama. Subsequent checking against the outcrop helps to identify errors and focuses attention on critical details. This can be particularly useful where local lateral variability is apparent and where a one-dimensional approach is inadequate to fully record the sequence. Where the exposure has promontories and recesses or is covered by vegetation, it is important to draw scaled diagrams which generalise the geometry of the outcrop.

Thickness of sedimentary units. Measurement of thicknesses of beds, using Standard International Units, should take place near to the vertical line selected for sampling the succession. This line may be determined by the outcrop, as in a stream section, although in other outcrops the decision may be either selective or simply arbitrary. Limits of units will normally be clear where bounding planes are sharp, but subjective decisions must be made where contacts are gradational.

It is sometimes helpful to work on bed-thickness data statistically in the laboratory. Several workers have found a logarithmic normal distribution of bed thickness in vertical sequences, and practically all bed-thickness frequency distributions are biased towards the thinner beds. Means, modes and measures of variability of bed thickness can be calculated. Studies of the lateral variation of grain size and bed thickness are valuable and the systematic gathering and plotting of data in the manner of Figure 2.5 should permit this.

Lateral variations. The geometry of a bed may be determined by tracing it laterally, thereby testing the principle of original continuity. If beds terminate, they do so in one of four ways:

(a) by convergence and merging of their upper bedding surface and sole;
(b) by lateral gradation of the composition of the bed so that the bounding surfaces die away;
(c) by meeting a cross-cutting feature such as a channel;
(d) by meeting a cross-cutting feature such as a fault or an unconformity. Some types of unconformity,

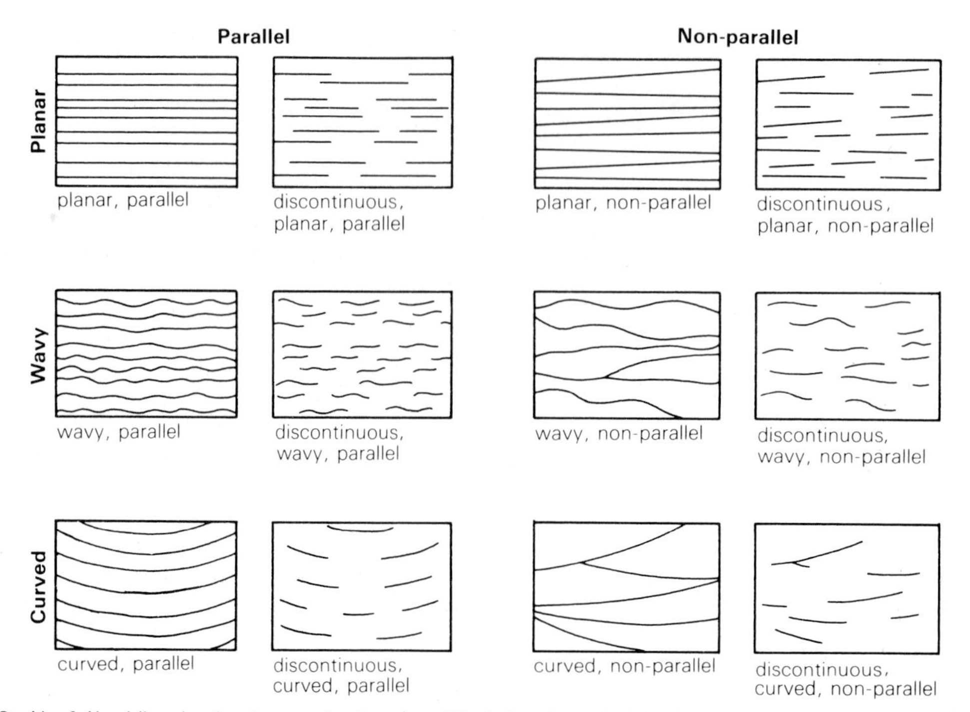

Figure 2.6 Useful bedding–lamination terminology (modified after Campbell 1967 and Reineck & Singh 1973).

which represent a considerable time gap, are not easy to distinguish from a syndepositional channel structure in a single outcrop.

Lateral inspection will sometimes reveal that in case (a), and partly in case (b), beds lap onto and drape previous structures (for example an organic mound or bioherm), although the angle of drape may be accentuated by postdepositional compaction. The upper bounding surface and sole may be parallel or non-parallel, continuous or discontinuous, and they can both be planar, wavy or curved (Fig. 2.6).

Features within the bed or bedset. Vertical variations within each bed are often due to changing composition, texture or internal structure. Beds and bedsets may be: (a) homogeneous or heterogeneous; (b) rhythmic; and (c) gradational. Their lithology may vary from homogeneous (e.g. well-sorted sandstone or siltstone) to heterogeneous (e.g. silty-mudstone, pebbly sandstones). Homogeneous and apparently structureless beds sometimes reveal unsuspected internal structure if special techniques (e.g. X-radiography) are applied to slabbed specimens in the laboratory. Some beds are heterogeneous due to sorting into layers, often showing repeated interlamination of material of contrasting composition or grain size, e.g. silt and sand, silt and mud (Fig. 2.2). Systematic variation of composition and grain size together from, say, sand to silt with interlamination of mud in the upper parts of the bed, is common in some beds deposited by episodic, decelerating currents.

The nature of the bounding surfaces (bed contacts). In recording a succession of beds, special attention should be paid to the nature of the **bounding surfaces**, also known as the **bed contacts**. Bed contacts can be gradational or sharp as illustrated in Figure 2.2. They may be marked by subtle or abrupt changes of composition and colour, texture and structure. Sharp changes may be non-erosional or erosional, the latter being marked by cross-cutting relationships, as at the base of a channel, or where there are downward projections on the sole of the overlying bed.

Attitude of beds. Sometimes bedding relationships at outcrop and on the scale of seismic sections are discordant. General terms **offlap**, **onlap** and **downlap** can be used to describe different types of relationships (Fig. 2.7) which can result from progradation, migration and infill at a wide range of scales.

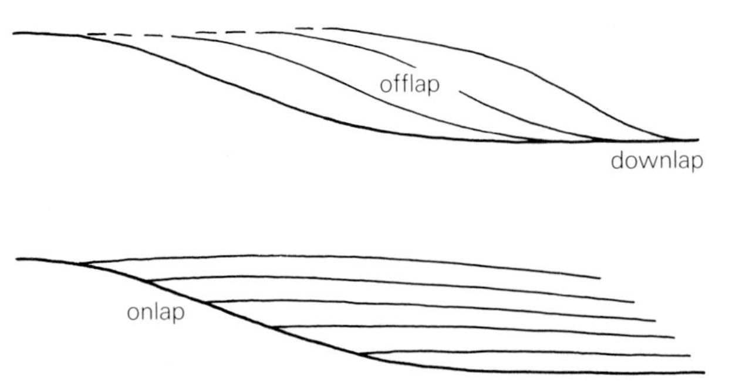

Figure 2.7 The terminology of discordant bedding relationships. These terms are used over a wide range of scales from that of a small outcrop to that of a reflection seismic section.

2.1.7 *Sequences and patterns: the classification of bedding*

Patterns of sedimentary units in groups of beds and bedsets may be discerned where thickness, grain size, composition or structure varies systematically (e.g. Fig. 2.2). The following kinds of pattern of vertical change are commonly recorded:

(a) Grain size of units fining upwards from coarse sandstones to siltstones and mudstones (as in some meandering-river sequences).
(b) Grain size of units coarsening upwards from shale to coarse sandstone (as in some deltaic sequences).
(c) Patterns of units such as shale – sandstone – coal –limestone – shale – sandstone, etc. (as in other kinds of deltaic sequences).
(d) Patterns involving units of carbonate–sulphate, hydrous and anhydrous sulphate, sulphate and halides (as in chemically precipitated evaporites).
(e) Patterns of bed thickness change (as in thickening – thinning – upward sequences).

A way of understanding and learning all this terminology for thickness, shape and types of beds is to practise making oral descriptions in front of others, preferably in the field, but also in the laboratory where much can be learned from writing descriptions of photographs (e.g. Fig. 2.9). Much will be gained if measurements, diagrams and photographs are brought back to the laboratory, and analysed, synthesised and plotted out as graphic logs (see Ch. 10). Laboratory examinations of rocks and photographs will lead to refinements of field descriptions and to further classification and analysis.

2.2 The significance of bedding

2.2.1 *Introduction*

Under present-day conditions an individual bed is observed to be deposited under either essentially constant physical and chemical conditions by uniform processes or by systematic changes of process. Bed contacts, true bedding planes and bounding planes represent changes in conditions to non-deposition, erosion or completely different conditions. Some contacts, however, record gradational changes. Laminae are produced as a result of minor fluctuations in the constant conditions over the same or a smaller area than the bed. Simple bedsets or cosets are the result of repetitions of genetically related variations in conditions. Composite bedsets represent repeated alternations of two sets of conditions. The preservational potential of beds and bedsets, i.e. their chance of becoming part of the geological record, is not necessarily related to the length of time over which a set of conditions operates in order to generate them. Conditions of net subsidence must prevail and erosional activity in the area must be relatively subordinate for accumulation to occur. Periods of erosion may be sufficiently long for major time gaps (represented by unconformities) to be apparent in the sequence, some of which are caused by tectonic events.

2.2.2 *Interpretation of the basic processes of sedimentation*

Several types of process contribute towards establishing the characteristics of a bed. These are: physical, chemical, biological and diagenetic.

Physical processes. Most sedimentary rocks result from deposition of material transported as individual grains in suspension or near the bed (i.e. in traction) by water flows of low sediment concentration, although some are derived from denser flows or, at the extreme, from mudflows. The nature of the transporting process and its intensity are controlled by physical parameters such as grain size and current strength, wave intensity and viscosity of the transporting medium and, in some cases, the flow depth. A change in any of these parameters might result in a change in the nature of the deposit and in the initiation of a new bed.

Where explosive volcanic processes (Ch. 8) operate and relief in the area is considerable, the gross shape of beds or groups of beds will reflect different modes of activity (Fig. 2.9). Such deposits are characterised by great lateral variability and discordant contacts.

Figure 2.8 Write a description of the bedding depicted in the photographs using the terminologies set down in Figures 2.2, 2.3, 2.6 and 2.7.

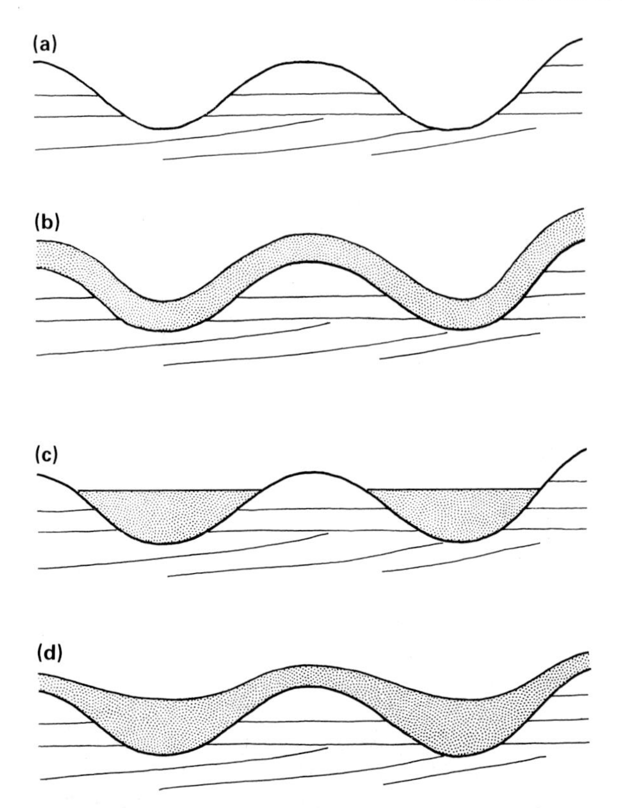

Figure 2.9 Variations in gross bed thickness produced by different types of volcanic process. (a) Pre-eruptive rock succession and topography. (b) Pyroclastic airfall deposits mantle topography. (c) Pyroclastic flow deposit infills the lower parts of the topography. (d) Pyroclastic surge deposits thicken into topographic lows (after Wright *et al.* 1980).

Chemical processes. Much material, particularly in sea water and in certain lakes, is carried in solution. In favourable conditions brought about by changes in temperature, pressure of carbon dioxide, or concentration of ions, these solutes may be precipitated as minerals, either directly on the floor of the basin or as loose particles (crystals) in suspension. Such precipitates are susceptible to reworking by physical processes.

Biological processes. The greater part of the calcium carbonate, which composes limestones and present-day carbonate sediments, results from the activities of organisms (both animal and plant) which precipitate calcium carbonate as part of their metabolic processes. Other organisms secrete silica or phosphate which may also contribute to sediments. Changes in the dominant organisms may produce changes in characteristics of sediment and may lead to the generation of beds. A bed may be formed from the hard parts of organisms more or less remaining *in situ,* but the skeletal material is most commonly redistributed by physical processes. Changes in organic activity commonly reflect changes in physical and/or chemical processes. Organisms, particularly those which burrow within sediment in order to feed, can also be effective in destroying bedding and lamination.

Diagenetic processes. The final appearance of a bed in the stratigraphic column results not only from the conditions of its deposition but also from its subsequent history during burial. The processes of postdepositional change (diagenesis) may vary with, for example, changing pore-water chemistry and, as a result, distinct 'beds' may be due to diagenetic differences rather than to changing processes during deposition.

In reality many beds result from combinations of these various types of process. For example, biological activity may depend on chemical and physical conditions. Similarly, diagenetic processes may vary in different host beds, possibly reflecting original textural differences in the sediment due to changing physical conditions of deposition. Study of bedding, therefore, should aim to interpret the full assemblage of processes which were active in generating successive beds.

2.2.3 *Changes of process and bed generation*

When we see a succession of beds in a vertical sequence and recognise that each bed records a change of process, we have the starting point for understanding something about the depositional environment. Before we are able to use the vertical succession, however, we must try to establish just how the vertical changes came about. Essentially there are two main processes of change which commonly act together but generally operate on different scales.

Changes due to lateral migration. When we look at environments of sediment accumulation at the present day we commonly observe that different processes operate in different parts of the environment and give rise to different deposits. If we were to make maps of the distribution of these deposits over a long enough period of time, it might be possible to detect changes. An area that once accumulated sand, for example, might later be a site of mud deposition and, if net sediment accumulation prevails, we might expect to find mud overlying sand in a trench dug at the site. In other words lateral migration of sub-environments under stable conditions can create changes (i.e. beds) in the vertical sequence. This idea which is Walther's principle of the succession of facies (see Section 1.3) is developed further in Chapter 10.

Changes due to temporal fluctuations. In contrast to the steady-state, lateral migration mechanism, many beds reflect changes due to external effects. On a lake floor, for example, coarser silts and sands may reflect periods of high river discharge, and muds may be deposited during quieter periods. With time this results in the interlamination of sand, silt and mud. Similarly, on a shallow sea floor, grain sizes and sedimentary structures may fluctuate and thus reflect periods of stormy and calmer weather. In both these examples the environment is not changing, but rather the vertical sequence results from the natural variation of process within that environment.

In some settings it is reasonable to distinguish *normal* deposits from *catastrophic* deposits. The classical example of such a setting is a deep-sea continental margin where, under normal conditions, fine-grained sediment accumulates from suspension. Occasionally, however, this process is punctuated by a catastrophic density current generated on the continental slope or shelf edge. This event deposits a sheet of sand in a matter of hours but may not recur for hundreds or even thousands of years. In the resultant sequence, however, the sand beds may be more abundant than the normal interbedded muds. Short-lived, catastrophic events can therefore make a contribution to the rock record out of all proportion to their duration.

2.2.4 The significance of sharp and gradational changes between beds: the importance of bedding planes

It is implicit in much of the previous section that some junctions between beds will be gradational whilst others will be sharp. Indeed the recording of boundaries between beds is at least as important as the recording of the features of the beds themselves, if we are to use our observations for environmental reconstruction. A gradational boundary may well suggest that a lateral migration of processes was taking place under steady-state conditions. Sharp junctions between contrasting beds might lead us to suggest catastrophic events. Where large-scale relief due to channelling is observed, this may mark the beginning of a radically different pattern of sedimentation.

Finally, it is worth thinking of bedding in relation to time. It is possible to recognise sedimentary deposits that must have taken a short time to accumulate: vertical tree stumps covered and infilled by muds after they had rotted but before they had fallen; dinosaurs covered by sediment before they had decomposed or oxidised; beds deposited from floods, dust storms and ash clouds. These beds often form part of thick, uniform sequences which, when dated, appear to have accumulated over considerable periods of geological time, often millions of years. This paradox of relatively swiftly deposited beds, making up sequences representing many millions of years, has led some geologists to coin somewhat intuitive statements such as '98% of geological time must be represented by the bedding planes', thus highlighting the importance of trying to recognise significant time gaps in seemingly continuous sedimentary sequences. The nature and distribution of certain trace fossils and some early diagenetic concretions (palaeosols and hardgrounds) (see Ch. 9) may sometimes enable us to do this with confidence.

2.2.5 Unconformities

An **unconformity** is a break in a stratigraphic sequence resulting from a change in conditions that caused deposition to cease for a considerable time. Various types of unconformity have been identified (Fig. 2.10), and they are easiest to recognise and define where sedimentary rocks succeed igneous or metamorphic rocks (**non-conformity**) or where they are laid upon previously folded strata (**angular unconformity**). Sometimes, however, angular unconformity may be suspected, albeit falsely, where flat bedding overlies very large-scale cross bedding, but checking the conformable attitude of the beds underlying the cross-bed set will soon settle the issue. Time gaps are less easily defined, indeed are often not recognised, without fossils. They may be inferred where a local succession appears conformable, or where local successions contain two sets of conformable beds separated by mere changes of sediment type (**disconformity**) or even where the sediment types are virtually the same (**paraconformity**). It must be emphasised that disconformity and paraconformity can be satisfactorily recognised only where there is proven absence of parts of a fossil sequence. Many successions, in which preservation potential of fossils is not high, as in red beds, lack the means to be analysed in this way. The largest erosional structures, for example fossil channels cut by rivers into alluvial fans or floodplains, or fossil tidal-channels cut into mudflats, can easily be confused with true disconformity. In the absence of fossils, a detailed sedimentological analysis may be necessary before the time significance of any gap becomes apparent. Hence seemingly normal successions may

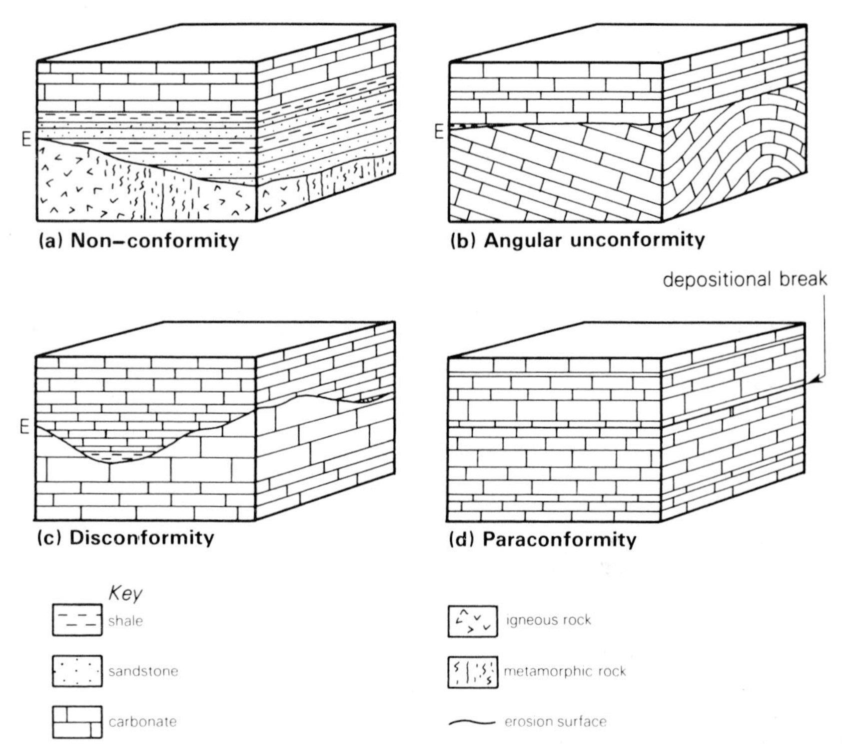

Figure 2.10 Four types of unconformity: (a) non-conformity, (b) angular unconformity, (c) disconformity and (d) paraconformity (after Dunbar & Rogers 1957). Note that the terminology of Figure 2.7 can be applied to beds above and below some unconformities.

contain many time-gaps. Students are therefore encouraged to observe and record sedimentary features and relationships above and below an 'unconformity' as carefully as possible and to generate and discuss as many working hypotheses as possible.

Short breaks in the record have been called **diastems**; they are time gaps so short that they are not reflected in evolutionary changes in the fossils associated with the sediments. It is useful to regard potential disconformities and paraconformities as diastems until proved otherwise. Bedding planes showing erosional contacts between beds may be the sites of major diastems, though it is frequently impossible to gauge how much of the sequence has been removed or the length of the time gap except in favourable circumstances.

2.2.6 *Preservation potential of bedding*

The preservation of a group of beds is due to the products of a set of depositional events escaping reworking or partial removal and being captured by net subsidence. This is a chance occurrence in many environments, particularly where powerful agents of erosion operate, for example in shallow marine environments, in floodplains where rivers are constantly shifting their channels, or where strong winds are eroding down to a water table. Rapid subsidence, coupled with the diminution of the agents of reworking and erosion, greatly enhance the chances of preservation.

Reference and study list

References marked with an asterisk * are suitable for 16–19 year old advanced level students in schools and colleges in the UK as well as for undergraduates.

Field experience

EXCURSIONS
Virtually any geological excursion early in a course will involve the observation and recognition of beds and bedding planes and the measurement of their thickness and their attitude in three-dimensional space.

FIELD GUIDES
The authors of guides rarely suggest what to do about studying bedding and students may have to rely on general advice in books such as Bates and Kirkaldy, Compton, Lahee, Moseley and Pugh, and in pamphlets such as that by Ensom. More likely, students will have to depend heavily on being taught these things by their tutors. Advanced level students in Britain may have recourse to the guidance and recommendations provided by examination boards.

No particular texts are recommended, but a list of books and pamphlets with elementary advice on the early stages of fieldwork is given below:

Barnes, J. W. 1981. *Basic geological mapping*. Milton Keynes: Open University Press, for Geological Society of London.

*Bates, P. E. D. and J. F. Kirkaldy 1976. *Field geology in colour*. London: Batsford.

Compton, R. R. 1961. *Manual of field geology*. New York: Wiley.

*Ensom, P. C. 1979. *The geological field notebook*. Dorchester: Dorset County Museum.

*Himus, G. N. and G. S. Sweeting 1972. *Elements of field geology*, 3rd edn. London: University Tutorial Press.

Lahee, F. 1961. *Field geology*, 6th edn. New York: McGraw-Hill.

Moseley, F. 1981. *Methods in field geology*. Oxford: W. H. Freeman.

*Open University 1975. 2nd level earth science summer school, geology programme. Milton Keynes: Open University. [Graphic logging system displayed.]

Pugh, J. C. 1975. *Surveying for field scientists*. London: Methuen.

Thompson, D. B. 1975. Types of geological fieldwork in relation to the objectives of teaching science. *Geology* **6**, 52–61.

Tucker, M. E. 1982. *The field description of sedimentary rocks*. Geological Society of London Handbook. Milton Keynes: Open University Press.

Laboratory experience

McKee, E. D. 1957. Flume experiments on the production of stratification and cross-stratification. *J. Sed. Petrol.* **27**, 129–34.

Open University 1980. Film: *Borehole logging*. Colour sound. Milton Keynes: Open University Enterprises.

Essential and further reading

A selection from:

*Allen, J. R. L. 1975. *Physical geology*. London: Allen & Unwin.

Blatt, H., G. V. Middleton and R. C. Murray 1980. *Origin of sedimentary rocks*, 2nd edn. Englewood Cliffs, NJ: Prentice-Hall. [Pages 5–6 give a brief introduction to the description of outcrop sections and the sampling and numerical analysis of data; see pp. 132–5 for a discussion of bedding.]

Broadhurst, F. M. and I. M. Simpson 1967. Sedimentary infilling of fossils and cavities in limestone at Treak Cliff, Derbyshire. *Geol Mag.* **104**, 443–8.

Campbell, C. V. 1967. Laminae, lamina set, bed and bedset. *Sedimentology* **8**, 7–26.

Donovan, D. T. 1966. *Stratigraphy: An introduction to the principles*. New York: Halsted. [Chs 1–3 deal with the principles of establishing the successions; Ch. 6 deals with breaks in the succession.]

*Dunbar, C. O. and J. Rodgers 1957. *Principles of stratigraphy*. New York: Wiley. [Part II is concerned with stratification (pp. 97–134); Part IV with local sections (pp. 257–70) and correlation (pp. 271–88).]

Fairbridge, R. W. and J. Bourgeois (eds) 1978. *Encyclopaedia of sedimentology*. Stroudsburg, Pa: Dowden, Hutchinson & Ross. [A section on bedding genesis by A. V. Jopling (pp. 47–55) is very helpful.]

Friedman, G. M. and J. E. Sanders 1978. *Principles of sedimentology*. New York: Wiley. [Pages 401–14 deal with field study of strata; pp. 427–9 are concerned with establishing lateral relationships.]

Ingram, R. L. 1954. Terminology for the thickness of stratification and parting units in sedimentary rocks. *Bull. Geol Soc. Am.* **65**, 937–8.

McKee, E. D. and G. W. Weir 1953. Terminology for stratification and cross-stratification in sedimentary rocks. *Bull. Geol Soc. Am.* **64**, 381–90.

Pettijohn, F. J. 1975. *Sedimentary rocks*, 3rd edn. New York: Harper & Row.

Reineck, H. E. and I. B. Singh 1980. *Depositional sedimentary environments*, 2nd edn Berlin: Springer-Verlag. [Especially pp. 82–113.]

Shrock, R. R. 1948. *Sequence in layered rocks*. New York: McGraw-Hill. [A great deal about the characteristics of beds and bedding is introduced incidentally while discussing way-up phenomena.]

Simpson, J. 1985. Stylolite-controlled layering in a homogeneous limestone: pseudo-bedding produced by burial diagenesis. *Sedimentology* **32**, 495–505.

Extra references relating to the figure and table captions

Griffiths, J. C. 1961. Measurements of the properties of sediments. *J. Geol.* **69**, 487–98.

Pettijohn, F. J. and P. E. Potter 1964. *Atlas and glossary of primary sedimentary structures*. New York: Springer-Verlag.

Pettijohn, F. J., P. E. Potter and R. Siever 1972. *Sand and sandstone*. New York; Springer-Verlag.

Wolfenden, E. B. 1958. Palaeoecology of the Carboniferous reef complex and shelf limestones in north-west Derbyshire, England. *Bull. Geol Soc. Am.* **69**, 871–98.

Wright, J. V., A. L. Smith and S. Self 1980. A working terminology of pyroclastic deposits. *J. Volcan. Geotherm. Res.* **8**, 315–36.

3 Basic properties of fluids, flows and sediment

3.1 Introduction

In order to understand the processes that produce many of the sedimentary structures observed in the geological record, we need to know something about the physical properties and mechanics of the fluids that erode, transport and deposit the sediment. Most of these processes result directly from movement of a fluid, commonly water, but also air and ice. Exceptions are sediments emplaced by the direct action of gravity on loose particles, usually on a slope. During gravity emplacement, fluids may be important as lubricants or supporters of the moving grains, and the moving mass of grains may itself show behaviour similar to that of a fluid.

We also need to know something of the physical properties of sedimentary particles themselves, both as individuals and as populations. The variation of size, shape and density found in natural sedimentary particles clearly influences their response to the flows that erode, move and deposit them.

In this chapter, therefore, we examine some of the properties of fluids and see how these influence the way in which the fluids move. We also consider the physical properties of sediments and show how particles and fluids interact during certain sedimentary processes.

This chapter may seem rather theoretical, but it mainly describes phenomena of common experience. Many of the features can be illustrated by simple experiment and by experience of everyday events. Try wherever possible to develop a feel for the physical reality of the various processes described. We indicate where we think experiments and observations of this type are helpful, but with a little imagination it may be possible to model features of fluids and flows other than those we suggest.

3.2 Properties of low-viscosity fluids and flows

3.2.1 Basic properties of fluids

The three simple fluids that account for virtually all sediment movement on the surface of the Earth are water, air and ice. Ice is only considered as a fluid when its behaviour is observed on a long timescale. In addition, mixtures of sediment and water, such as slurries and mudflows, can be considered to behave as fluids.

Consider initially only water and air. It is immediately obvious that these two media differ very significantly in certain physical properties. The two properties that most strongly influence the flow of these media and the way in which they react with sedimentary particles are **density** and **viscosity**.

The fluid density (ρ_f) determines the magnitude of forces such as stress which act within the fluid and on the bed, particularly when the fluid moves down a slope under gravity. Density also determines the way in which waves are propagated through the fluid and it controls the buoyant forces acting on sedimentary particles immersed in the fluid and determines their effective density ($\rho_s - \rho_f$) (where ρ_s is the density of the solid particle). For example, quartz grains in water have an effective density of 1.65 gm cm^{-3} compared with 2.65 in air, a difference which strongly influences the ability of the different media to move the grains.

The viscosity (μ) describes the ability of the fluid to flow. It is defined as the ratio of the shear stress (τ, shearing force/unit area) to the rate of deformation (du/dy) sustained by that shear across the fluid:

$$\mu = \frac{\tau}{du/dy} \qquad (3.1)$$

For any fluid, the viscosity is not constant but varies with temperature (cf. for example, hot and cold syrup).

We can visualise flow by means of a simple model where a fluid is trapped between two parallel plates moving relative to one another. The fluid is envisaged as a stack of sheets parallel to the plates. These sheets move relative to one another at a uniform rate so that a straight line drawn perpendicular to the plates will then deform into a straight, inclined line, leaning in the direction of shear (Fig. 3.1). The viscosity reflects the force needed to produce a particular rate of deformation or sliding of the imaginary sheets. Increased viscosity demands a greater shear stress to produce the same rate of deformation.

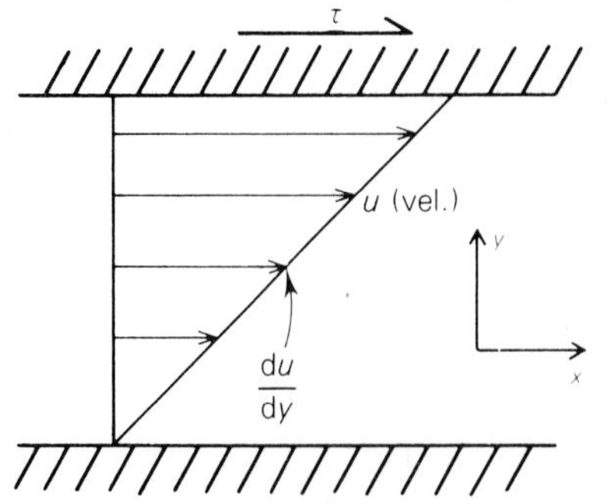

Figure 3.1 Definition diagram for viscosity. Two rigid, parallel plates enclose the fluid. A shear stress, τ, acting parallel to the sheets sets up the steady-state velocity profile shown by the inclined line, where the length of arrow is proportional to velocity, u.

As density and viscosity both play an important role in determining fluid behaviour, it is usual to combine them into a single term, the so-called kinematic viscosity (ν):

$$\nu = \mu/\rho_f \tag{3.2}$$

3.2.2 *Laminar and turbulent flow*

Some of the basic features of flow can be investigated by means of a simple experiment. Inject a thin stream of dye into a very slowly moving unidirectional flow of a viscous fluid, such as glycerine, in a narrow channel and carefully observe the form of the dye downstream of the injection point. Repeat the procedure at progressively increasing flow speeds, or with different fluids of lower viscosity. You will notice that, with low speeds and high viscosity, the dye persists as a fairly coherent and reasonably straight stream, whereas with increased velocity or decreased viscosity the stream breaks down and moves as a series of deforming masses within which there are components of movement perpendicular to the overall flow direction (Fig. 3.2).

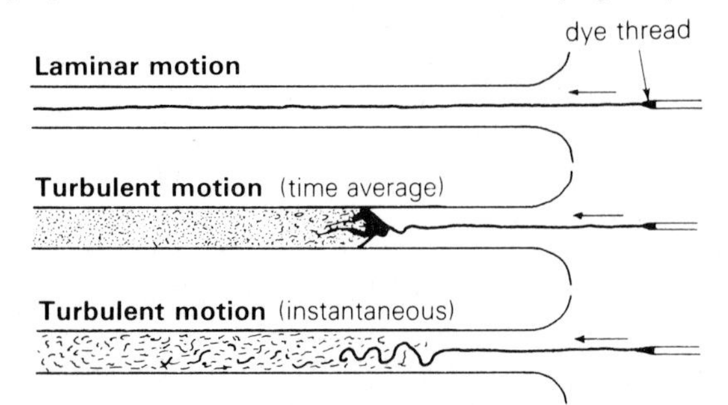

Figure 3.2 The Reynolds experiment to illustrate the difference between laminar and turbulent flow. Dye injected into the flow from a point source behaves in different ways (after Allen 1968).

With low velocity and high viscosity, the flow corresponds to the model outlined in Section 3.2.1 and the flow is said to be **laminar**. With more rapid flow or a lower fluid viscosity, the fluid can no longer be visualised as a series of parallel sheets or filaments but clearly has some form of random motion superimposed upon the unidirectional flow. This secondary motion is the very important phenomenon of **turbulence.**

3.2.3 *Turbulence*

Some appreciation of the nature of turbulence is vital to our understanding of many of the sedimentary structures described later. The tubulence seen in the flow of water in a smooth-sided channel is a random movement of parcels of fluid superimposed upon the overall flow. If we slow down the flow sufficiently or increase the viscosity of the fluid, it is possible to eliminate this random motion and achieve conditions of laminar flow, but in virtually all natural conditions involving air or water turbulent flow is the norm. Velocity measured at a point in a laminar flow is constant through time, whereas velocity at a point in turbulent flow will vary considerably about a 'time-averaged' value. This distinction between the two flow types suggests that it should be possible to use some variable to predict the boundary conditions separating them. The factors that control the level of turbulence are usually combined to derive a **Reynolds number** (Re) for the flow. This is a dimensionless number which expresses the ratio between the inertial forces related to the scale and velocity of the flow – which will tend to promote turbulence – and the viscous forces – which tend to suppress turbulence:

$$Re = \frac{\bar{U}L\rho}{\mu} = \frac{\bar{U}L}{\nu} \tag{3.3}$$

where $\bar{U}$ is the mean velocity of the flow and L is some length which characterises the scale of the flow (e.g. depth of flow). Being dimensionless the Reynolds number is of great value when comparing different examples or designing scaled-down models. The transition from laminar to turbulent flow takes place at a critical value of the Reynolds number which will depend in each case upon the boundaries of the flow (e.g. channel, etc.).

The existence of turbulence has important effects on flow properties. Because the movement of eddies in a turbulent flow absorbs energy, a greater shear stress is required to maintain a particular velocity gradient in turbulent flow than in laminar flow. Equation 3.1 has to be modified to account for the turbulence:

$$\tau = (\mu + \eta)\,{}^{d\bar{u}}/_{dy} \tag{3.4}$$

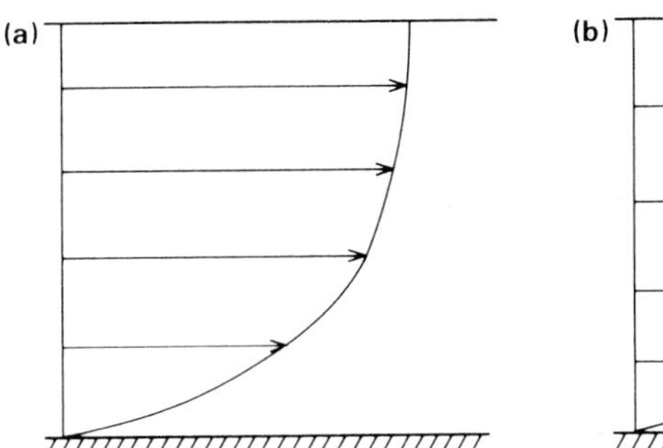

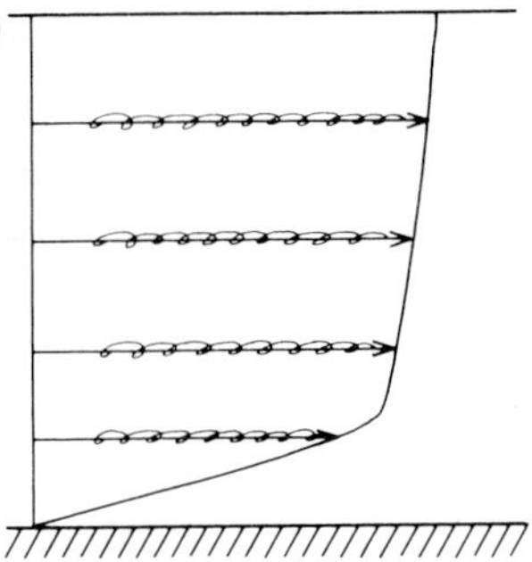

Figure 3.3 Velocity profiles for (a) laminar and (b) turbulent flows in an open channel of great width. In the profile for turbulent flow, the velocities are time-averaged values. Only close to the bed does a near-laminar pattern of movement occur.

where η is the so-called **eddy viscosity**, an additional term which accounts for the extra shear needed to maintain turbulence and $\bar{u}$ is the time-averaged velocity. Eddy viscosity is not, however, a constant for the fluid but depends upon the level of turbulence in the flow; in other words it depends upon the Reynolds number. This makes calculation of shear stresses in a turbulent flow rather complicated.

Another consequence of turbulence is that the velocity profile through a turbulent flow has a different shape from that through a laminar flow. While the profile of laminar flow (Fig. 3.3a) is a realistic representation of the velocities at any instant, the profile for turbulent flow (Fig. 3.3b) is averaged over time to eliminate the fluctuations due to turbulence. For the same reason, the time-averaged velocity $\bar{u}$, rather than instantaneous velocity u, is used in Equation 3.4. Instantaneous values of velocity in turbulent flows have components of direction and magnitude superimposed on the time-averaged velocity. The variations of velocity are commonly of the same order as the time-averaged value itself. This phenomenon accounts for the irregular buffeting experienced in trying to wade a rapidly flowing stream or when standing in a strong wind. One example of turbulence can commonly be seen on the water surface of rivers, particularly during floods: 'boils' can be seen rising to the surface, particularly in subdued light and in rain, when reflections are reduced (Fig. 3.4).

Where currents are strong enough to move sediments, turbulence will be well developed and will influence the way in which the grains move. Turbulence is the crucial mechanism in the transport of sediment in suspension when the upward components of the turbulent motion support the suspended grains (see 3.9.1). This implies that turbulence cannot be

Figure 3.4 (a) 'Boils' on the surface of a rapidly flowing river, showing the upwards movement of turbulent cells to the surface. (b) Small-scale turbulent mixing between clear and turbid water at the junction of two streams. Note the discrete cells of water, which move at a high angle to the flow direction.

entirely random and that there must be a net upward energy flux.

So far we have discussed turbulence in terms of secondary movements distributed throughout the unidirectional flow. However, there are also more localised eddies which are associated with the shape of the boundaries of the flow. Obstructions and irregularities fixed on the margins of flows generate eddies, the shape and organisation of which closely relate to the shape of the obstruction and to the prevailing flow conditions. Sometimes a 'captive' body of fluid rotates in the lee of the obstruction, whereas in other cases a spiral eddy is shed back into the main flow. In such cases, the flow is said to separate from the boundary at a **separation point** or **line** and to reattach itself downstream at a **reattachment point** or **line** (Fig. 3.5).

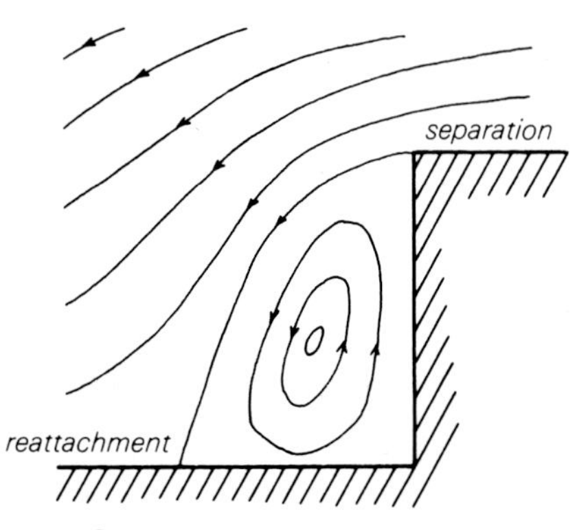

Figure 3.5 Separation and reattachment of flow at a negative step on the perimeter of a flow. A cell of rotating fluid is trapped within the separation eddy or 'bubble' (after Allen 1968).

You can learn much about separated flows and the structure of eddies in water by simple experiments in laboratory channels or in small, natural, clear-water streams. The pattern of water movement can be seen from the movement of any small suspended particles, but the best method for visualisation is to inject dye into the flow at selected points through a fine tube. A solution of potassium permanganate serves very well for this purpose. Place obstructions of different shapes and sizes on the bed of the channel and carefully explore the local pattern of water movement. Try to determine the points or lines where flow separates from and reattaches to the bed. See if you can determine the volumes and shapes of the eddies and describe the patterns of water movement within them (cf. with Fig. 3.6). If experimental facilities are not available, you can still learn much from carefully watching the patterns of water movement around bridge piers or large boulders in rivers or the movement of smoke or dry leaves on windy days. You need to develop a feel for the three-dimensional shape and organisation of eddies in relation to the obstacles that create them in order to understand erosional structures and depositional bedforms seen on present-day surfaces and in rocks at outcrop.

3.2.4 *Bed roughness*

Obstacles on the boundary of a flow generate eddies that influence the general level of turbulence. The larger and more abundant the obstacles, the more turbulence is generated and the more energy is absorbed, causing a retardation of the flow. This introduces the idea of **bed roughness** which expresses the frictional effect which the boundary of the flow, for example a river bed, has on the flow. Roughness is made up of two components when the boundary consists of loose moveable grains. The grains themselves constitute one component (**grain roughness**) and their frictional effect relates to grain size. If the sediment is poorly sorted, however, large grains may be enveloped in finer material and their frictional effect reduced. The grain relief is the critical factor. The second component is the bedforms into which the sediment may be moulded by the flow (**form roughness**) (Fig. 3.7). These bedforms depend very much upon the conditions of flow which, in turn, depend to some extent on the bed roughness. The equilibria established between the bedforms and the flow are, therefore, highly sensitive (see Ch. 6).

When the relief at the boundary of a flow is very small, the roughness elements do not generate eddies and the bed may be described as **smooth**. The critical relief for this condition to apply is determined by the nature of the prevailing flow. Directly above the smooth bed and beneath the fully turbulent flow there is a thin layer within which the flow is much less turbulent, the so-called **viscous sub-layer**, whose thickness depends upon the depth, velocity and viscosity of the total flow. If the bed relief exceeds this thickness, no sub-layer can exist. The sub-layer is therefore only important with fine bed material. Recently it has been shown that the viscous sub-layer does not exhibit laminar flow, as had been thought previously, but that it is characterised by **streaks** of faster- and slower-moving fluid, aligned parallel to the flow. These periodically 'burst' into the overriding turbulent flow. Streaks may be important in the initiation of grain movement and the formation of ripples, in the

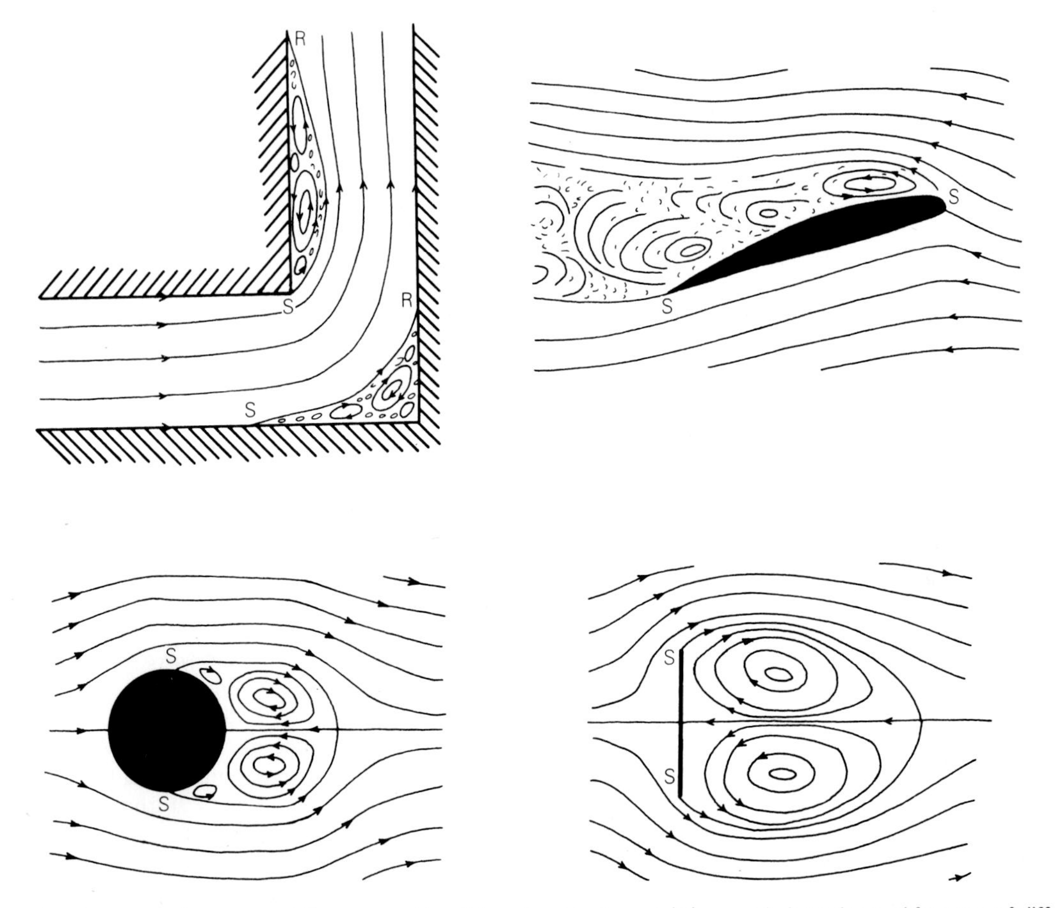

Figure 3.6 Examples of patterns of flow separation (S) and reattachment (R) around obstacles and features of different shapes (after Allen 1968).

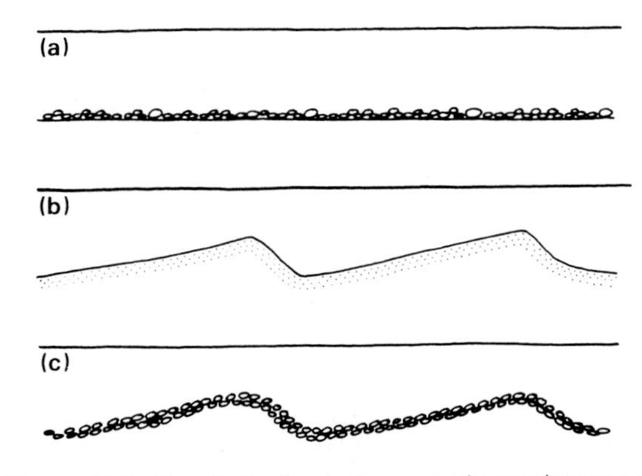

Figure 3.7 The distinction between grain roughness and form roughness: (a) purely grain roughness, (b) purely form roughness over a smooth artificial bedform, and (c) grain and form roughness over a natural bedform.

production of some current lineation and in the development of lamination in fine-grained sediments deposited from suspension (see Chs 5 & 6).

3.2.5 Boundary shear stress

The behaviour of sediment on a bed below a flow is determined largely by the force that the flow is able to exert on the bed. The boundary shear stress (force/unit area parallel to the bed) is a function of depth (h), slope (S) and the nature of the fluid, and indirectly a function of velocity of flow. Calculation of the boundary shear stress (τ_0) is complex. It depends upon the Reynolds number, the frictional characteristics of the bed and the shape of the velocity profile of the flow close to the bed.

A simple approximation of the boundary shear stress for an open channel of large width can be obtained from the idealised situation in Figure 3.8. For calculating the shear exerted by wind, this method is clearly inapplicable, since depth is indeterminate, as it also is in very deep water. Details of the shape of the velocity profile close to the bed are needed to estimate boundary shear stress more accurately.

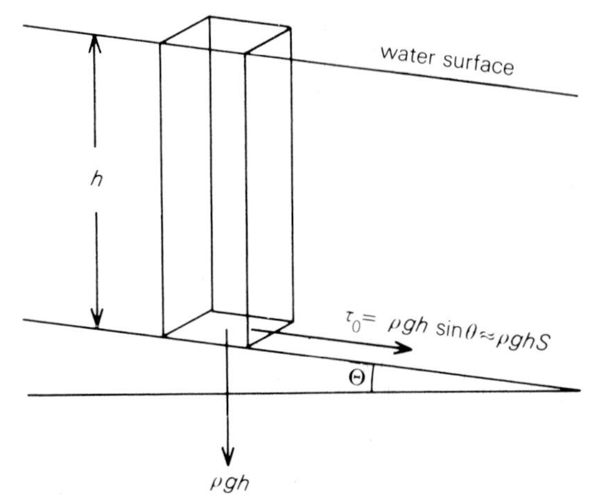

Figure 3.8 Definition diagram for the calculation of bed shear stress for water flowing downslope as an open-channel flow. Forces acting on unit area of the bed are indicated.

3.2.6 The role of gravity: rapid and tranquil flows

In addition to the controls exerted by viscous and inertial forces on the character of a flow through their influence on turbulence, controls by gravitational forces are also important. In particular, gravity, being a body force acting on the fluid as a whole, influences the way in which the fluid transmits surface waves. The speed at which a wave can be propagated in shallow water is given by the equation:

$$c = \sqrt{gh} \tag{3.5}$$

where c is the wave speed (celerity) and h is the water depth.

It is clear that for flowing water, there will be a velocity above which it will no longer be possible to propagate waves in an upstream direction. This critical velocity separates two distinct types of flow, **tranquil flow** and **rapid flow**. The distinction is commonly drawn by reference to another dimensionless number, the Froude number (Fr), given by the ratio of inertial to gravitational forces in the flow:

$$Fr = \bar{U}/\sqrt{gh} \tag{3.6}$$

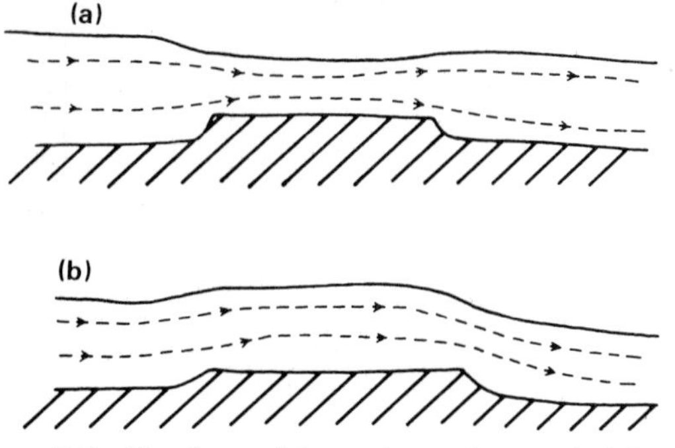

Figure 3.9 The form of the water surface and of time-averaged flow lines over a positive obstruction on the bed under conditions of (a) tranquil and (b) rapid flow.

For $Fr > 1$, we have rapid flow conditions in which it is not possible for waves to be propagated upstream and for $Fr < 1$, we have tranquil flow where this is possible.

The Froude number and this distinction of flow types apply only to liquids. In air, analogies are provided by the Mach number and by subsonic and supersonic velocities, although then the wave motion involved is compressional and not gravitational.

In tranquil flow the water surface is rather irregular as cells of turbulence move freely. In rapid flow the water surface looks more glassy and the flow appears rather 'streaked out' with turbulence somewhat suppressed. When these two types of flow encounter obstacles at their base, they react differently (Fig. 3.9).

Try to recognise which type of flow occurs in small streams, in rivers, in rainwater flow in gutters, or in laboratory channels. Rapid flow will be most likely where the slope is high. It is quite common to see sharp transitions between the two states, when rapid flow

Figure 3.10 A 'hydraulic jump' from shallow rapid flow to tranquil flow conditions. The sudden transition in both depth and velocity is accompanied by an upstream breaking wave.

passes down stream into tranquil flow. The resulting breaking wave or **hydraulic jump**, which commonly moves in an upstream direction, marks the sudden increase in depth and reduction of velocity (Fig. 3.10). A small scale version of this phenomenon can be produced by directing a jet of water vertically downwards on to a flat, smooth horizontal surface. This distinction of flow type is independent of any sediment in the system. It is however related in a general way to the existence of upper and lower **flow regimes** defined by bedforms developed on a sand bed (see Ch. 6).

3.3 Waves

So far we have considered only unidirectional currents in fluids, but water in particular also transfers energy through the movement of waves. Waves at their simplest involve localised vertical and horizontal movement of water without any net displacement of water taking place. Most of us have at some time watched waves at the seashore or thrown stones into ponds and watched the concentric pattern of waves that result. The behaviour of waves in shallow water can be studied by means of a small laboratory wave tank (see end of this chapter). With simple waves, try to measure the length, height, speed and period of the waves, and the water depth. How do these properties relate to one another?

Some basic terminology used in the description of water waves is shown in Figure 3.11. In addition, any wave is characterised by its period T, the time between the movement of successive wave crests past a point, and wavelength L, the distance apart of successive wave crests. From this it follows that:

$$c = \frac{L}{T} \tag{3.7}$$

However, it is important to know how wave speed c (or propagation velocity or celerity), is controlled by other properties of the water body. Wave theory is mathematically complex and for our purposes it is sufficient to present two results. The theory recognises two distinct types of waves which depend on the ratio of wave height to water depth. Shallow-water waves have lengths of at least twenty times the water depth, whereas deep-water waves have lengths of less than four times the water depth. In consequence, there are also intermediate forms. For the deep (d) and shallow (s) extremes, wave speed is derived in the following ways:

$$c_s = \sqrt{gh} = 3.1\,h^{\frac{1}{2}} \tag{3.8}$$

$$c_d = \frac{gT}{2\pi} = 1.55T \tag{3.9}$$

Here c_s and c_d are in metres per second, g (acceleration due to gravity) is in metres per (second)^2, h (the water depth) is measured in metres, and T in seconds. For intermediate wave types, the relationships are rather more complex. The most obvious point about these relationships is that for deep-water waves, the wave speed is independent of depth, indicating that wave behaviour is not influenced by the bed. For shallow waves, water depth is the prime control and the waves can be said to 'feel' the bottom and react to it.

Associated with these differences in speed are differences in the pattern of water movement associated with the passage of a wave. Armed with some beads, some of which just float and others which just sink, it should be possible to visualise the pattern of water movement as a wave passes a point. The beads will have an ellipitical

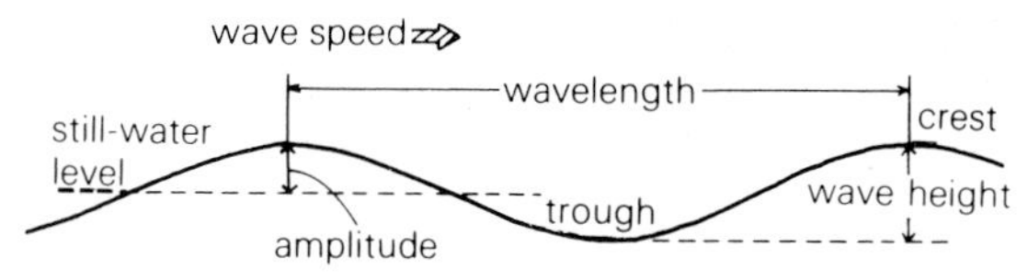

Figure 3.11 Definition diagram for the main physical properties of simple water surface waves.

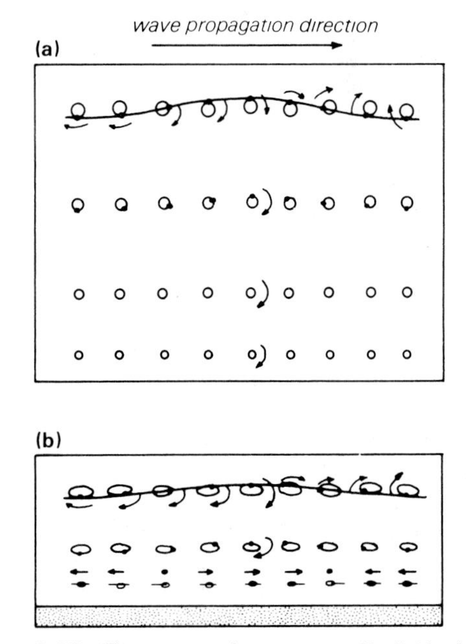

Figure 3.12 The pattern of movement of individual water particles associated with the passage of a surface wave for (a) deep water waves and (b) shallow water waves.

pattern of movement near the water surface, but close to the bed the sinking beads may show a more or less linear pattern of to-and-fro movement. This so-called 'orbital' motion characterises all the molecules of water, but the form of the orbit changes with depth and position in the water column. With deep-water waves, circular orbitals decrease in diameter with depth until movement dies out (Fig. 3.12). With shallow water waves, the orbitals change their shape from nearly circular at the surface through elliptical to a linear (forwards and backwards) movement close to the bed. Such oscillatory motion is important in the movement of sediment and in the development of wave ripples (see 6.1.5).

As waves approach the shore, they become steeper and eventually become unstable and break. The style of breaking varies with the steepness of the beach. With steep beaches, waves **plunge** strongly close to the shore giving a short run-up (**swash**) and strong **backwash**. With more gentle slopes, breaking occurs further offshore and waves move as **surges**, often over long distances, and the strength of the backwash is less.

The view of wave motion presented above is a simplification of what happens in most natural settings. Several groups of waves of different length, and even different direction, commonly coexist and resolve into interference patterns in both the water movement and the sediment response.

3.4 Properties of sediments moved by flows

Grains that are moved by flows have their own physical properties and these influence the way in which they respond to flows. The most important are size, shape and density. With respect to the grain size only a few points need emphasising. If all sedimentary particles were spheres, cubes or some simple geometrical shape, then a dimension such as diameter or side length would be appropriate. However, natural sedimentary particles are much more irregular. Grain size is usually measured by sieving through meshes of known spacing and by using a series of graduated sieves it is possible to derive, for any sediment, the percentage of grains (by weight) falling between any two sieve sizes. Sieving, in effect, measures the smallest cross-sectional area of a particle and reflects its intermediate axis. What we would really like is some measure of size which reflects the behaviour of the particle in fluid flows, and there is no reason why least cross-sectional area or intermediate axis should provide this. These two parameters can be identical for particles showing a whole range of shapes.

A simple experiment shows how shape influences the behaviour of a grain in a fluid. Take two identical pieces of paper, screw up one of them into a ball and allow both to fall to the ground through the air. The unfolded one falls more slowly and with a pronounced side to side motion while the ball falls more or less directly, although both 'particles' are of the same volume and mass. More realistic and controlled experiments along the same lines should also be made using real sedimentary particles. Try to obtain three or four glass cylinders (1000 ml measuring cylinders are ideal). The cylinders should be filled with water at different temperatures, with glycerine or left empty (i.e. full of air). Using a selection of sands of varying size, shape and mineralogy, see how the speed with which grains fall through these various fluids is influenced by grain diameter, grain density, grain shape and fluid viscosity.

The behaviour of any particle will be controlled by some combination of these variables, but at present it is not possible to combine measurable physical parameters in such a way that they describe the hydrodynamic behaviour of a particle. Instead it is more productive and less time-consuming to investigate the hydrodynamic behaviour directly and to measure some value which reflects the combination of the physical parameters. The usual hydrodynamic parameter is the **fall velocity** of the particle, that is the constant velocity with which it falls through a column of water at a fixed temperature (i.e. fixed viscosity), after an initial phase of acceleration.

We can illustrate the nature of this equilibrium by reference to a small spherical particle falling through a column of still water (Fig. 3.13). At the fall velocity v_0, the constant speed at which the sphere falls after its initial acceleration, the gravitational forces acting downwards on the sphere are balanced by the viscous drag which the fluid exerts on it, i.e.

$$\frac{4}{3}\pi\left(\frac{d}{2}\right)^3 g(\rho_s-\rho_l)=C_d\,\pi\left(\frac{d}{2}\right)^2\rho_l\frac{V_0^2}{2} \qquad (3.10)$$

and so

$$V_0^2=\frac{4gd}{3C_d}\left(\frac{\rho_s-\rho_l}{\rho_l}\right) \qquad (3.11)$$

where C_d is a drag coefficient which depends upon particle shape and particle Reynolds number Re $(=V_0d/v)$

where d is particle diameter). For low particle concentrations and low Reynolds numbers:

$$C_d = \frac{24}{Re} = \frac{24}{V_0 d/\nu} \quad (3.12)$$

Thus we can write

$$V_0 = \frac{1}{18}\frac{(\rho_s - \rho_l)\, gd^2}{\mu} \quad (3.13)$$

which is the Stokes Law of settling. In other words, the fall velocity is proportional to the square of the grain diameter, and is therefore a reflection of grain size.

The Stokes Law of settling only applies, however, for particles of small grain size (low Reynolds number) where laminar flow around the particle can be assumed. For larger particles, turbulence is generated and the equations must be modified. In general terms, the velocity is proportional to reducing powers of grain diameter with increasing turbulence. Also Equations 3.10–13 only apply to isolated or highly dispersed grains. With high grain concentrations there is interference between the falling grains, giving slower settling rates. However, in spite of these drawbacks, the idea of using fall velocity as a way of describing grain size is useful and goes some way towards resolving the otherwise intractable relationships between size, shape, density and hydraulic behaviour.

It is a relatively simple experimental procedure to measure the fall velocity of any particle, whatever its shape, and from this measured velocity to calculate the diameter of the sphere that would fall at the same speed as that particle. If we standardise the density (ρ_s) to that of quartz, it is possible to express the effective size of any grain, of whatever shape or density, in terms of the diameter of the equivalent quartz sphere (**hydraulic equivalence**).

In terms of sediment response to flow, it is usually the nature of the grains in bulk that is important, rather than properties of the individual grains. The various measures used to characterise grain-size populations (e.g. median, mean, sorting, skewness, etc.) have been reviewed at length in many standard texts on sedimentary petrography. These measures can often be useful in describing a sediment, but despite much effort, as yet, they tell us remarkably little about processes and environments of deposition. Recent work on cumulative curves suggests that it is possible to recognise subpopulations within natural sediment and

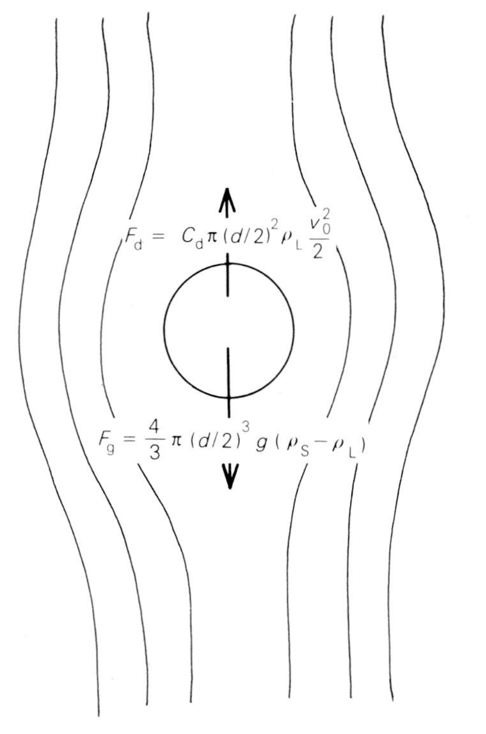

Figure 3.13 The forces acting on a spherical particle falling through a fluid. The streamlines indicate that flow around the particle is laminar and therefore either the viscosity of the fluid is high or the particle is small.

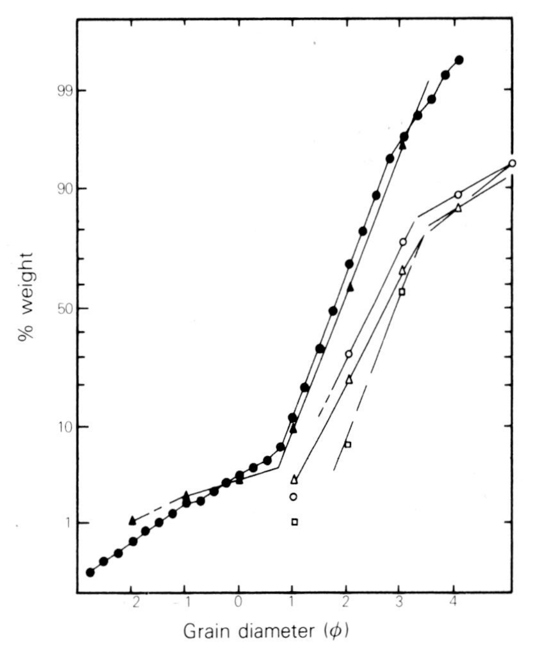

Figure 3.14 Cumulative grain-size curves, plotted on probability paper showing the straight-line segments thought to be due to suspended sediment, intermittently suspended bedload and traction load. Symbols relate to particular samples of sieved sand (after Middleton 1976).

that these may correspond to different modes of transport such as suspension, intermittently suspended load and traction bedload (Fig. 3.14).

3.5 Erosion

The behaviour of grains when subjected to shear by a current is important for understanding sedimentary structures. At some point, with increasing boundary shear stress, grains begin to move, and **erosion** is then said to be taking place. Under what conditions do grains actually begin to move?

Consider the Hjulström-Sundborg curve which plots grain size against critical erosion velocity (Fig. 3.15). From this diagram one can read off the velocity at which grains of a given composition and size begin to move if the flow velocity above the bed is gradually increased. However, we should not only try to understand the gross relationship between grain size and erosion velocity but also to think in detail about what is going on at the bed when grains are set in motion.

If we look at a rather idealised situation it is possible to isolate some of the factors involved. We can then go on to examine the rather more complex reality of erosion of natural materials.

Consider a roughly equidimensional particle resting in a surface made up of similar particles (Fig. 3.16). When a fluid moves over this surface, four types of force act upon the particle: (a) the weight of the particle, (b) frictional forces between adjacent particles, (c) hydraulic lift forces, and (d) the tangential shear couple. Types (a) and (b) are forces that resist motion, whereas (c) and (d) are forces that encourage motion. Let us briefly consider each of these in turn:

(a) The weight of the particle acting vertically downwards (F_g) will act as a moment trying to rotate the grain about the contact point with its underlying downstream neighbour or neighbours.
(b) Frictional forces between particles resist sliding motion and relate to the roughness of the particle surfaces and to the electrochemical forces between particles, which are only important with very small particle sizes.
(c) Hydraulic lift forces, F_l, are a consequence of the flow accelerating over an upward protrusion on the bed. Low pressure above the particle and hydrostatic pressure below combine to give an upward-directed force, as on an aerofoil.
(d) The boundary shear stress (τ_0) of the flow acts on the exposed particle and can be envisaged as a horizontal force through the particle centre. Its effect is to cause the particle to rotate about the point of contact with the underlying downstream particle. The force on the particle (F_d) depends both upon the boundary shear stress and upon the degree of exposure of the grain to the shear stress:

$$F_d = \frac{\tau_0}{N} \quad (3.14)$$

where N is the number of exposed grains per unit area.

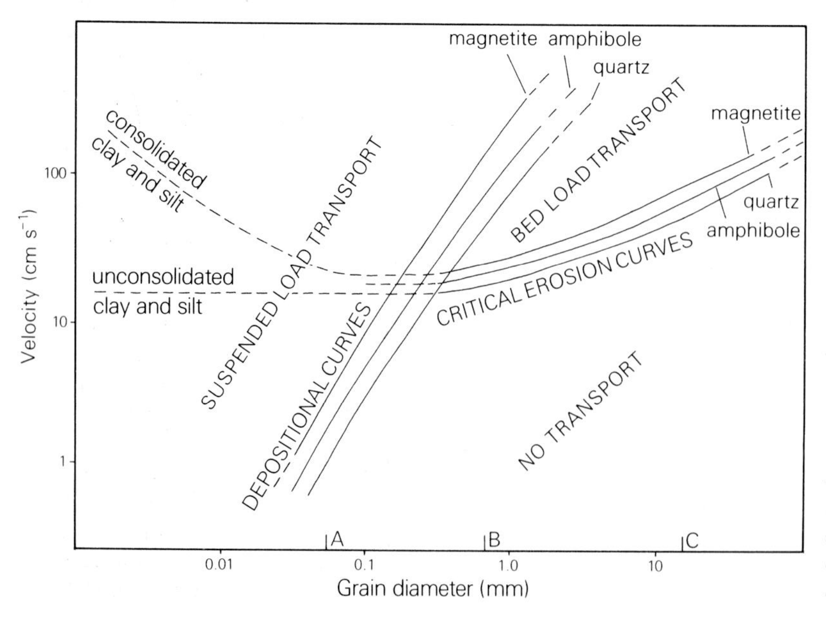

Figure 3.15 The Hjulström–Sundborg plot of grain size against flow velocity 1 m above the bed for initiation of movement of particles of different densities. The nature of the transport mechanism, once erosion has taken place, is also shown (after Sundborg 1956 and Ljunggren & Sundborg 1968). As an example, quartz particles of grain size *B* would begin to move as bedload at the critical erosion velocity of *c.* 12 cm s^{-1} and would go into suspension at a velocity of 1 m s^{-1}. On deceleration, the particles would settle from suspension at a similar velocity and cease movement close to but below the critical erosion velocity. Try to predict the threshold conditions for changes of behaviour of particles of grain sizes *A* and *C* for similar cycles of acceleration and deceleration.

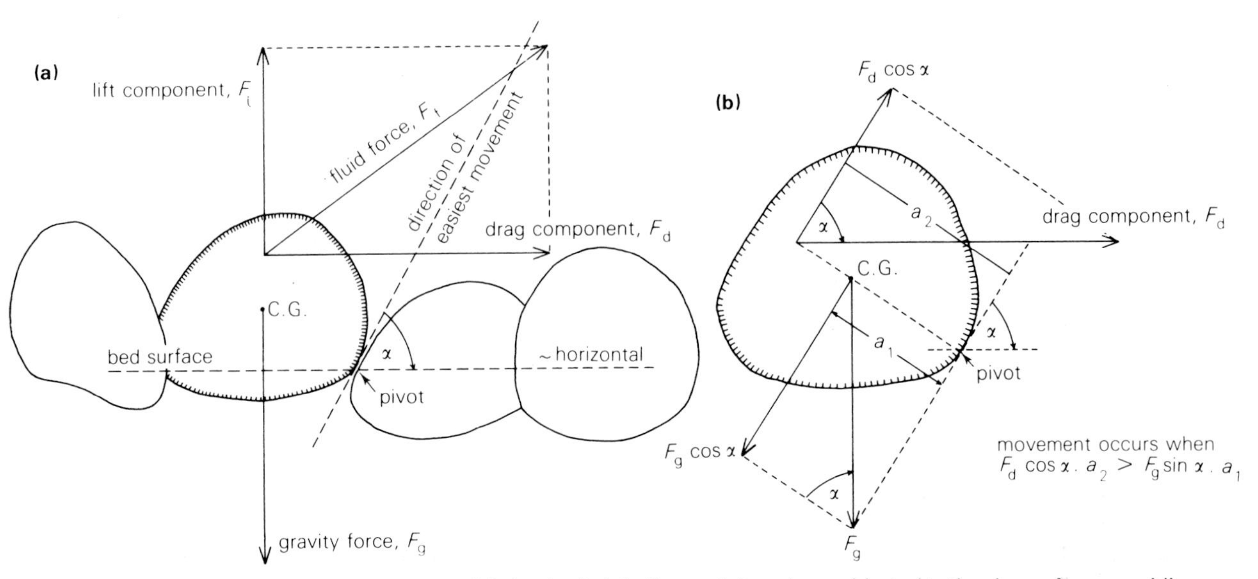

Figure 3.16 (a) The forces acting on a particle in a bed of similar particles when subjected to the shear of an over-riding current and (b) the movement forces which must be balanced for incipient motion; movement begins when $F_g a_1 \sin \alpha = F_d a_2 \cos \alpha$ (after Middleton & Southard 1977).

If, as a gross simplification, frictional forces and hydraulic lift forces are ignored, then for grain movement to occur:

$$F_d a_2 \cos \alpha \geqslant F_g a_1 \sin \alpha \qquad (3.15)$$

In natural settings many other factors complicate matters. First, natural grains are not all equidimensional, nor are they of uniform size. Irregularly shaped grains have a stability that varies with their orientation with respect to the current. Naturally deposited grains tend to rest where drag and lift forces are at a minimum. Flattened grains are most stable when they are inclined upstream at angles in the range 10–20°, as seen in the phenomenon of 'imbrication' of flat pebbles on beaches and river beds (see 7.4.4). Elongated grains are most stable if their long axes are parallel to the current; when aligned transverse to the current, they roll relatively easily. Measured experimental values of critical boundary shear stress tend therefore to diverge somewhat from the values predicted by simple models of erosion.

The packing of grains often relates to the overall grain-size distribution. Well sorted sediments show a pattern of behaviour more closely related to the ideal than do poorly sorted ones, where large grains are only partly exposed to the prevailing shear stress because of partial burial. Simple relationships between grain size and critical boundary shear stress or critical erosion velocity do not apply over a full range of grain sizes as shown by the Hjulström-Sundborg diagram (Fig. 3.15). For grains over about 0.6 mm diameter there is a gradual increase in critical shear stress and velocity while below that grain size these parameters tend to increase with diminishing grain size (Fig. 3.15). This rather unexpected result has been attributed to the increasing importance of intergranular forces in fine sediments, especially those which have been allowed to settle and compact under gravity for considerable periods. With decreasing grain size, the ratio of surface area to volume increases and, in consequence, the surface forces become proportionally greater. The result is that grains show **cohesive** behaviour. Where grains are easily moved as individuals they are said to be non-cohesive. In addition to raising the critical shear stress, cohesion also enables muds and silty sediments to remain stable on high-angle slopes, sometimes even going beyond the vertical.

The protrusion of grains into the flow, giving lift forces and controlling the presence or absence of a viscous sub-layer (see 3.2.4) will also help to determine just when a grain moves. An added complication is that the natural turbulence of any flow may give fluctuations in the instantaneous values of boundary shear stress. Some eddies may cause boundary shear stress and hydraulic lift forces to be temporarily large enough to entrain a particle which would not move under time-average conditions. Effects of this type make the satisfactory definition of the onset of grain movement difficult.

The streaks of high-velocity flow in the viscous sublayer may also localise the initial grain movement in fine-grained sediment. Wave action, when coexistent with a current, will give pulses of increased and diminished boundary shear stress and these can lead more readily to movement than will a steady unidirectional flow.

In natural settings, cohesive sediment in the size range of mud and silt may diverge considerably from the behaviour suggested by the Hjulström–Sundborg curves by virtue of the erosion commonly taking place by removal of aggregates of grains (mud and silt clasts) rather than of individual grains. Also, any material already moving in the flow may help to promote or accelerate erosion by means of a 'sand blast' effect when moving grains hit the bed.

The onset of grain movement due to wind is broadly similar to that with water. However, the occurrence of particles already in motion lowers the critical erosion velocity quite significantly. A sand bed which is stable in winds of a certain speed can be set in motion if a few grains are thrown (seeded) on to the bed. The grain impacts trigger new grain movements and set off a chain reaction of movement down wind. The movement quickly ceases when seeding ends. There are therefore two critical shear stresses for wind erosion, an **impact threshold** where seeding is essential and a higher threshold above which movement takes place without any seeding of bed movement.

3.6 Modes of sediment transport

Having reviewed some of the factors involved in the initiation of grain movement, it is now necessary to outline the various ways in which movement is sustained. These fall into two main groups: **suspension** and **bedload transport.**

3.6.1 Suspension

Sediment carried in the fluid without coming into contact with the bed is supported by the fluid turbulence of the flow. The sediment moves at roughly the same rate as the fluid and the movement results from a balance between downward gravitational forces on the grains and upward forces derived from the fluid turbulence. This suggests that the turbulent flow has a net upward energy flux. In theory, any grain size of material can be carried in suspension if currents are strong enough, but in most natural situations it is usually the finer material of silt and mud grade that moves in this way. Indeed, below the grain size of about fine silt, grains, when eroded, go directly into suspension without an intermediate phase of bedload movement (Fig. 3.15). As the level of turbulence is increased, the suspended-sediment carrying capacity and competence (maximum potentially transportable grain size) of the flow increase. Increased load increases the viscosity and density of the flow with the result that larger grains can be more readily moved in suspension. This process is, however, self-limiting as increased sediment concentration has a damping effect upon the turbulence due to the increase in viscosity until conditions approach those of a mudflow (see 3.7.1).

3.6.2 Bedload transport

The movement of grains in continuous or intermittent contact with the bed may be by saltation, rolling or creep. **Saltation** describes the jumping and bouncing motion of grains close to the bed during vigorous bedload movement. Grains follow asymmetrical trajectories, which are commonly complicated in water by random, turbulence-induced fluctuations. There are gradations between true saltation and suspension as turbulence becomes more vigorous. As descending grains hit the bed, they either bounce back into the flow, dislodge stationary grains on the bed and help to set them in motion, or simply have their kinetic energy dispersed into the bed. In air, collisions on impact are more vigorous than in water because of the lower viscosity and higher immersed weights of the grains. Mobilisation of grains resting on the bed is then an important process. In water, a general damping of impacts takes place and hydraulic lift forces are probably more important in starting grain movement. Where a grain collides with the bed and does not bounce, its kinetic energy may be dispersed amongst several grains resting on the bed. Some of these may, as a result, be pushed a short distance down current or down wind. This is the phenomenon of **creep** and it can account for up to 25% of total bedload movement during wind transport. **'Rolling'** occurs when rather large or elongate clasts are set in motion. It will be favoured if a larger grain is moving over a relatively flat surface of smaller grains. There will be a much greater chance of a grain coming to rest if it is surrounded by, and it rests upon, grains of a similar size to itself. All the modes of bedload transport can coexist to a greater or lesser extent. They will usually be associated with the development of **bedforms** on the sediment surface. These bedforms commonly occur as repetitive patterns on the bed at a variety of scales. When fully developed, they reflect equilibrium between the strength of flow and the grain size of the sediment. These important sedimentary structures are described in Chapters 6 and 7.

3.7 Sediment gravity flows

Mixtures of sediment and water, under appropriate conditions, are able to move as mass flows through the action of gravity. Mass flows include a variety of processes, distinguished theoretically as distinct **mechanisms** whereby sedimentary particles are supported *within* the flows (Fig. 3.17). In this section we outline some of the main types, but you should realise that natural flows are complex, that more than one support mechanism usually operates and that the dominant mechanism may change during the duration of a flow. Flows which have a **yield strength** (see below) are considered as **debris flows**, of which **mudflows** have **cohesive strength** while **grainflows** have **frictional strength**. While the fluid phase between particles is commonly water, in some grain flows (see 3.7.2) it may be air.

3.7.1 Mudflows

Sediment–water mixtures with high sediment concentrations tend to have high apparent viscosities as a result of inter-particle interaction. Mud is particularly important as it provides the material with cohesion. Such natural mixtures are very variable in terms of both the sediment: water ratio and also the grain sizes of the sediment particles. Their viscosities also vary along with these factors, higher water contents and finer grain sizes often tending to promote lower overall viscosity. Such mixtures, which are capable of flowing if present on sloping surfaces, are termed **mudflows**. The movement of such a mixture is determined by the relationship between the applied shear stress (downslope component of gravity) and the **shear or yield strength** and viscosity of the mixture. The yield strength is defined as the applied shear stress below which the mixture is stable. Once this is exceeded, sediment masses tend to flow in a more or less laminar fashion, but if the applied shear stress is high, or if the flow becomes diluted by additional water, then turbulence may develop. Large-scale turbulent churning is most likely to occur in a sub-aerial setting, whilst in a sub-aqueous setting full turbulence may develop as the flow is transformed into a turbidity current (see 3.7.3) Once the applied shear stress falls below the shear strength of the mixture, the flow will cease to move (it 'freezes'). The existence of shear strength, whether cohesive or frictional in nature, gives the mixture the properties of a rheological **plastic**. A general equation for describing the rate of shear strain in such flows can be written as:

$$\frac{du}{dy} = \frac{1}{\mu}(\tau - \tau_{crit})^k \qquad (3.16)$$

where μ is the apparent viscosity of the mixture, τ is the shear stress, and τ_{crit} is the critical shear stress, yield strength or plastic limit: k is a coefficient, which describes the stress–strain relationship during deformation. For most debris flows, k is close to 1, approximating ideal

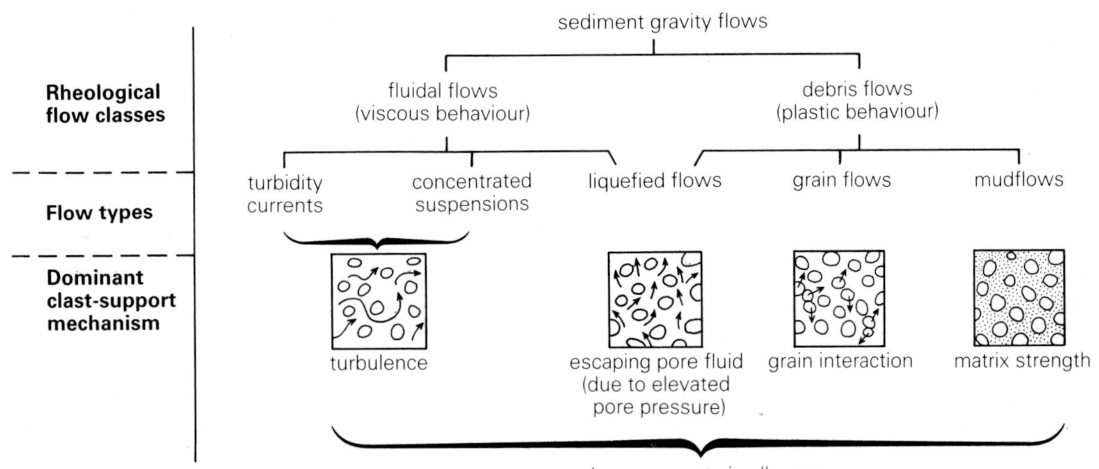

Figure 3.17 Various types of sediment gravity flow where different types of interaction between the water and the sedimentary particles create the mobility necessary for movement (Middleton & Hampton 1973).

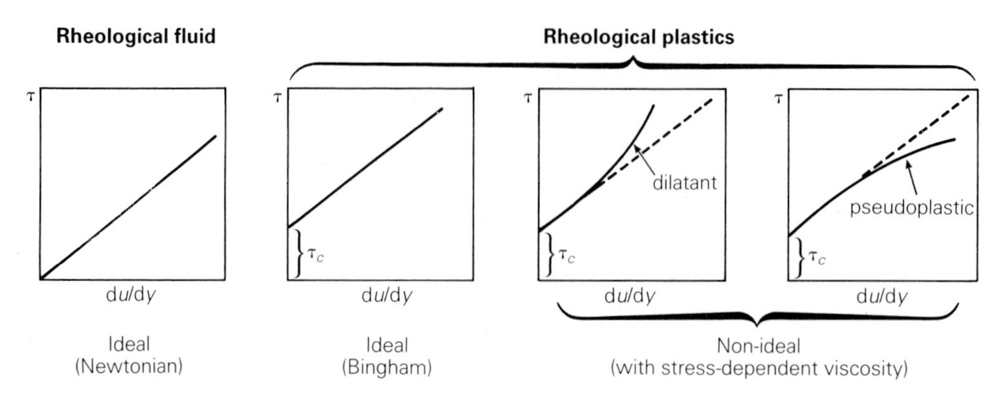

Figure 3.18 Graphs to show the relationship between applied shear stress (τ) and the strain rate (du/dy) for flows with different rheological properties (after Nemec & Steel 1984).

(Bingham) plastic behaviour, but deviations from this value occur and their effects are illustrated in Figure 3.18.

Sediment–water mixtures clearly have a density greater than that of pure water. Larger clasts, surrounded by a sediment–water slurry involving finer particles, will experience greater buoyant support than in clear water. **Buoyancy** therefore, as well as viscous strength, is an important means of supporting clasts in a mudflow and inhibiting their sinking through the flow.

In order to investigate some of these features in at least a qualitative way, make some experimental mudflows in the laboratory. Mix clay and sand with water to give different viscosities and see how these flow on slopes of differing inclination and roughness, both subaerially and under water. The volumes, velocities, thicknesses and degree of internal deformation should all be noted. To preserve a record of internal deformation, make flows of plaster of Paris injected with spots of dye or add plaster of Paris to the matrix of a sandy flow. Careful slicing of the solidified mass can help you understand the internal deformation which acted.

In general, the deposits of mudflows are characterised by a lack of stratification and of sorting of particles within them. This is because all particles come to rest at the same time when the flow stops. As we will see in later chapters, selective transport and deposition of grains of different sizes are much more effective in producing grading, stratification and lamination.

3.7.2 Grain flows

If you watch a truck tipping a load of dry sand or if you walk on the steep side of a sand dune you will notice that the sand grains move as a flowing layer, a **grain flow**. This type of debris flow is cohesionless and it is the frictional interaction of the constituent particles which is important. You will also have noticed that a pile of loose sand is stable only at less than a certain angle of slope. Attempts to steepen the slope trigger movement after the angle has been increased by a few degrees. The angle is thereby reduced to one at which the slope is again stable. This type of movement and the existence of a particular **angle of rest** are important in the deposition of inclined sediment laminae on the lee faces of ripples and larger bedforms in both air and water (see Ch. 6).

In order the develop a feel for this type of sediment behaviour, you should carry out some simple experiments. First of all, measure the angle of rest of sediments of different grain sizes, grain size sorting and particle shapes under different conditions. Methods of doing this are outlined in some of the references given at the end of this chapter. See how the size and shape characteristics influence the angle. Try the experiments in air with dry sediment and, if possible, repeat the measurements under water. A second experiment is to steepen the slope of a pile of sand to see what angular difference exists between the angle of rest and the angle of slip. How does this angle vary with sediment type and conditions?

When the angle of rest is exceeded, internal shear stresses due to the downslope gravitational component overcome the internal shear strength of the sand caused by intergranular friction. Once the surface layer begins to move, individual grains must start to move relative to one another within the layer. For movement to be maintained, space must be created between particles; in other words, there must be an expansion of the layer normal to its surface (dilatant flow). Such expansion is created and maintained by vigorous intergranular shear and grain-to-grain collision (Fig. 3.17) which can be expressed as a **dispersive pressure**, the magnitude of which depends on the magnitude of the shear. When the shear falls below a critical value, the flowing layer will 'freeze' as it collapses upon itself and grains resume a closer packing. A consequence of this type of behaviour is that larger particles will be forced upwards in the flowing layer. An everyday experience of this is the fact that shaking a bowl of sugar will bring any larger lumps to the surface. In addition, during the vigorous particle movements, smaller particles may filter downwards through the dispersed layer, a process known as **kinetic sieving**. The effect of these two processes, either individually or combined, is to promote the development of an **inverse grading** in the layer (see 6.8.2 & 7.4.3).

3.7.3 Density and turbidity currents

Some simple experiments are valuable for understanding something of these types of current. Ideally, construct or obtain a narrow but deep glass- or clear plastic-sided tank at least 1 m long and arrange a vertically sliding, fairly watertight gate near one end to create a small compartment. Such an apparatus is shown in Figure 3.19. Fill the whole tank with water and add dye to the water in the small compartment. Pull up the gate and observe how the coloured and the clear water interact. Now add a quantity of salt to the small chamber, add dye and repeat the experiment several times using different salt concentrations. What is the nature of the interaction now? How does the salinity of the introduced water influence the results?

By using a small immersion heater or by allowing ice cubes to melt in the small compartment, repeat this experiment so that the released water is either warmer or colder than that in the main tank. How do the temperature differences influence the interaction of the water?

In a second series of experiments, add finely powdered, clay-sized mineral grains to the water in the compartment and stir them up into a suspension before releasing it into the main tank. These experiments can be varied by using different concentrations of suspended material, by using minerals of different densities (e.g. kaolinite, calcite and barite) and by setting the tank at slopes of different magnitudes. Try substituting cold milk for the sediment suspension. If a tank is unavailable and cannot be improvised, you may still be able to appreciate the effects under consideration by stirring up the mud at the edge of a pond and watching the behaviour of the resultant suspension.

With clear water of uniform temperature, mixing of the water masses was probably rather sluggish and evenly distributed. With warmer inflow, the coloured warm water probably rose to the surface of the main tank and moved along the tank as a floating layer, an **overflow**. With colder inflow the reverse is the case, the cold layer moving along the bottom beneath the clear water of the main tank as an **underflow**. The experiments with salt solutions and with suspended sediment should also have produced underflows. The thickness of a particular flow and the speed at which it moves along the floor of the tank should vary with the temperature contrast, the strength of the solution, the concentration of the suspension and the density of its mineral matter. From this it should be obvious that it is the excess density of the introduced water mass which controls the behaviour of the underflow. Both types of underflow are examples of **density currents**, the one due to suspended sediment being a **turbidity current**, a particularly important type of sediment gravity flow.

In nature, density currents and particularly turbidity currents play an important role in the transport and deposition of sediment. Clear-water density currents most commonly develop where cold river water enters a warmer lake. The higher-density cold water plunges below the lower-density warm water and may cause sediment transport down the sloping lake floor. If the river water is also charged with sediment, the flow may be a turbidity current. Large turbidity currents, commonly originating by slumping of poorly consolidated material

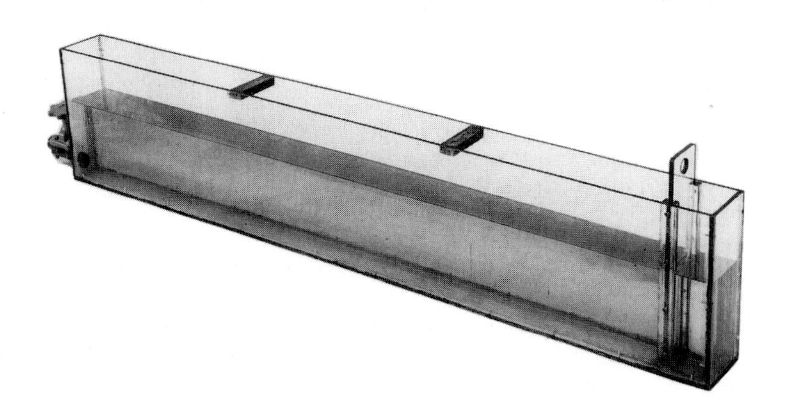

Figure 3.19 A simple apparatus for studying the influence of density on the mixing of different water masses. Mixing is achieved by lifting the lock gate between the two compartments.

on the continental slope are thought to be important in transporting sand to the ocean floor, usually via submarine canyons. Such slumps, as they accelerate down slope, mix with increasing volumes of sea water and become more dilute until the sediment is fully suspended, and the flow is a true turbidity current.

A turbidity current is driven by gravity acting on the excess density of the suspension over that of the surrounding clear water. The grain support mechanism of the suspension is the turbulence generated by the flow, and clearly all the properties of the flow are highly interdependent (Fig. 3.17). Alteration to any one property will trigger changes in other properties. Turbidity currents, which carry large volumes of sand out onto the deep ocean floor, deposit their load when a reduction in slope causes deceleration which in turn reduces turbulence and thereby the capacity of the flow to carry sediment. The currents also exert shear stresses on the bed and these may move, as bedload, sediment already deposited from suspension by the current (see 6.8.4).

Where bodies of water (the tank in our experiments) are density stratified for reasons of temperature or salinity, certain introduced currents may have densities which fall between the extremes of the stratified column. In such cases **interflows** develop which can, perhaps, also carry fine material in suspension for long distances.

3.8 Pyroclastic processes

These processes originate in magmatic explosions in both sub-aerial and sub-aqueous settings. Erupted material may be transported as flows, surges and falls, and the processes involve both elements of the normal sediment transport modes and of sediment gravity flow, supplemented by the effects of volcanic gases and the generation of steam. **Pyroclastic flows** and **surges** are end-members of a spectrum of gas-rich gravity flows which range from concentrated laminar and plug flows to dilute turbulent currents. **Pyroclastic falls** involve the settling of subaerially erupted particles through air and through water which may be flowing or static. Volcanic particles are ejected into the air as an **eruption column** and transported as an **ash plume** which may involve lateral expansion, lateral explosions, winds, convection currents, waves and tides. In all these processes, turbulence of either air or water is the dominant means of clast support. (See Fig. 2.9.)

Reference and study list

References marked with an asterisk * are suitable for 16–19 year old advanced level students in schools and colleges in the UK as well as for undergraduates.

Field experience

It is important in the field and the laboratory to generate a feel for the interaction of the many variables involved in sedimentary processes: different media (air and water in particular), mass, weight, density, effective density, temperature, size and shape of grains, laminar and turbulent flow; viscosity, eddy viscosity; bed roughness (grain and bedform roughness); shear stress, dispersive pressure; temperature; velocity, critical erosion velocity, celerity of waves; angle of slip and angle of rest. It is also important to develop an appreciation of certain dimensionless relationships, Reynolds number and Froude number.

In this chapter we have suggested that your field programme should include investigations, planned by both your tutors and yourself, to observe and record some or all of the following processes in their natural settings.

RIVERS, STREAMS AND ESTUARINE CHANNELS
Organised turbulence (around obstacles and bedforms; flow separation and attachment points; captive eddies); less organised turbulence (e.g. boils); tranquil and rapid flow, hydraulic jumps and streaking.

GUTTERS AND BEACHES
Flow in very shallow currents; streaking in the viscous sublayer.

GRAINFLOWS, MUDFLOWS, DEBRIS FLOWS, SOIL CREEP (SOLIFLUCTION LOBES) AND AVALANCHES
Features on embankments of roads or cliffs in clays or tills commonly show the products of many of these processes. After wet weather, it may prove possible to observe and document active processes, especially those operating at intermediate speeds. Documentation of slow processes (e.g. soil creep) require sustained periods of observation, while direct experience of rapid processes such as avalanches will usually be by accident rather than design. In observing both products and processes try to identify features of laminar flow and plastic and brittle deformation (e.g. crevasses, joints and rotational shear).

FLOW OF WIND OVER DUNES OR OBSTACLES
Organised or less organised turbulence (see Ch. 6); suspension, saltation, creep, rolling; deflation.

DENSITY AND TURBIDITY CURRENTS
Flows of mud at the edge of ponds; cold air or water beneath warm; salt water under fresh water.

WAVES IN PONDS, LAKES AND AT THE SEASHORE
Wavelength, wave height; celerity; wave base; breaking waves; swash and backwash.

Laboratory experience

*Farrow, G. 1976. Film: *Living sediments: a prelude to palaeoecology*. 16 mm, colour, sound; *c*. 25 min. Glasgow: Audiovisual Aids Unit, University of Strathclyde. [A film concerning sediments, organisms and trace fossils, in which there are good shots of standing and breaking waves in a channel, and of megaripples on adjacent sand banks.]
Friend, P. F. 1982. Film: *Current and wave ripples*. 16 mm,

colour, sound; *c*. 8 min. Department of Earth Sciences, University of Cambridge. [Detailed processes seen in flume experiments.]

*La Chapelle, E. 1970. Film: *Time lapse studies of glacier flow*. 16 mm, colour, silent with explanatory titles; 14 min. Seattle: University of Washington Press.

United States Geological Survey 1963. Film: *Flow in alluvial channels*. 16 mm, colour, sound; *c*. 30 min. Bandelier Productions for US Geol. Surv. [A film which demonstrates the basic properties of flows and the response of bedforms to them. The use of dye injection is demonstrated.]

*Walton, E. K. and J. B. Wilson 1969. Film: *Shoreline sediments*. 16 mm, colour, sound; 29 min. Edinburgh: Audiovisual Aids Unit, Department of Genetics, Edinburgh University. [Useful short sections on wave processes and flumes.]

Allen, J. R. L. 1968. *Current ripples: their relations to patterns of water and sediment motion*. Amsterdam: North Holland.

Allen, J. R. L. 1985. *Experiments in physical sedimentology*. London: Allen & Unwin. [A set of simple experiments using very basic, everyday equipment.]

Bagnold, R. A. 1941 and 1954. *The physics of blown sand and desert dunes*. London: Methuen. [Pages 25–31 are concerned with the construction of a wind tunnel. Much of the rest of the book is concerned with the results of experiments in the tunnel.]

*Harrison, S. S. 1967. Low-cost flume construction. *J. Geol Education* **15**, 105–8.

*Jopling, A. V. 1965. Angle of repose box suitable for either class use or research. *J. Geol Education* **13**, 143–4.

*King, C. H. J. 1980. Experimental sedimentology for advanced level students using a motorised wave tank. *Geol. Teaching* **5** (2), 44–52. [Fourteen experiments and practical schedules using the wave tank described by Whitfield (1979).]

Logan, A. 1975. A modified Bagnold-type wind tunnel for laboratory use. *J. Geol Education* **23**, 114–15.

Middleton, G. V. 1966–7. Experiments on density and turbidity currents. Parts I–III. *Can. J. Earth Sci*. **3**, 523–46, 627–37; **4**, 475–505.

Middleton, G. V. 1970. Experimental studies related to problems of flysch sedimentation. In *Flysch sedimentology in North America*, J. Lajoie (ed.), 253–72. Geol. Assoc. Canada, Spec. Pap. 7. [The 'kinetic sieving' mechanism of smaller grains working themselves downwards in a flow, possibly giving rise to inverse grading.]

*Open University 1980. *Summer school laboratory notebook no. 4. Earth sciences*, 22–4, 60. Milton Keynes: Open University. [Foundation Science Course S101. Sieving: grain-size distributions (median, mean, mode skewness, kurtosis).]

*Whitfield, W. B. 1979. The development and educational uses of a motorised wave tank. *Geology Teaching* **4** (2), 64–8. [The basic description of the tank plus suggestions for experimentation.]

Williams G. P. 1977. *Aids in designing laboratory flumes*. Washington, DC: U.S. Geol Surv. Water Resources Division.

Yoxall, W. H. 1983. *Dynamic models in earth-science instruction*. Cambridge: Cambridge University Press. [Ch 7 and 8 involve the flume and the wave tank.]

Essential reading

A selection from:

Allen, J. R. L. 1985. *Principles of physical sedimentology*. London: Allen & Unwin. [A wide-ranging account of many of the important processes of sediment transport and deposition.]

Blatt, H., G. V. Middleton and R. C. Murray 1980. *Origin of sedimentary rocks*, 2nd edn. Englewood Cliffs, NJ: Prentice-Hall. [Chs 3 and 4 especially.]

Derbyshire, E., K. J. Gregory and J. R. Hails 1979. *Geomorphological processes*. London: Butterworth.

Drewry, D. 1986. Glacial Geologic Processes. London: Edward Arnold.

Embleton, C. and J. Thornes (eds) 1979. *Process in geomorphology*. London: Edward Arnold.

Komar, P. D. 1976. *Beach processes and sedimentation*. Englewood Cliffs, NJ: Prentice-Hall.

Kuenen, P. H. and C. I. Migliorini 1950. Turbidity currents as a cause of graded bedding. *J. Geol*. **58**, 91–127. [A series of experiments in a small laboratory tank, contributing enormously to the growth of a theory.]

Middleton, G. V. and M. A. Hampton 1973. Sediment gravity flows: mechanics of flow and deposition. In *Turbidites and deep water sedimentation*, G. V. Middleton and A. H. Bouma (eds), 1–35. Los Angeles: Pacific Section, SEPM.

Middleton, G. V. and J. B. Southard 1984. *Mechanics of sediment movement*. Short Course Notes No. 3. Providence: Eastern Section, SEPM.

Nemec, W. and R. J. Steel 1984. Alluvial and coastal conglomerates: their significant features and some comments on gravelly mass-flow deposits. In *Sedimentology of gravels and conglomerates*, E. H. Koster and R. J. Steel (eds), 1–31. Can. Soc. Petrol. Geol. Mem. 10.

Patterson, W. S. B. 1969. *The physics of glaciers*. Oxford: Pergamon Press. [The basic text on the mechanics of the flow of ice.]

Visher, G. S. 1969. Grain size distributions and depositional processes. *J. Sed. Petrol*. **39**, 1074–106. [A stimulating but controversial attempt to relate parts of grain-size distribution curves to depositional processes.]

Further reading

Allen, J. R. L. 1982. *Sedimentary structures: their character and physical basis*. Developments in Sedimentology 30. Amsterdam: Elsevier. [Two volumes in hardback and a single volume in softback: the most comprehensive account of physically produced sedimentary structures available, and likely to remain the prime source of information for a long time.]

Bagnold, R. A. 1973. The nature of saltation and of 'bed-load' transport in water. *Proc. R. Soc. Lond*. **332A**, 473–504.

Grass, A. J. 1971. Structural features of turbulent flow over smooth and rough boundaries. *J. Fluid Mech*. **50**, 233–55.

Hails, J. R. and A. P. Carr (eds) 1975. *Nearshore sediment dynamics and sedimentation*. New York: Wiley.

Hampton, M. A. 1972. The role of subaqueous debris flow in generating turbidity currents. *J. Sed. Petrol*. **42**, 775–93.

Hampton, M. A. 1975. Competence of fine-grained debris flows. *J. Sed. Petrol.* **45**, 834–44.

Johnson, A. M. 1970. *Physical processes in geology*. San Francisco: Freeman, Cooper.

Kawamura, R. 1951. Study of sand movement by wind. *Inst. Sci. Tech., Tokyo Rep.* **5** (3–4), 95–112.

Leeder, M. R. 1983. On the interactions between turbulent flow, sediment transport and bedform mechanics in channelised flows. In *Modern and ancient fluvial systems*, J. D. Collinson and J. Lewin (eds), 5–18. Int. Assoc. Sedimentol. Sp. Publ. 6.

Lowe, D. R. 1976. Subaqueous liquified and fluidised sediment flows and their deposits. *Sedimentology* **23**, 285–308.

Middleton, G. V. 1976. Hydraulic interpretation of sand size distribution. *J. Geol.* **84**, 505–26.

Raudkivi, A. J. 1976. *Loose boundary hydraulics*, 2nd edn. Oxford: Pergamon Press.

Yalin, M. S. 1977. *Mechanics of sediment transport*, 2nd edn. Oxford: Pergamon Press.

Extra references relating to figure captions

Ljunggren, P. and Å. Sundborg 1968. Some aspects of fluvial sediments and fluvial morphology. II. A study of some heavy mineral deposits in the valley of the river Lule Alv. *Geogr. Ann.* **50A**, 121–35.

Sundborg, Å. 1956. The river Klaralven: a study of fluvial processes. *Geogr. Ann.* **38**, 127–316.

4 Erosional structures

4.1 Introduction

Most areas of present-day sediment accumulation reflect complex interactions between erosion, transport and deposition. Even with net long-term accumulation, deposition may be interrupted in the short-term by periods of erosion. Similarly, most ancient sequences are not the products of steady, continuous deposition but result from alternating periods of deposition, non-deposition and erosion. This chapter deals with the various features which indicate that erosion has taken place.

As with most structures (Ch. 5–7), the chances of an erosional structure being preserved in the rock record are very small. For erosional structures to be preserved the sediment has to be sufficiently cohesive and strong to maintain the erosional relief until it is buried, probably almost immediately, by a contrasting sediment. Erosional structures are almost always recognised on bedding surfaces as relief on the base of the bed immediately overlying the erosion surface. Erosion is also recognised in vertical section by truncation of bedding or lamination in the sediment below the erosion surface. If erosion has been widespread, however, it may have taken place without the preservation of relief and its recognition may then depend upon indirect evidence. Even where relief is observed, it does not follow that it reflects the total amount of erosion. Widespread erosion of a large thickness of sediment may result in only small-scale features. The observed relief can therefore only reflect the minimum thickness of sediment removed.

Many erosional structures are valuable indicators of 'way-up' and of palaeocurrent direction. They can, therefore, play an important role in both structural and palaeogeographical analysis, as well as giving an insight into the processes prevailing during sediment accumulation.

Classification of erosional structures has to be arbitrary as different types grade into one another. The scheme adopted here is based on both descriptive and genetic criteria. Three broad categories are recognised, within which further subdivision is possible:

(a) sole marks on the bases of coarser beds in interbedded sequences;
(b) small structures seen on modern sediment surfaces and more rarely on upper bedding surfaces in ancient strata;
(c) large structures normally recognised in vertical section in ancient sediments (i.e. channels and slump scars).

4.2 Sole marks

4.2.1 Preservation

Sole marks comprise a varied group of structures found as casts on the bases of coarser-grained beds interbedded with mudstones. The coarser-grained sediments are commonly sandstones but exceptionally may be limestones or conglomerates. The sole marks result from the erosion of cohesive, fine-grained sediments which go directly into suspension on erosion. On erosion, coarser-grained, non-cohesive sediment moves first as bedload and the resulting depositional bedforms (see Ch. 6) prevent the development of erosional structures. The cohesive strength of the fine-grained sediment also allows the details of the erosional relief to be maintained until it is buried by coarser-grained material (Fig. 4.1). Resumption of deposition of fine-grained sediment, similar to that eroded, would not normally provide the lithological contrast needed to pick out the structures on weathering.

Erosion and deposition can often be phases of the same current separated by only a short period of time. Subsequent lithification renders the coarse-grained sediment more resistant to eventual weathering than the finer material, with the result that the fine-grained sediment is preferentially removed to expose the cast on the sandstone base. It is very important to understand this mode of preservation and to recognise that the structures observed are negative impressions of the erosional relief.

Sole marks are characteristically the products of environments with episodic sedimentation. Prolonged deposition of muds is punctuated by sudden influxes of coarser sediment in events comprising an early erosive phase and an immediately succeeding depositional phase. One of the most common events generating such beds is the turbidity current. It was previously thought

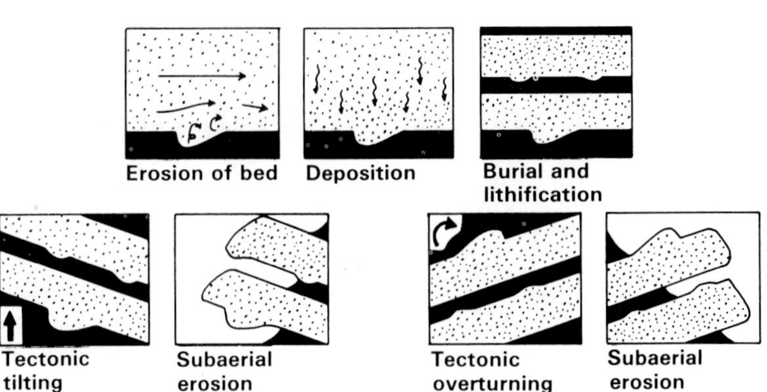

Figure 4.1 Stages in the development of a sole mark and its potential use as an indicator of 'way-up' (after Ricci-Lucchi 1970).

that sole marks were diagnostic of turbidites; but storm surges in shallow seas, sheet floods in semi-arid environments, and crevasse surges into floodplains all have the necessary properties and may produce sole marks. Interpretation should always be restricted to the process rather than to the type of event or the environment until the full context of the structure is understood.

We now consider the varieties of sole marks and the processes that give rise to them. They are divided here into two broad classes which differ principally in the way the structures are generated: structures due to turbulent scour (scour marks); and structures due to objects moved by the current (tool marks).

4.2.2 Scour marks

Scour marks are distinguished by their generally smooth shape and often by their rather streamlined appearance. They may occur as isolated casts or in patterns covering the bedding surface. A variety of shapes occurs, amongst which it is possible to recognise groups which can be similarly named and described together. Four main groups seem important: obstacle scours, flutes, longitudinal scours and gutter casts.

Figure 4.2 Obstacle scour around a pebble in the base of a sandstone bed. Several smaller pebbles also show their own scours. Current from top left to bottom right. Hecho Group, Eocene, Pyrenees, Spain.

Obstacle scours. Large clasts such as pebbles, fragments of wood and the more robust fossils sometimes occur on the bases of sandstone beds and are associated with distinctive ridges of sandstone. The ridges are commonly crescentic or horseshoe-shaped, partially encircling the larger clast with the tails dying away in one direction (Fig. 4.2). These ridges are, of course, casts of troughs developed around the large clasts. In order to understand the development of these structures, try to visit a sandy beach or find a small stream with a sand bed over which there is quite a strong flow. Place a pebble or some other obstruction on the bed and see what happens when the stream flow or the backwash of waves goes over the bed. A crescentic scour trough will commonly develop around the obstruction with the deepest part of the trough on the upstream side of the obstacle and the tail pointing downstream. The scour trough is caused by the accelerated flow around the obstacle and the details of its shape relate to the pattern of eddies generated by this acceleration. The eddies are directed vigorously downwards onto the bed on the upstream side of the obstacle and spiral eddies shed on either side of the obstacle, die out down stream producing the tails (Fig. 4.3). The

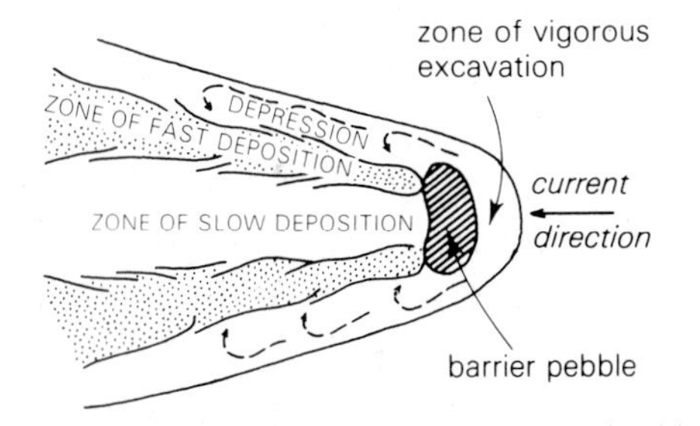

Figure 4.3 The pattern of eddies associated with an obstacle on the bed and its relationship to the formation of obstacle scours (after Sengupta 1966).

structure of the localised captive eddies can be picked out around obstructions by means of a stream of injected dye (see 3.2.3).

The structures are not very common as sole marks but, when present, they provide a good indication of current direction and 'way-up'. In some cases, probable obstacle scours occur although the obstacle itself has been removed. In such cases, a horseshoe-shaped ridge occurs in isolation and the nature of the obstacle is left to the imagination of the observer.

Flutes. Flutes are similar to obstacle scours and they occur both as isolated features and as patterns or groups. Individually they are variable in shape and size, but on any one surface they tend to be rather similar. Flutes are characterised by a rounded, although sometimes tightly curved, 'nose' at one end. The deepest part (i.e. maximum relief) occurs close to the nose, from which point the mark flares away and dies out. On any bedding surface the 'noses' of all flutes will point in the same general direction. Flutes range commonly in length from 5 cm to 50 cm, in width from 1 cm to 20 cm and in depth commonly up to 10 cm. In shape they range from highly elongate forms, gradational with longitudinal scours, to very wide forms with gently curved noses which can be called **transverse scours** (Figs 4.4, 4.5f & g). Some flutes have highly twisted shapes (Fig. 4.5d) particularly close to the nose, whereas others have very simple streamlined shapes (Fig. 4.5b & c). The sides of some flutes are unusual in showing a pattern of small-scale steps instead of the more usual smooth form. These steps can be generally

Figure 4.4 Varieties of flutes on the bases of sandstone beds. Try to judge the direction of the palaeocurrent in each case. (Photograph (b) courtesy G. Kelling.)

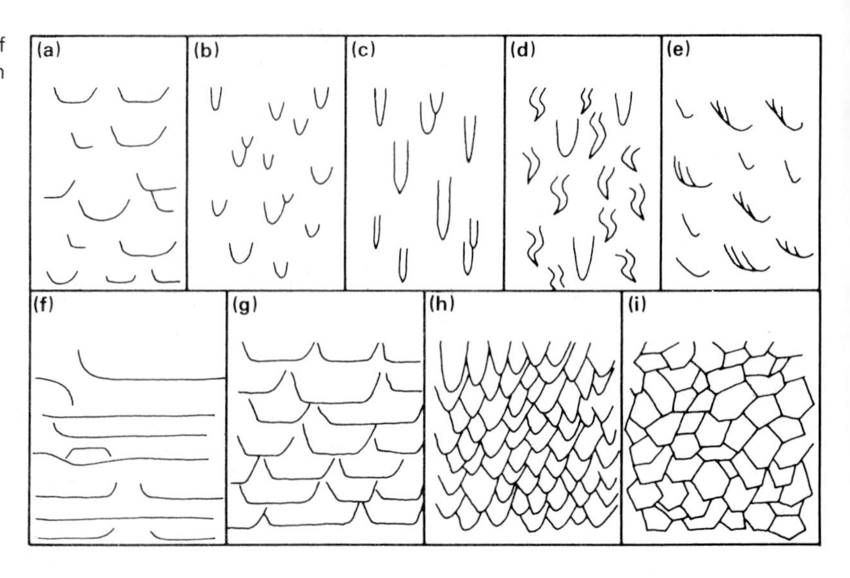

Figure 4.5 Schematic shapes of flutes in plan view. Current from bottom to top in each case (after Allen 1971).

(a)

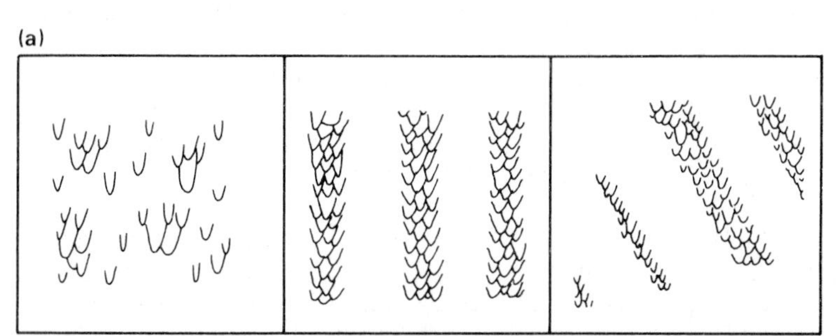

(b)

Figure 4.6 Patterns of distribution of flute casts on bedding surfaces: (a) heterogeneous patterns of distribution (after Allen 1971); (b) an example of the more common case of homogeneous distribution. Bude Formation, Upper Carboniferous, Cornwall.

related to the lamination or thin bedding in the underlying sediments where slight differences in grain size cause differences in the resistance to erosion.

In describing flutes, always include measurements of their dimensions, orientation and the direction in which they point, and a description of their overall shape. Also note if the flutes are distributed in a pattern on a bedding surface. Linear patterns may have the flutes arranged longitudinally or *'en echelon'* and other surfaces may be totally covered with the flutes in 'fish-scale' pattern (Figs 4.5h & 4.6).

Flutes can be produced experimentally when water flows over a surface of cohesive sediment or over slightly soluble substrate. Small bumps and depressions on the bed will cause acceleration of the flow which gives rise to flow separation. The associated higher shear stresses lead to erosion which, in turn, emphasises the relief near the irregularity, and so on. The scale of separation and the erosional relief increase together and will continue to do so as long as suitable flow conditions are sustained. Eventually evidence of the

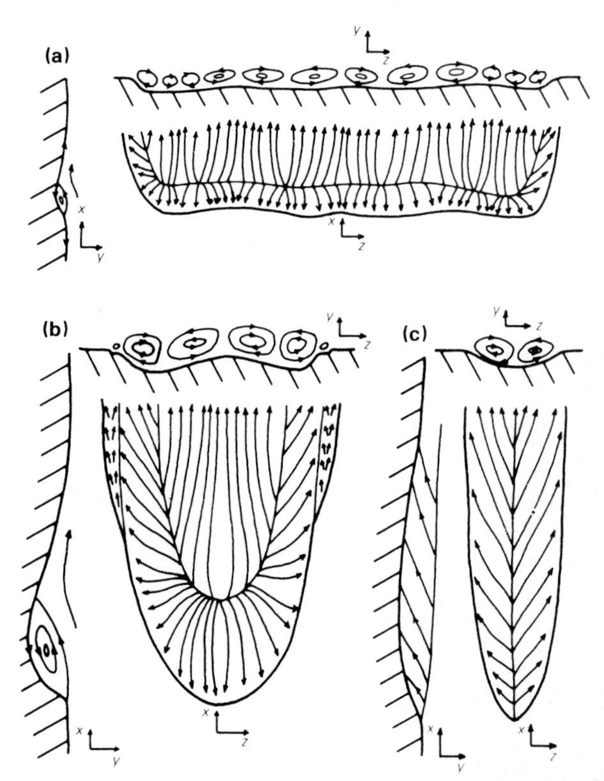

Figure 4.7 Simplified patterns of water motion (eddying) associated with erosional scours of different shapes: (a) transverse scour; (b) and (c) parabolic flutes of differing width. Flow from bottom to top in all cases. *x* direction is down stream, *z* is transverse to flow, *y* is vertical (after Allen 1971).

initial irregularity will be destroyed. Erosion is most concentrated near the nose of the flute from where it dies out down stream as the eddies are absorbed into the main flow (Fig. 4.7). The shape of the flute is intimately related to the structure of the eddies in the nose.

The shape and pattern of flutes bear quite a close relationship to the shape and distribution of the initial irregularities if the scour and growth of the flutes did not last very long. Where erosion was sustained, the flutes may reflect the strength and duration of the current that eroded them.

Not all flutes develop from initial irregularities of the bed. Some may develop from the lateral merging of longitudinal scours. As well as being a valuable indicator of 'way-up' in deformed sequences, flutes are amongst the most abundant and important indicators of palaeocurrent direction.

Superficial examination could cause transverse scours to be confused with straight or sinuously crested ripples, leading to misinterpretation of both the way-up and the palaeocurrent direction. If there is any uncertainty, try to resolve it by looking for related internal structure. Ripples normally show an internal cross lamination (see 6.1.4) whereas transverse scours usually have no related internal structure.

Longitudinal scours (longitudinal ridges and furrows). Longitudinal scours occur as patterns of closely spaced parallel ridges and furrows on the bases of sandstone beds. In transverse cross section the ridges are rather rounded and the intervening furrows are rather sharp, reflecting round-bottomed troughs and sharp ridges on the surface of the eroded mudstone (Fig. 4.8). The spacing of the ridges is commonly 0.5–1 cm with a relief of a few millimetres. Although the overall pattern is one of parallelism, ridges do end if traced far enough along their length. Some die out by coalescing with a neighbour, and others show rounded ends reminiscent of the noses of flutes. Some patterns are parallel and continuous whereas others are markedly dendritic. Wider ridges with distinct noses are gradational to flutes.

Longitudinal scours result from patterns of small-scale eddying close to the bed where spiral eddies have axes parallel to the current. Adjacent eddies have opposite senses of rotation so that flow at the bed has zones of alternate upward- and downward-directed components. The line along which descending vortex limbs impinge on the bed will be sites of high stress and rapid erosion, and beneath ascending limbs stress will be at a minimum and erosion at its lowest (Fig. 4.9). Once localised on scour features, the eddies become fixed and accentuate the relief. Where there are

Figure 4.8 Longitudinal scours on the bases of sandstone beds. In most cases it is only possible to tell the direction of flow and not its sense. Examples with 'noses' also enable the sense of movement to be judged. Mam Tor Formation, Upper Carboniferous, Derbyshire.

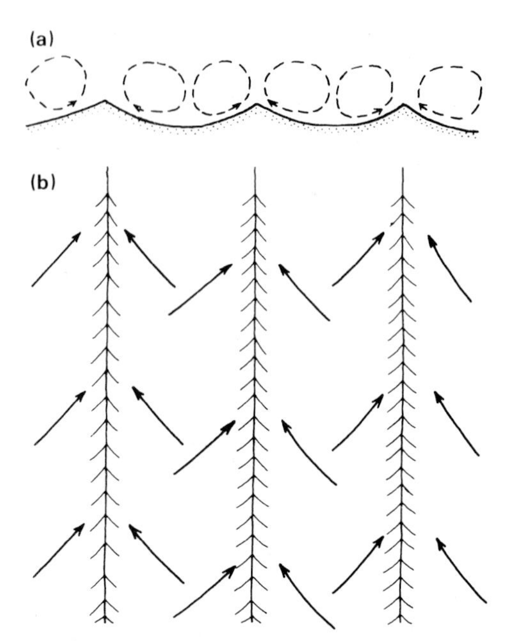

Figure 4.9 The pattern of water movement associated with the development of longitudinal scours; (a) in vertical section normal to flow; (b) in plan view on the bed (in part after Allen 1971).

Figure 4.10 (a) Sandstone-filled gutter casts within a sequence of mudstones and (b) the lower surface of the fill in a loose, overturned block. Campanuladal Formation, Proterozoic, North Greenland.

rounded noses, these are convex up stream and they probably reflect a pattern of eddying similar to that occurring in flutes, with flow separation and a local, transverse component to the axis of the eddy.

Longitudinal scours can be useful as indicators of 'way-up' and of palaeocurrent direction. However, only examples with flute-like noses can indicate both the palaeocurrent direction and its sense (i.e. upstream and downstream directions).

Gutter casts. These structures generally occur as isolated elongate ridges on the bases of sandstone or coarse-grained limestone beds. They protrude into the underlying finer-grained sediment from an otherwise rather flat bedding plane, and in vertical section they may show U- or V-shaped profiles (Fig. 4.10). These are generally symmetrical but, more rarely, may be asymmetrical with one side steeper than the other. They are commonly up to 10 cm wide and almost as deep. Where the coarse-grained sediment does not give a continuous bed above the erosion surface, the coarse infills are preserved as isolated bodies in the finer-grained sediment.

In plan, the casts are commonly slightly to moderately sinuous and they extend for several metres. Sometimes ends are seen and these may be quite steep, similar to flutes, or they may gradually die away. Other ridges are gently curved, when it is common for the margin on the outer wall to be steeper. Smaller features (commonly tool marks) may be superimposed on the walls and floors of the gutter casts, showing preferred orientation parallel to the elongation of the gutter cast.

Gutter casts are the product of fluid scour, possibly aided by the 'sand blast' effect of coarser grains carried by the flow. They appear to reflect a pattern of helical vortices with their horizontal axes parallel to the flow. Pairs of vortices are probably responsible, but on bends one may become dominant to give the oversteepening of the outer wall in a way analogous to river channel meanders.

4.2.3 *Tool marks*

Tool marks differ from scour marks in being produced by objects carried by the flow rather than by the flow itself. They also have rather more sharply defined shapes, and they often carry detailed patterns of small-scale relief. A simple morphological classification is:

Continuous { sharp and irregular profile: **grooves**; smooth and crenulated: **chevrons** }

Discontinuous { single: **prod marks, bounce marks**; repeated: **skip marks** }

Grooves. Grooves are elongate ridges on the bases of sandstone beds. They occur in isolation or in parallel groups. In transverse vertical section they show an irregular, sharply defined relief, usually related to smaller superimposed grooves and ridges (Fig. 4.11). The smaller features tend to trend parallel to the larger ones, but sometimes they are twisted to give a cork-screw effect. Ends of groove casts are seldom seen but they may be gradual or quite sharp. Rarely, a mudflake, a plant fragment or a fossil may be found at the end of the groove casts.

Most bedding planes with groove casts commonly show only one direction of groove. Carefully measure and record the directions of groove casts. On surfaces where it is possible to see more than one direction, record the various directions and also try to put the different directions into chronological order on the basis of cross-cutting relationships (e.g. Fig. 4.11b).

Groove casts result from the infilling of erosional relief gouged by some object, or tool, being dragged through the cohesive substrate by a current. More rarely, it is possible that grooves may result from the rolling of disc-shaped tools leaving tracks similar to those of wheels in soft sand or mud. The identity of the tool will usually be unknown. Where an object is found at the end of the groove, the nature of the tool and the sense of movement of the current are both established. With most grooves it is only possible to judge the direction of movement. Measurements should therefore be recorded as undirected lineations (e.g. 120–300°). The twisted appearance of some grooves reflects rotation of the tool as it was dragged along the bed.

Chevrons. Rarer than grooves, chevrons are linear zones of V-shaped crenulations which consistently close in one direction to produce a chevron pattern (Fig. 4.12). The individual linear zones are seldom more than 2 cm or 3 cm wide and the relief is generally less than 5 mm.

Drag a stick through soft mud or any very viscous liquid and a similar pattern is produced. Chevrons record small-scale folding or 'rucking-up' of the surface of rather weak but cohesive mud by the passage of a tool very close to the bed. The V-shaped ridges, which make up the chevron mark, close downstream thus giving a sense as well as a direction to palaeocurrent measurements. Like other sole marks, chevrons are good 'way-up' indicators.

Figure 4.11 Examples of groove marks on the bases of sandstone beds. Note the difficulty of establishing the sense of movement except where there are associated prod marks or where a groove ends. In (b) it is possible to detect successive erosive episodes. Which direction is earlier? Scale in (a) is 5 cm.

(a) Mam Tor Formation, Upper Carboniferous, Derbyshire; (b) Aberystwyth Grits, Silurian, Aberystwyth, Wales.

Figure 4.12 A chevron mark on the base of a sandstone bed. The structure records the 'rucking-up' of the surface mud layer by a passing tool. Both direction and sense of movement can be judged from the structure. Skipton Moor Grits, Upper Carboniferous, Yorkshire.

Prod marks and bounce marks. Sharply defined, discontinuous marks, often elongate and with a preferred orientation, occur on the lower surfaces of many sandstone beds. Some, called prod marks, are notably asymmetrical along their length, with one end being deep, blunt and well defined while the other end is gradational and slopes gently to the bedding surface (Fig. 4.13). Others, called bounce marks, are more symmetrical along their length, being gently sloping at both ends.

Both types vary in size from several centimetres wide and tens of centimetres long down to very delicate forms, less than 1 cm in length and 1–2 mm wide. Depths are roughly proportional to width, the smallest forms being only 1–2 mm deep. Larger examples may show the superimposition of delicate ribbed relief comparable to that seen on grooves.

In describing and recording these marks in the field try to measure their size and direction, making sure always to record the sense of any longitudinal asymmetry.

Prod and bounce marks record the impact of larger

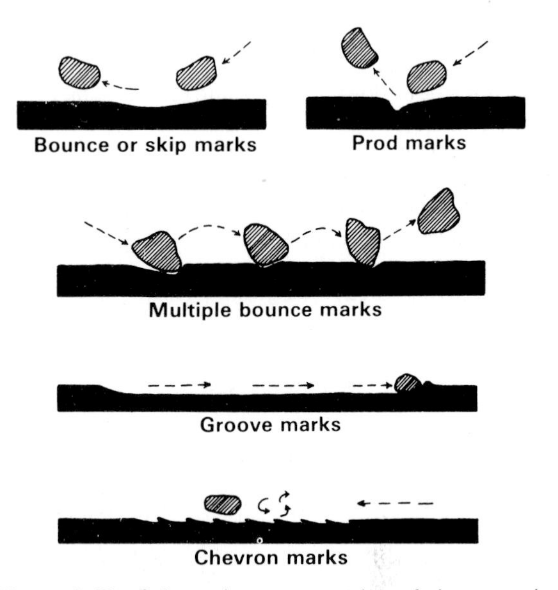

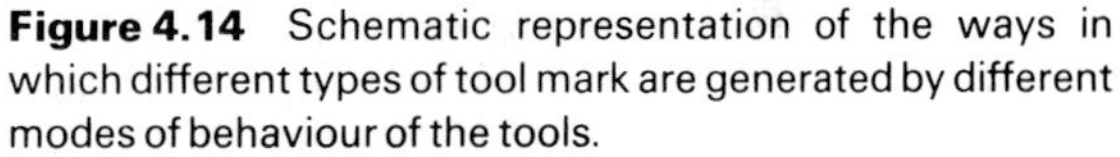

Figure 4.14 Schematic representation of the ways in which different types of tool mark are generated by different modes of behaviour of the tools.

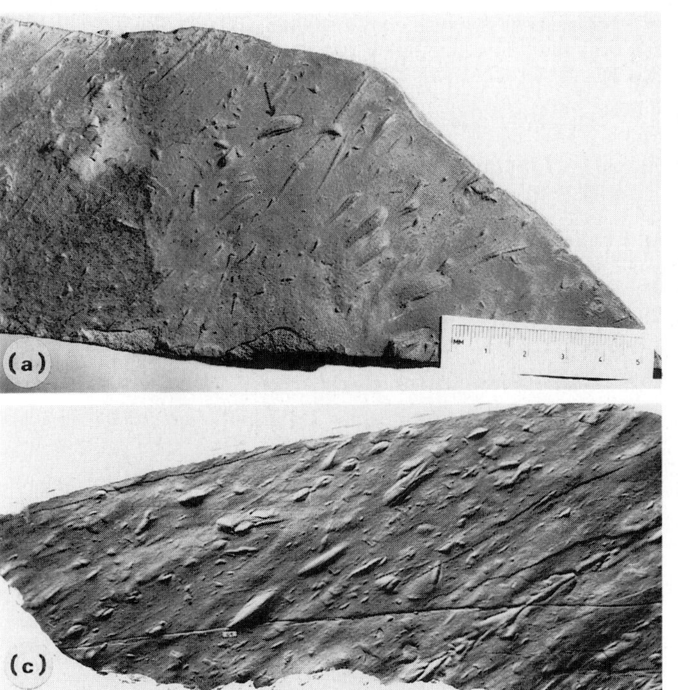

Figure 4.13 Sandstone bases showing a variety of tool marks including bounce and prod marks. Prod marks are more asymmetrical along their length with the blunt end down stream. Bounce marks are more symmetrical along their length. Try to assign examples to these classes, but do not expect them all to fall neatly into a class. Try to assess the direction of movement in each case. Scale bar in (b) is 5 cm. (a, b) Upper Carboniferous, Staffordshire; (c) Krosno Beds, Oligocene, Carpathians, Poland.

objects on the bed. With prod marks, the approach angle of these objects was rather large, so that on impact they dug deeply down into the mud before being pulled steeply out by the flow to leave a blunt 'nose' at the downstream end of the mark (Fig. 4.14). Note particularly that the asymmetry of prods is opposite to that of flutes whose 'noses' are at their upstream ends. Bounce marks reflect a lower approach angle of the tool, so that on impact it digs more gently into the mud before lifting off again (Fig. 4.14). No asymmetry is thereby produced and so only the direction of the current can be deduced. In most cases it is impossible to identify the tool though some exceptional examples are so distinctive that they can be related to, for example, fish vertebrae or ribbed shells.

Skip marks. These are a series of similar bounce marks arranged linearly, usually with rather even spacing. The individual marks need not be identical, but they should be similar enough to suggest that they were produced by the same tool (Fig. 4.15). In some cases skip marks may be very closely spaced and almost gradational with a groove.

Skip marks represent the repeated bouncing of the same tool above the bed. Differences in the shape of bounce marks can sometimes be related to rotation of the tool as it bounced along. Where the marks are almost continuous and approach grooves, the tool may have been disc-shaped and its behaviour analogous to that of a wheel rolling and bouncing down hill.

Distribution and association of tool marks. Tool marks generally occur in mixed assemblages of different types with grooves and prods being found together. Grooves, being longer and larger, generally give a better measure of direction; prods, by their asymmetry, provide the sense of the movement. In addition, tool marks are rarely seen on the same surfaces as scour marks. In many sequences, particularly of turbidites, tool marks are more common on the bases of thin sandstone beds, whereas scours occur on the bases of thicker beds. This perhaps suggests that tool marks represent superficial erosion by relatively weak currents, and scour marks record more wholesale removal of the bed, possibly cutting down into more cohesive mud.

4.3 Small-scale structures on modern and ancient upper surfaces

Erosional structures are fairly common on sandy and muddy sediment surfaces of modern depositional settings but are rarely preserved on upper bedding surfaces in rocks. Both water and wind can act erosively on sediment surfaces. Water erosion gives rise to obstacle and longitudinal scours and rill marks; wind erosion leads primarily to erosional remnants, but may also etch out internal depositional structures in the exposed sand.

4.3.1 Water erosion forms

Obstacle scours. The main features, nature and origin of obstacle scours, have been partly described with reference to sole marks. They occur on both sandy and muddy surfaces and usually take the form of horseshoe-shaped troughs around a pebble, a block of ice or a shell fragment, their size relating to the size of the obstacle.

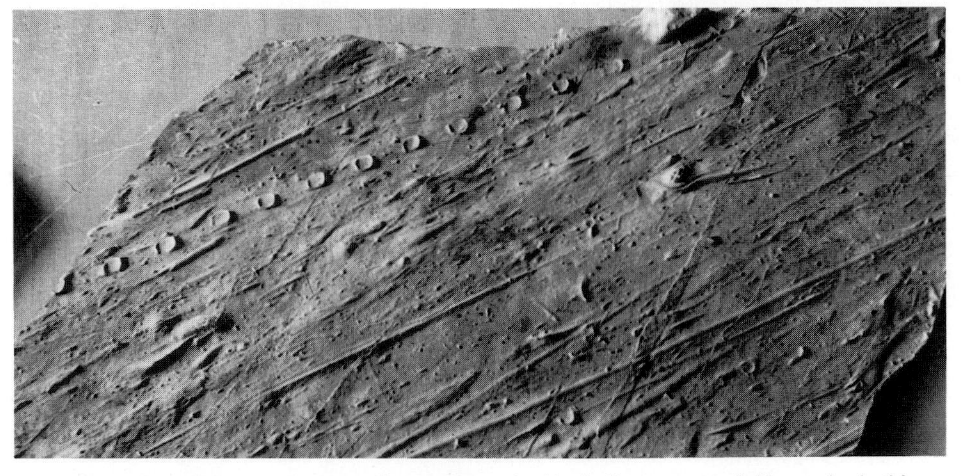

Figure 4.15 A variety of tool marks on the base of a sandstone bed including one set of skip marks, in this case probably made by the repeated bouncing of a fish vertebra. Krosno Beds, Oligocene, Carpathians, Poland.

The trough is usually deepest along the upstream side or around the flanks of the obstacle and dies away down stream. On sand the scour trough and the obstacle may locally perturb the ripple pattern. If the mineralogy of the sand is varied, the scour may be accentuated by mineral sorting.

Longitudinal ridges and furrows. On flat muddy areas, especially in the intertidal zone, patterns of longitudinal ridges and furrows of gentle relief and variable length and spacing occur. They are parallel to dominant currents and are probably related to spiral patterns of secondary circulation in the water (cf. Fig. 4.9).

Rill marks. Rill marks are small-scale, dendritic channels a few centimetres wide and are found on modern sand and silt surfaces, but rarely on bedding planes in ancient sediments (Fig. 4.16).

They result from the emergence of water at the exposed sediment surface following a fall in water stage. They occur most commonly on beaches and on slopes of larger tidal bedforms at low tide. They may also occur on the flanks of large bedforms in rivers. They are almost invariably destroyed by a rise of water level and they have a very low preservation potential.

4.3.2 *Wind erosion forms*

Strong winds blowing over damp or slightly cohesive sediment can lead to erosional forms reminiscent of flutes but showing a positive relief on the upper surface. A blunt nose points up wind with a tail streaking out down wind (Fig. 4.17). Often, on modern surfaces the erosional remnants are localised around pebbles or shell fragments. They are commonly up to a few centimetres wide and up to a few tens of centimetres long and they occur in groups rather than as isolated forms. They are rather uncommon in the rock record where they could be confused with flute marks, thereby providing ambiguous 'way-up' criteria.

Figure 4.17 (a) Wind erosional remnants on the surface of damp sand following a period of strong wind, from top left to bottom right. The blunt ends are up wind, and ridges tail away down wind. Tana Delta, Norway. (b) Supposed ancient examples on the upper surface of sandstone bed. What was the direction of the palaeowind? Independence Fjord Group, Proterozoic, North Greenland.

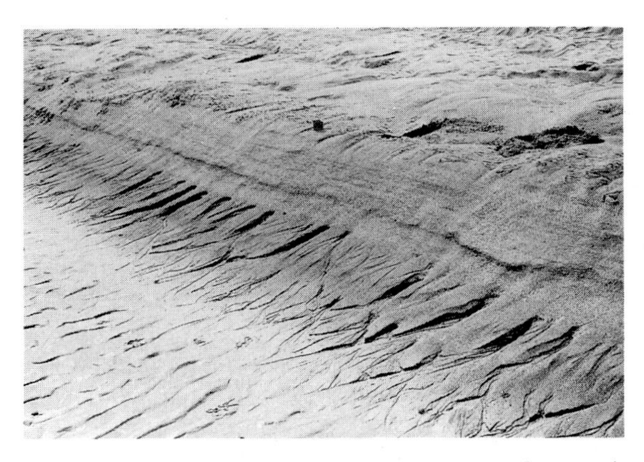

Figure 4.16 Rill marks on the lower part of a sandy channel margin in an intertidal setting. The pattern of small dendritic channels is cut by water emerging from within the sand during falling and low-water stage. Tana Delta, Norway.

4.4 Erosional features in vertical section

The recognition that erosion has taken place during the accumulation of a sediment sequence commonly depends on the occurrence of surfaces that truncate earlier lamination or bedding. On the larger scale, these features are best seen in vertical sections rather than on bedding planes, particularly if exposure is laterally extensive. Clearly, the chances of recognising large-scale erosional structures are much reduced in a narrow stream section or in a borehole.

4.4.1 *Down cutting relief*

A surface that sharply truncates earlier bedding or lamination will commonly be inclined to the depositional horizontal and may be part of a larger structure if traced laterally for a sufficient distance. The form of the larger structure will depend upon the way in which the erosion took place. There are two main processes to consider when looking at any suspected erosion surface:

(a) erosion by scour creating a feature elongated in the direction of fluid movement, e.g. channels or in aeolian sediments 'blowouts' (see 6.3.5);
(b) erosion by mass movement down a slope, creating a feature of less definite shape and orientation but commonly arcuate along the slope, i.e. a slump scar.

However, the two processes can occur together. For example, a river channel, eroded mainly by fluid scour, may have slump scars as smaller-scale features on its banks.

Erosional features, of whatever origin, occur over a wide range of scales, up to hundreds of metres deep and kilometres wide. These largest forms require exceptional exposure for them to be seen in one outcrop and normally their existence has to be inferred from mapping, from the comparison of appropriately spaced, laterally equivalent sections or from reflection seismic data.

The erosional features most commonly recognised at outcrop usually show small- to medium-scale relief but may have a wide range of shapes, orientations and subsidiary features, all of which may be important to their full understanding (Fig. 4.18). Rather than trying to impose some scheme of classification, we suggest that you find an example for yourself and then observe, measure and describe it with the following groups of questions in mind:

What is the overall three-dimensional shape of the erosion surface? Surfaces may have quite complex shapes with both flat and curved sectors and it is necessary to break the problem down into more specific questions.

Is the surface a continuously concave-upwards curve or does it have a distinct base and sides?
If it has sides, what is their maximum inclination?
If both sides can be seen are they similar? In other words, is the cross section symmetrical or asymmetrical? Commonly, only one margin of a channel-like surface is seen and one can then only guess at the nature of the unseen margin.
Is the apparent shape in the observed cross section the true shape or is it distorted by an oblique orientation of the exposure to the true cross section? Sections other than those perpendicular to a channel axis will have higher width:depth ratios than the true cross section. Some means of estimating the orientation of the channel axis is needed and this is discussed below.

What are the dimensions of the erosion surface? Is it possible to measure the depth and width of channel-like forms? Bear in mind the distortion of width when the exposure is oblique to the true cross section. In incomplete exposure it is still valuable to record maximum observed values. Observed relief may in some cases represent the full depth of the erosional form but, in other cases, it may be only a small fraction of total depth. Even a small and apparently complete channel could be superimposed on the floor of a much bigger form.

What is the orientation of the erosional form? If a channel form has been established, it will usually be important to know its orientation so that it can be used to establish a palaeogeography. If you can see a clear 'channel' shape in an exposure, this tells you that the axis of the channel makes a considerable angle with the face of the exposure. However, you should try to be more precise than that. Walking around outcrops, it is often possible to judge the orientations of small channels by eye with quite reasonable accuracy. This is much easier if steep channel sides are exposed. Their strike will commonly parallel the channel trend. Small-scale structures superimposed on the floor or walls will also give a clue. Erosional sole marks on a channel floor give a good indication of channel trend even though the natural fluctuation of flow direction in the channel will reduce the accuracy of any measurement. In rare cases of extensive bedding-plane exposure, such as on a wave-cut platform or in conditions of semi-arid or desert weathering it may be possible to trace the channel over long distances and thereby establish a more reliable trend. It may even be possible to judge its sinuosity. If possible, it is always worth looking down from cliff tops on to bedding surfaces if channel sand bodies are suspected (cf. Fig. 6.44).

Figure 4.18 Examples of different channel margins and channel forms at a medium scale. (a) Ordovician, Gaspé, Quebec, Canada. (b) Central Clare Group, Upper Carboniferous, Co. Clare, Eire. (c) Shale Grit, Upper Carboniferous, Derbyshire. (d) Montañana Group, Oligocene, Pyrenees, Spain.

Examination of aerial photographs is also valuable in this regard.

4.4.2 *Superimposed features on erosion surfaces*

Not all channel cross sections have simple shapes, and the sides, in particular, often show minor features of steps, terraces and even overhangs. There are probably two main controls on the development of these features: the nature of the substrate and the history of erosion and infilling.

Steps and terraces on a channel side often relate closely to the lithology of the eroded sediment. Beds of varying composition or grain size respond differently to flowing water; some will be more readily eroded than others. More cohesive, generally fine-grained sediments commonly form more resistant features, whereas coarser, less cohesive sediment is preferentially eroded. In extreme cases, overhangs may develop.

Many channel forms seen in the field may have undergone a series of erosional and depositional events. Figure 4.19 shows ways in which different sequences of 'cut and fill' can produce a similar channel shape. The sequence of events can only be deduced from observing the channel fill as well as the erosion surface. Erosion surfaces, which are obvious when there is a clear contrast between the lithologies of the substrate and the fill, will be less easily detected within the fill where the lithology is more homogeneous. An erosion surface within the fill may sometimes be a sharp parting and may often be associated with and accentuated by a thin conglomeratic layer of exotic or intraformational clasts.

4.4.3 *Problems and complications*

There are three aspects of larger-scale erosional features that need separate discussion:

(a) recognition of erosion where no erosional relief is seen;

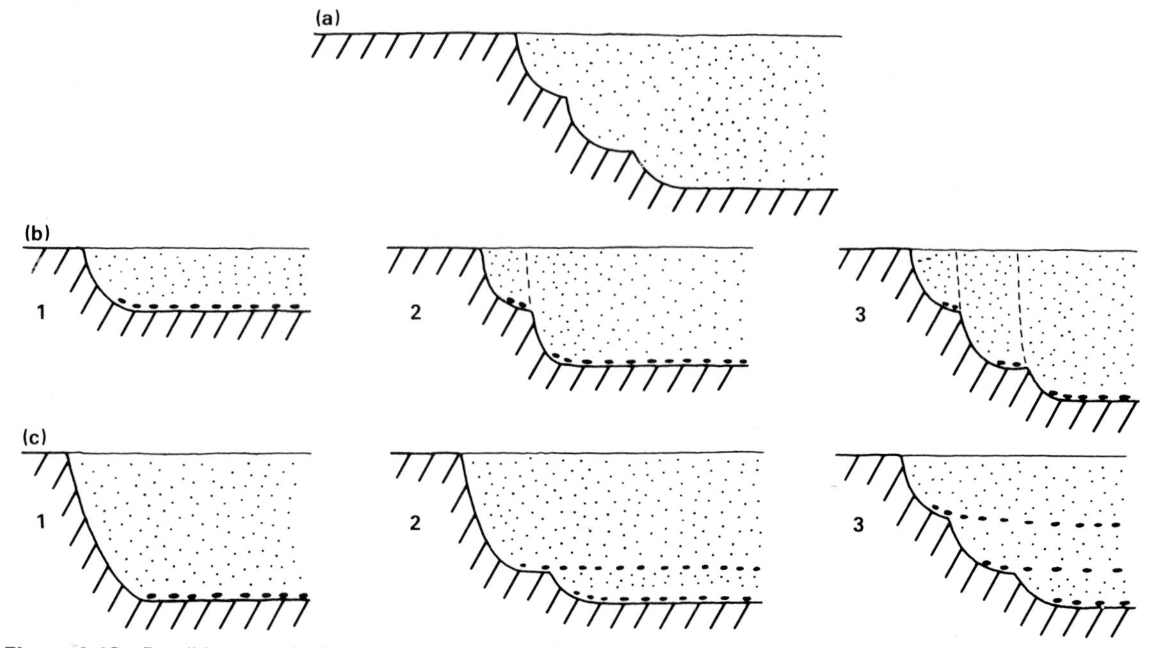

Figure 4.19 Possible stages in the erosion of a stepped channel margin by repeated episodes of cut and fill. (a) Shows the final channel shape. (b) and (c) Show different stages by which this form could be achieved. The ability to recognise erosion surfaces *within* the channel fill may be vital in understanding the full history of development of the channel.

(b) relationship between preserved 'channel' form and the instantaneous shape of the active channel;
(c) distinction between forms due to water scour (channels) and those formed by mass movement (slump scars).

Absence of distinct erosional relief. The absence of distinct erosional relief does not necessarily mean that no erosion has taken place or even that deposits are of non-channel origin. A very wide or very large channel may require an exceptionally large outcrop to establish its channel shape. An outcrop trending parallel to a channel axis will also prevent a channel-shaped cross section from being seen. Clearly we must be aware of clues that still may lead us to suggest a channel origin, even though none of them gives an entirely unambiguous answer.

A sandstone or conglomerate resting sharply on a unit of finer sediment will often have been deposited in a channel. This explanation is supported if a layer of coarse clasts occurs at or just above its base, particularly if the clasts are of intraformational origin. If the actual surface of contact is exposed, smaller-scale erosional structures such as flutes or grooves may give additional evidence of scour.

However, sheets of sand and conglomerate with slightly erosive bases and with coarse basal layers may also be deposited in non-channelised settings, for example by sheet floods or by large turbidity currents.

Preserved channel form and active channel shape. It should not be assumed that one must observe a channel shape in order to infer that a channelised flow was responsible for the deposition of a particular rock unit. Clearly, if a channel form is seen, a channelised flow was involved, but the absence of a channel form does not necessarily rule out the presence of a channel during deposition. In many channel deposits, preserved channel margins are relatively rare features. This somewhat paradoxical state of affairs is resolved if you consider the behaviour of an active channel.

For the cross section of a preserved channel form to be identical to that of the active channel, the channel must have been eroded and infilled without shifting its position. This leads to the preservation of narrow 'shoestring' sand bodies. If channels stay active for sustained periods, the chances are that they will migrate laterally, possibly shifting position by several channel widths while maintaining their cross-sectional shape. This results in an erosion surface which, although eventually ending in a channel margin, may be so laterally extensive that the

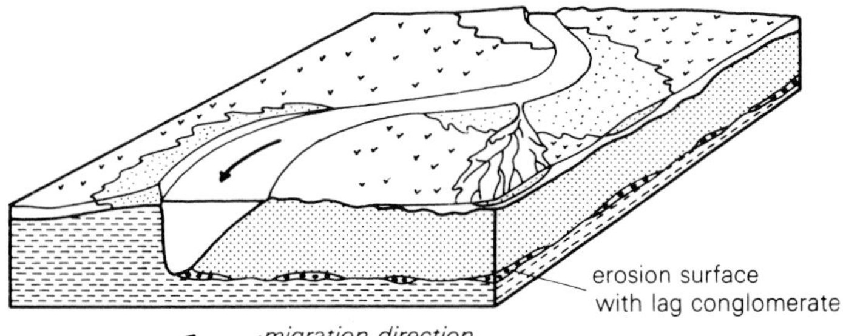

Figure 4.20 The lateral migration of a channel, in this example due to the development of meanders, may erode a horizontal erosion surface with little or no relief. The presence of a channel can only be inferred unless its final position, prior to its abandonment, is observed at outcrop. See Figure 10.8a for an example of the likely preserved sequence. (Based on Allen 1964.)

chances of seeing the margin are small. The most common examples of this behaviour are provided by meandering river or tidal channels (Fig. 4.20). Channels in deep-sea settings, such as submarine fans, show much less tendency to migrate laterally.

In a case where the recognition of channelling depends upon criteria such as those outlined above, determination of the instantaneous shape of the active channel may be difficult or impossible. In some cases, internal features of the channel deposits above the erosion surface such as 'epsilon cross bedding' may suggest the shape of the channel cross section (see 6.2.10).

The distinction between channel and slump scars. Not all erosional cross-cutting surfaces in rocks are the margins of channels scoured by water. Many are the product of large-scale slumping on sub-aqueous slopes which leaves behind a slump scar. The ability to distinguish slump scars from channels is important for interpreting processes and environments of deposition and for predicting the probable extent of an erosion surface and its relationship to the palaeoslope.

Slump scars commonly have their maximum horizontal extent perpendicular to the downslope direction, whereas channels cut by currents are elongate down the slope (Fig. 4.21). In addition, all but the most strongly arcuate slump scars are single-sided; channels have two sides. However, partial preservation and poor exposure make it important to have other criteria for distinguishing channels and slump scars.

In vertical section slump scars are usually smoothly curving, concave-upwards surfaces whose inclination may vary from near-vertical to near-horizontal. Sediments below the surface may show small, normal faults with a similar orientation to the surface, suggesting local horizontal extension. The surfaces lack small-scale superimposed sole marks, such as flutes and grooves, and the steps and terraces that are common in many scoured channels. However, the most important criterion is the similarity of the sediment above and below the surfaces of slump scars. Slumps originate because of instability of sediment on a slope and they commonly move off spontaneously without any external trigger. Depositional conditions in the area are unaltered, and later sediments drape and gradually eliminate the topography of the slump scar. In contrast, the cutting of a channel implies the action of strong currents, and these will be reflected not only in the erosional surface but also in coarser

Figure 4.21 Pattern of major and minor slump scars on the slope of the present-day Mississippi delta. Note that most of the slump scars are subparallel to the bathymetric contours. (Based on Roberts *et al.* 1976.)

sediments laid down above that surface. These coarser sediments may include intraformational conglomerates, which are not readily produced by slumping, and also depositional structures that reflect high-energy currents. However, if a channel is suddenly abandoned, fine sediments may infill it and make it difficult to distinguish from fine-grained bank material. With some slump scars the mass of slumped sediment may not have moved far and a deformed slump mass may be found in association with the scar. In other cases, a series of subparallel slip surfaces may occur with slices of slightly shifted but otherwise undisturbed sediment between them.

Slumping and channel scouring do, in addition, quite commonly coexist. The undercutting of river banks by scour commonly leads to blocks or masses of bank material slumping into the channel and creating slump scars in the process. The toe of the slip surface may extend below the floor of the channel. Abandonment of such a channel soon after a slump may lead to the preservation in the rock record of slump scars at the channel margin and of slumped, rotated blocks in or below the channel fill.

Reference and study list

References marked with an asterisk * are suitable for 16–19 year old advanced level students in schools and colleges in the UK as well as for undergraduates.

Field experience

Present-day features: your field programme should include investigations of areas of interaction of erosion, transport and deposition which leave records of erosional features intact. Scour marks, such as obstacle scours caused by water or wind, are observable on beaches, shallow sandy stream beds, estuarine sandflats or aeolian inter-dune areas. Flute marks, longitudinal ridges and furrows may be seen on cohesive mudflats. Tool marks (grooves, chevrons and prod marks) may be made by dragging an object such as a stick across a mudflat or a drying pond. Rill marks and dendritic channels are best seen on beaches. Larger channel forms are easiest to examine on alluvial fans and in fluvial and intertidal areas. Aerial photographs may prove to be useful in tracing larger channels in all kinds of settings.

In the geological record, sole structures are most frequently displayed in the field in turbidite facies. Channels are most commonly observed in rocks from alluvial fan, fluvial, intertidal, deltaic and some turbidite situations. Slump scars may be discernible in deltaic and turbidite facies. Larger erosional structures should be sought by measuring sections, recognising marker beds and a datum level, and plotting successions in the laboratory, and hence discovering that parts of a succession are missing.

Laboratory experience

In flumes and wind tunnels it is easy to observe the erosional features produced by pressure changes and eddies around obstacles placed in the flow. In addition, useful experience may be gained from the following:

Dzulynski, S. and F. Simpson, 1966. Experiments on interfacial current markings. *Geol. Romana* **5**, 197–214.

Farrow, G. 1976. Film: *Living sediments: prelude to palaeo-ecology*. Colour, sound; 25 min. Glasgow: Audio Visual Aids Unit, University of Strathclyde. [Good pictures of intertidal channels and diagrams of their migration.]

*Walton, E. K. and J. B. Wilson 1969. Film: *Shoreline sediments*. Colour, sound; 29 min. Edinburgh: Audiovisual Aids Unit, Department of Genetics, Edinburgh University. [Brief coverage of aqueous and aeolian obstacle scours and tool marks, and rill and swash marks.]

Essential reading

A selection from:

*Allen, J. R. L. 1970. *Physical processes of sedimentation*. London: Allen & Unwin. [Chapter 2 contains a simple introduction to sole structures. Larger channels are described in several later chapters (4–6).]

Allen, J. R. L. 1982. *Sedimentary structures; their character and physical basis*. Developments in Sedimentology 30. Amsterdam: Elsevier. [Chapter 7 of volume B deals with flutes and related structures.]

Clari, G. and G. Ghibaudo 1979. Multiple slump scars in the Tortonian type area (Piedmont Basin, northwestern Italy). *Sedimentology* **26**, 719–30.

Collinson, J. D. 1970. Deep channels, massive beds and turbidity current genesis in the Central Pennine Basin. *Proc. Yorks. Geol Soc.* **37**, 495–519.

Dzulynski, S. and J. E. Sanders 1962. Current marks on firm mud bottoms. *Trans Conn. Acad. Arts Sci.* **42**, 57–96. [Twenty-two plates and 37 pp. text in illustration of sole marks on undersides of sandstones.]

*Pettijohn, F. J. and P. E. Potter 1964. *Atlas and glossary of primary sedimentary structures*. Berlin: Springer-Verlag. [A compendium of photographs and a glossary in four languages. Sole structures common.]

Reading, H. G. (ed.) 1986. *Sedimentary environments and facies*, 2nd edn. Oxford: Blackwell Scientific. [Many chapters contain discussion of large scale erosional structures, especially Chs 3 and 6–13.]

Reineck, H. E. and I. B. Singh 1980. *Depositional sedimentary environments*, 2nd edn. Berlin: Springer-Verlag. [A very balanced discussion of the origin of many erosional structures in many environments.]

*Ricci-Lucchi, F. 1970. *Sedimentografia*. Bologna: Zanichelli. [Written in Italian with English–Italian index. Many sole marks in flysch sandstone–mudstone successions.]

Walker, R. G. 1966. Deep channels in turbidite-bearing formations. *Bull. Am. Assoc. Petrolm Geol.* **50**, 1899–917.

Whitaker, J. H. McD. 1974. 'Guttercasts', a new name for scour and fill structures: with examples from the Llandoverian of Ringerike and Malmøya, Southern Norway. *Norsk. Geol. Tidsskr.* **53**, 403–17.

Further reading

Allen, J. R. L. 1962. Intraformational conglomerates and scoured surfaces in the Lower Old Red Sandstone of the Anglo-Welsh Cuvette. *Liverpool & Manchester Geol. J.* **3**, 1–20.

Collinson, J. D. and J. Lewin (eds) 1983. *Modern and ancient fluvial systems*. Int. Assoc. Sediment. Spec. Publ. 6. Blackwell Scientific. [Several papers deal with channel forms in alluvial deposits, particularly useful are Lewis & Lewin, Ramos & Sopena and Okolo.]

Dzulynski, S. and E. K. Walton 1965. *Sedimentary features of flysch and greywackes*. Amsterdam: Elsevier.

Ginsburg, R. N. (ed.) 1975. *Tidal deposits*. Berlin: Springer-Verlag. [Small erosional structures at bed contacts and larger structures in intertidal and subtidal deposits in mud, sand and carbonate rocks are well described in many papers.]

Gubler, Y. D. *et al.* 1966. *Essai de nomenclature et characterisation des principales structures sedimentaires*. Paris: Editions Technip. [186 pp. of sedimentary structures.]

Karcz, I. 1968. Fluvial obstacle marks from the wadis of the Negev (southern Israel). *J. Sed. Petrol.* **38**, 1000–12.

Laird, M. G. 1978. Rotational slumps and slump scars in Silurian rocks, western Ireland. *Sedimentology* **10**, 111–20.

Lanteume, M. 1967. *Figures sedimentaires du flysch. Grés d'Annot: du synclinal de Peira-Cava*. Paris: Editions CNRS. [Many photographs of sole marks in turbidites with captions in English, French, German, Italian and Spanish.]

Miall, A. D. (ed.) 1978. *Fluvial sedimentology*. Can. Soc. Petrol. Geol. Mem. 5. [Several useful papers deal with erosional features at the scale of channels, particularly by Barwis, Cotter, Nami & Leeder, Puigdefabregas & van Vliet and Rust.]

Plint, A. G. 1986. Slump blocks, intraformational conglomerates and associated structures in Pennsylvanian fluvial strata of eastern Canada. *Sedimentology* **33**, 387–99.

Extra references relating to figure captions

Allen, J. R. L. 1964. Studies in fluviatile sedimentation: six cyclothems from the Old Red Sandstone, Anglo-Welsh Basin. *Sedimentology* **3**, 163–98.

Roberts, H. H., D. W. Cratsley and T. Whelant 1976. *Stability of Mississippi delta sediments as evaluated by analysis of structural features in sediment borings*. Offshore Technol. Conf. Paper No. OTC 2425.

Sengupta, S. 1966. Studies on orientation and imbrication of pebbles with respect to cross-stratification. *J. Sed. Petrol.* **36**, 362–9.

5 Depositional structures in muds, mudstones and shales

5.1 Introduction

The terminology of siliciclastic sediments of grain size finer than sand is rather confusing. A range of terms has been used in overlapping and sometimes ambiguous ways. It is not our business to become deeply involved in questions of definition, and our usage will conform to the following loosely defined terms:

mud and mudstone Unconsolidated and lithified (respectively) sediment in which grains of sand size (4ϕ or larger) are absent or form an insignificant part. Where coarser grains are conspicuous the terms can be suitably qualified (e.g. sandy mud, pebbly mudstone). These terms include the more precisely defined terms **silt, siltstone, clay** and **claystone** and are useful as field terms because of the difficulties of accurately judging the grain size of fine-grained sediments in the field.

silt and siltstone These are rather more narrowly defined terms for sediments containing a dominance of grains in the range 4ϕ to 8ϕ. If you put such sediments between the teeth they feel gritty, but grains are not generally visible to the naked eye.

clay and claystone Unconsolidated and lithified (respectively) sediment where the dominant grain size is less than 8ϕ. Such sediments feel smooth and greasy to the touch, even between the teeth. Although many clays and claystones contain a high proportion of clay minerals (i.e. hydrated aluminosilicates), grain size rather than mineralogy is the basis of the definition.

shale A widely and often loosely used field term for mudstone which shows a conspicuous fissility on weathering. It is somewhat unsatisfactory in that the weathering state must play a part in its recognition and it cannot be consistently used in comparing rock at outcrop with, say, that of a borehole core.

Muds and mudstones are exceedingly abundant in both modern depositional environments and the rock record, accounting for about 60% of the latter. They are derived as the products of chemical weathering of many unstable source rocks, e.g. basic igneous rocks, and from extreme physical attrition. The fine-grained debris, produced by chemical weathering of silicate minerals other than quartz, comprises mainly clay minerals and chlorite, whereas physically derived sediment, for example in glacial 'rock flour', has a mineral content dependent upon the rocks of the source area.

While most mudstones are the result of deposition following a phase of transport, some may result from *in-situ* weathering of unstable source material. The resultant soil profiles (palaeosols), when found within a rock sequence, will commonly be associated with significant depositional breaks or even unconformities.

In addition, fine-grained sediments are generated directly by explosive volcanic activity resulting in both airfall and water-lain tuffs, which may be subsequently reworked by currents or as mass flows. Such volcanic deposits are often recognised by their contrasting colour and weathering state. Confirmation of volcanic origin commonly requires laboratory analysis of clay minerals. High volcanic eruption columns (tens of km in Plinian eruptions) give very widespread sheets of ash through pyroclastic fall. Widespread distribution is achieved by winds in the upper atmosphere after settling from the stratosphere. Material is transported worldwide, with the paradox that the most powerful processes give rise to extremely thin horizons in the geological record. However, fine ash may fall close to the volcanic centre as a result of a weak explosion or due to rain flushing the eruption cloud. In the latter case, fine ash may occur as accretionary lapilli. Bed thickness will be controlled by the pattern of rainfall rather than by distance from the vent.

Many muds and mudstones are also rich in organic matter, which occurs as either finely divided organic debris or as organic molecules chemically attached to the clay mineral particles.

It is much more difficult to interpret the physical conditions of deposition of muds and mudstones than those of coarser-grained sediments (Chs 6 & 7). There are two main reasons for this. First, the range of physical processes that operate during deposition of muds is much more restricted than for coarser-grained sediments. Secondly, fine-grained sediments, particularly those rich in clay minerals and organic matter, have a much higher initial porosity than most coarse-grained sediments and this makes them more susceptible to compaction on burial. This has the effect of distorting and compressing any structures, sometimes to the point where they are completely obliterated. The amount of compaction will vary with the composition of the sediment and with its burial history. Although some carbonate muds appear to have suffered little compaction, it is not uncommon for some clay- or organic-rich mudstones to have been compacted to a quarter or even an eighth of their depositional thickness. This effect can be observed by study of the internal structure of concretions that formed early in the post-depositional history of the mud (see 9.3.1). Early-formed carbonate-cemented concretions sometimes preserve relatively uncompacted depositional structures and uncrushed fossils. If concretions occur in a mudstone sequence, always try to see if their internal structure offers any help in understanding the deposition of the mud (Fig. 5.1).

Tectonic movements have much more drastic effects on fine-grained sediments than on coarser ones. During folding, fine-grained sediments generally behave in an incompetent manner and also readily develop cleavage, thus obscuring and distorting any original structures (Fig. 5.2). Sedimentologically useful structures will be much more commonly found on cleavage planes than in sections perpendicular to them. The distortion of structures of known or assumed original shapes in the cleaved mudstone may be used to estimate tectonic strain. Elliptical reduction spots are commonly assumed to have been circular (i.e. spherical) prior to distortion.

5.2 Structures and lamination

5.2.1 Detection of lamination

Cut and varnished slabs of fresh rock or naturally polished sections on the coastal foreshore or in stream beds provide the best opportunities of seeing structures in mudstones. Structures are usually of small scale and are described in terms of different types of lamination. These types are intergradational, but we describe them below under headings that give some indication of potentially useful criteria. The detection of grain-size differences in fine-grained sediments is usually based on colour differences as the grains themselves are not normally visible. As a general rule, lighter colours indicate coarser-grained sediment, but there are cases where the opposite is true.

5.2.2 Very fine lamination and fissility

Very fine parallel lamination, which leads to fissility on weathering, is usually confined to claystones or to micaceous siltstones. On freshly cut surfaces perpendicular to the lamination it is usually impossible to see any colour banding which may reflect grain-size differences. The surfaces parallel to the fissility are commonly smooth and flat. When describing these mudstones it is helpful to try to judge whether the rock will only split down to layers of a particular thickness or whether, if you had the equipment and patience, it would be possible to go on splitting it indefinitely. If there seems to be a limiting thickness, it should be measured and recorded even though it must be accepted that fissility is a function of weathering history as well as an intrinsic property of the rock. In splitting the rock, try to see if the surfaces of splitting correspond to mica- or organic-rich layers. The term **paper-laminated** is sometimes used to describe shales which can be split apparently indefinitely.

The lack of any obvious grain-size differences in very-fine-grained fissile claystones suggests that grain orientation is responsible for the fissility. Clay minerals, chlorites and micas commonly occur as platy grains which, on compaction, are squeezed into a parallel orientation. Fine clay particles are carried in suspension by water and it requires a reduction in the level of turbulence for the grains to be deposited. This is usually achieved when a flow carrying a load of suspended sediment slows down on entering a larger body of quiet water. In many cases the settling of clays from suspension is aided by a change in the salinity of the water as it enters the depositional basin, such as when a river enters the sea. In estuaries and other marginal marine settings, the salinity of the water allows the small clay particles to form aggregates known as **flocs** by a process of **flocculation**. Flocs are much larger than their constituent particles and they tend to settle out more quickly. The extent of flocculation is a function of particle concentration, fluid turbulence and of the chemistry of the particles and the receiving basin. Turbulent conditions tend to break down the flocs.

5.2.3 Fine lamination with grain-size differences

Close examination of artificially or naturally polished surfaces of some mudstones often reveals colour banding of paler and darker layers of the order of 1 mm or less in thickness (Fig. 5.3). This normally reflects slight grain-size differences which can sometimes be detected by close examination with a hand lens. If you can see a

Figure 5.1 Cut and etched section of a carbonate-cemented concretion collected from a sequence of compacted mudstones. Within the concretion the original lamination is in an uncompacted state and goniatite (G) and bivalve (B) shells occur in their uncrushed three-dimensional form. Upper Carboniferous, Staffordshire.

Figure 5.2 Cleaved mudstone in which a tectonic foliation has been superimposed on the original depositional fabric. The depositional lamination is picked out by slight colour/grain-size differences. Light meter for scale. Late Precambrian, Finnmark, Norway.

Figure 5.3 Finely laminated homogeneous mudstone where the lamination and parting is probably caused by slight grain-size differences (not seen) and by preferred orientation of platy grains. Lens cap is 5 cm diameter. Upper Carboniferous, Middle Shales, Dyfed, S. Wales.

difference in grain size, the coarser layers at least must fall within the silt size class, and in such a case the individual layers are likely to be only a few grains thick. Try always to record the thickness of the laminae and to judge their lateral continuity and parallelism. With such thin layers, thickness is perhaps best indicated by quoting an average, calculated by counting the number of layers in a known thickness, rather than by measuring thicknesses of individual laminae. Parallelism and continuity of lamination can be quite variable. Some examples show extreme continuity and others have laminae that pinch out laterally.

Once it is established that laminae are defined by grain-size differences, it follows that the process responsible for deposition must have fluctuated, although it is difficult to estimate the time scale of this.

Two possible depositional processes must be considered. One is that the sediment settles from suspension from the whole water column or from lighter turbid water floating near the surface. Fluctuation of the supply of suspended sediment will give rise to the lamination. The second is that the coarser layers are the product of weak, dilute density currents flowing close to the bed while the finer layers record the background sedimentation. When the laminae are very thin, other features of the overall sequence must be assessed. For example, if lamination occurs in a sequence which has evidence of larger-scale density currents in the form of turbidite sandstones, the inference that the lamination is due to density currents may be reasonable.

Some fine lamination could also be the product of shorter-term fluctuations in more sustained currents. Sweep and burst processes in a viscous sub-layer (see 3.2.4) may sort sediment into coarser and finer layers, particularly in the silt-size range.

5.2.4 *Thicker lamination or thin bedding with gradational boundaries*

Many mudstones have a distinct 'striped' appearance of alternating lighter and darker layers from a millimetre up to a few centimetres thick. Such layering is usually dominated by silt-grade material but darker, finer-grained layers may have a substantial clay content, whereas paler, coarser-grained layers may contain some fine sand. It is important in such cases to try to check whether the beds or laminae are controlled mainly by overall grain size or whether they reflect fluctuations in a component such as comminuted plant debris.

These units are usually parallel-sided with gradational boundaries (Fig. 5.4). Layer thickness can be estimated by counting layers over a measured interval or, in more detail, by measuring each layer. The second approach is important if the coarser and finer layers differ in thickness and also if it is suspected that some overall trend of thickness change occurs throughout the vertical sequence.

These mudstones reflect fluctuations in the suspended sediment supply on a time scale too large to be attributed to sweep and burst mechanisms or to other short-period fluctuations in an otherwise constant flow. Seasonal or other climatic factors may control sediment discharge to deltas and river basins where these sediments are common. The gradational contacts

Figure 5.4 Striped siltstone where the parallel lamination or thin bedding results from gradational grain-size changes suggesting long term fluctuations in sediment load. Upper Carboniferous, Middle Shales, Dyfed, S. Wales.

suggest gradually increasing and waning high discharge episodes rather than sudden incursions, for example of slump-triggered turbidity currents. However, it is again impossible to tell if suspended sediment settled from the whole water column, from a floating plume or from a fluctuating but perhaps permanent density underflow.

5.2.5 *Thin bedding with sharp-based, graded beds*

Mudstones with sharply differentiated dark and pale layers are commonly characterised by the coarser-grained, paler layers having sharp bases and gradational tops (Fig. 5.5). The thicknesses of both the dark and pale layers are more varied than in striped mudstones with gradational boundaries. In these mudstones, beds tend to be laterally continuous and the coarser layers may show grading of grain size. Even if this cannot be directly observed, it can be inferred from the gradational top of the bed. Bases of the coarser layers may sometimes be slightly irregular with relief of a few millimetres.

The sharp base and clear definition of the coarser layers suggest that they represent relatively sudden events superimposed upon the background of quieter, more constant sedimentation of the finer-grained, darker layers. The internal grading and the gradational tops of coarse-grained layers suggest a waning of the suspended load during the more active episodes. The small-scale irregular morphology on their bases is probably due to loading of the silts into soft, waterlogged clays (see 9.2.1).

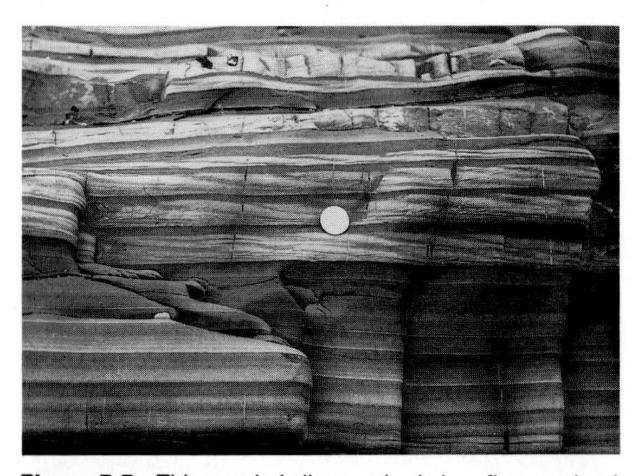

Figure 5.5 Thin, graded siltstone beds in a finer-grained background in the lower part. Even though grain size is not directly visible, the grading may be inferred from the contrast between the sharp bases of the coarser grained beds and their more gradational tops. Fine sandstone beds above show ripple drift cross lamination (see Sec. 6.1.4). Upper Carboniferous, Bude Formation, Cornwall.

5.2.6 *Structureless mudstones*

Some mudstones show no obvious lamination, bedding or fissility irrespective of their weathering state. In some cases a rather blocky pattern of fracture is evident, but in others the sediment is completely massive and homogeneous, even to the point of breaking with a conchoidal fracture. Often there is little to describe in these rocks, but it is still worth looking carefully at them as their lack of lamination may be due to one of several possible causes. It may reflect an original lack of depositional layering (i.e. continual steady deposition) or it could be due to later destruction of layering. On fresh surfaces of some apparently homogeneous mudstones it may be possible to see mud clasts in a mud matrix.

Lack of original layering in water-lain mudstones may be due to a very homogeneous and possibly rather rapid depositional process or to a lack of platy grains. Rapid deposition of muds from suspension is probably not uncommon, but direct evidence for it in the rock record is quite rare. Preservation of tree trunks in an upright growth position in some coal measure mudstones is one of the more compelling pieces of evidence. Sub-aerially deposited muds also commonly lack lamination. Thick accumulations of wind-blown silt (**loess**), which typify many proglacial areas, are examples of this. A mud which has been deposited as a mudflow may also lack structure if the large clasts which typically characterise a mudflow were unavailable in the source area (see 7.4.3).

Original layering may be destroyed later by a variety of agencies such as burrowing organisms, plant roots and soil-forming processes. It is often possible to find recognisable burrow forms (see 9.4) or the remains of rootlets as in mudstone seat-earths (fireclays). Vertical colour variation or the development of concretions may be due to soil-forming processes.

Large-scale post-depositional movement of thick mud beds is common, particularly in rapidly deposited sequences. Muds buried under younger sediments commonly flow both vertically and laterally to give diapiric structures (see Ch. 9), and this can lead to the development of massive, blocky or scaly fabrics.

Reference and study list

References marked with an asterisk * are suitable for 16–19 year old advanced level students in schools and colleges in the UK as well as for undergraduates.

Field experience

Many of the features that develop on cohesive mud surfaces have been dealt with in Chapter 4.

PRESENT-DAY ENVIRONMENTS
Study of present-day areas of deposition of mud can often be unrewarding, for suspended clay often obscures the depositional surface when the area is covered by water. After-the-event observation of muddy intertidal areas is often useful, but the dangers inherent in attempting to traverse mudflats and the physical effort of squelching through them are never to be forgotten. Muddy density flows can often be generated at the edges of still, clear ponds by disturbing small masses of sediment at their edges. Mudflows can often be seen in cliffs and excavations in muddy sediments.

ANCIENT SEQUENCES
Vast thicknesses of shales and mudstones from many environmental origins are encountered during many field excursions. Interpretation of the processes of origin of these rocks is often limited. The ascribing of the rocks to a palaeo-environment often rests upon the evidence of body fossils, upon features in mudstones described in other chapters (e.g. post-depositional structures, Ch. 8) and upon the structures and features of the sediments interbedded with the mudstones.

Laboratory experience

Useful experience may be gained by introducing mud suspensions and small amounts of somewhat coarser grain sizes into long settling tubes, so as to produce laminations and normally graded laminations. Attention should be given to controlling and measuring the effects of variables: the medium (air, water or even glycerine), grain size, temperature, viscosity, relative density of medium, relative density of grain, velocity of fall, etc.

Milner, H. B. 1922. *Sedimentary petrography* (2 vols). London: Allen & Unwin. [Techniques for using sedimentation tubes.]

*Open University 1978. *Summer School S100 Laboratory notebook No. 4. Earth sciences*. Milton Keynes: Open University [Sedimentation from suspension into air, water and glycerine (pp. 16–21, 25–7).]

Essential reading

A selection from:

Boswell, P. G. H. 1961. *Muddy sediments*. Cambridge: Heffer.

Cook, H. E. and P. Enos (eds) 1977. *Deep water carbonate environments*. Tulsa: SEPM Sp. Publ. 25.

Hsu, K. J. and H. C. Jenkyns (eds) 1974. *Pelagic sediments on land and under the sea*. Int. Assoc. Sedimentol. Sp. Publ. 1. Blackwell Scientific.

Kranck, K. 1975. Sediment deposition from flocculated suspensions. *Sedimentology* **22**, 111–23.

Kulm, L. S., R. C. Roush, P. H. Harlett, R. H. Neudecki, D. M. Chambers and E. J. Runge 1975. Oregon continental shelf sedimentation: interrelationships of facies distribution and sedimentary processes. *J. Geol.* **83**, 145–74. [Discussion of the processes of transport of silt and clay, and the effects of bioturbation.]

O'Brien, N. R., K. Nakazawa and S. Tokuhashiu 1980. Use of clay fabric to distinguish turbiditic and hemipelagic siltstones and silts. *Sedimentology* **27**, 47–61.

Picard, M. D. 1971. Classification of fine-grained sedimentary rocks. *J. Sed. Petrol.* **41**, 179–95.

Potter, P. E., J. B. Maynard and W. A. Pryor 1980. *Sedimentology of shale*. Berlin: Springer-Verlag. [A comprehensive source of information about shale with a considerable section on sedimentary structures (pp. 17–38), many of which, however, fall outside the scope of the present chapter.]

Reineck, H. E. and I. B. Singh 1980. *Depositional sedimentary environments*, 2nd edn. Berlin: Springer-Verlag. [Many features produced in mudrocks from intertidal areas.]

Ricci-Lucchi, F. 1970 *Sedimentografia*. Bologna: Zanichelli. [Written in Italian with English–Italian glossary. Many structures from mudrocks.]

Further reading

Byers, C. W. 1974. Shale fissility: relation to bioturbation. *Sedimentology* **21**, 479–84.

Jopling, A. V. and B. C. McDonald (eds) 1975. *Glaciofluvial and glaciolacustrine sedimentation*. Tulsa: SEPM Sp. Publ. 23. [Several papers on varves: see particularly Gustavson, Gustavson, Ashley & Boothroyd and Ashley.]

Lambert, A. and K. J. Hsu 1979. Non-annual cycles of varve-like sedimentation in Walensee, Switzerland. *Sedimentology* **26**, 453–61.

McKee, E. D. and M. Goldberg 1969. Experiments on formation of contorted structures in mud. *Bull. Geol Soc. Am.* **80**, 231–44.

Millot, G. 1970. *Geology of clays*. Berlin: Springer-Verlag.

O'Brien, N. R. 1970. The fabric of shale – an electron microscope study. *Sedimentology* **15**, 229–46.

Odom, I. E. 1967. Clay fabric and its relation to structural properties in mid-continent Pennsylvanian sediments. *J. Sed. Petrol.* **37**, 610–23.

Schluchter, Ch. (ed.) 1979. *Moraines and varves: origin, genesis, classification*. Rotterdam: Balkema. [A good collection of papers on varved sediments.]

Smalley, I. J. (ed.) 1975. *Loess: lithology and genesis*. Stroudsburg, Pa: Dowden, Hutchinson & Ross.

Spears, D. A. 1976. The fissility of some Carboniferous shales. *Sedimentology* **23**, 721–5.

Stow, D. A. V. and D. J. W. Piper (eds) 1984. *Fine grained sediments; deep-water processes and facies*. London: Sp. Publ. Geol. Soc. Lond. 15.

Warme, J. E., R. G. Douglas and E. L. Winterer (eds) 1981. *The Deep Sea Drilling Project; a decade of progress*. SEPM Sp. Publ. 32.

6 Depositional structures of sands and sandstones

Structures developed in siliciclastic or carbonate sands and in sandstones and calcarenites reflect a variety of transport processes and they are our clearest indicators of the types and strengths of currents that move sediment. The transporting medium may be water or air. Deposition of sand is generally due to the cessation of an episode of bedload transport, to accumulation during steady flow with excess sediment supply, or to deposition from suspension from powerful, decelerating currents. After deposition from suspension, sand may continue to move as bedload before it finally comes to rest. To classify structures in sand we have adopted a scheme which is partly descriptive and partly interpretative. Sand-size sediment of pyroclastic origin may also form many of the structures described in this chapter. Most pyroclastic deposits are of such a grain size as to be more appropriately described in Chapter 7.

6.1 Ripples and cross lamination

6.1.1 Introduction

Ripples are more or less regularly spaced undulations on a sand surface or on a bedding plane of sandstone. Their spacing (wavelength) is usually less than 50 cm and relief seldom exceeds 3 cm. Above these dimensions bed undulations are referred to as **dunes** or **sandwaves** (see 6.2). Ripples show a wide variety of shapes many of which relate to particular sedimentary processes and hence are used to interpret conditions of deposition.

Cross lamination is the pattern of internal lamination which develops within sand as a result of the migration of ripples. It can be seen both on bedding planes and on vertical surfaces. Patterns of cross lamination are often specific to particular types of ripple and so can also be used in interpretation.

6.1.2 Material

Although ripples and cross lamination are principally features of sand-grade sediment, they may also occur in coarse silts. They are most common in fine- to medium-grained sand and are rare in material coarser than coarse sand, except where they are due to wave action or to strong winds.

6.1.3 Ripple morphology

Ripples are described in terms of their appearance in both profile and plan view (Fig. 6.1). The important distinction between symmetrical and asymmetrical ripples is based on their profile perpendicular to the crestline. Although there is some truth in the generalisation that ripples with symmetrical profiles are the product of wave action and those with strongly asymmetrical profiles are due to current activity, the reality is rather more complex. At least as important in interpretation is the ripple pattern in plan view and the shape and continuity of ripple crestlines in particular. A whole range of patterns is seen in the rock record and on present-day beaches, river beds and tidal flats.

Detailed measurement and recording of ripple morphology can be very informative and should always be attempted in any serious study. Basic features can be measured and the values combined to yield indices that point towards the dominant process, even if interpretation can be ambiguous (Fig. 6.2).

The relationship between profile symmetry and crestline continuity and curvature is complicated. Whereas symmetrical ripples commonly have straight and rather continuous crests, not all straight or continuously crested ripples are symmetrical. Some straight-crested ripples may show a marked asymmetry (Fig. 6.6).

Ripples with highly sinuous crests (Fig. 6.2) and those with a strongly three-dimensional shape, such as linguoid ripples (Fig. 6.3) usually have asymmetrical profiles. They have steeper, concave-upwards lee faces and more gently sloping convex-upwards stoss sides. Such ripples result from currents flowing in one direction only. There is, however, a continuum of asymmetrical, current ripples ranging from straight-crested through sinuous-crested to linguoid in shape. Associated ridges and hollows on the stoss sides of ripples are aligned roughly parallel to flow. Scour pits occur in the lee of ripples, commonly in front of a downstream concavity in the crestline or downstream of the gap between two linguoid ripples.

On some beaches, ripple forms with low relief have

(a) Vertical profiles

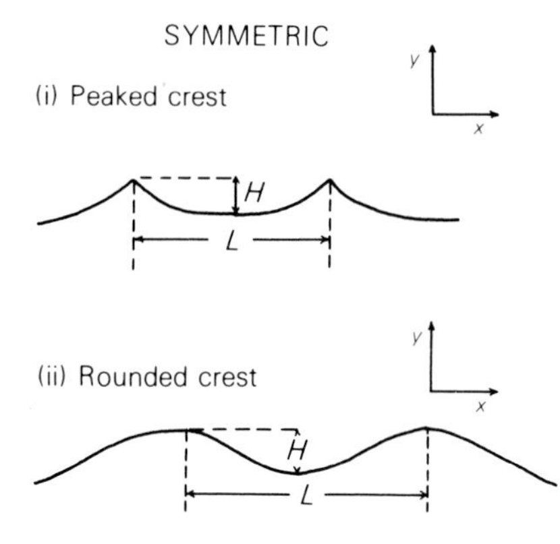

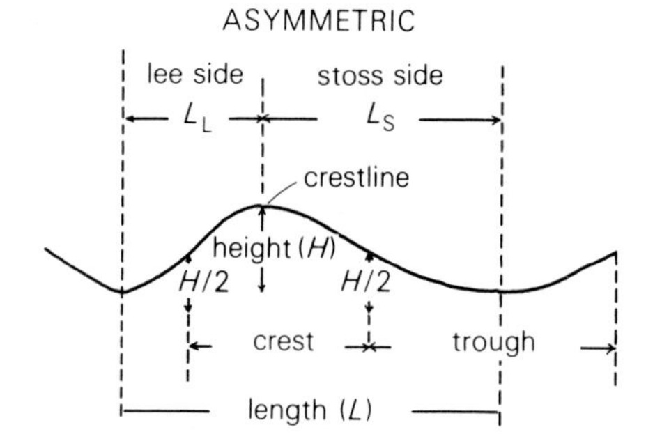

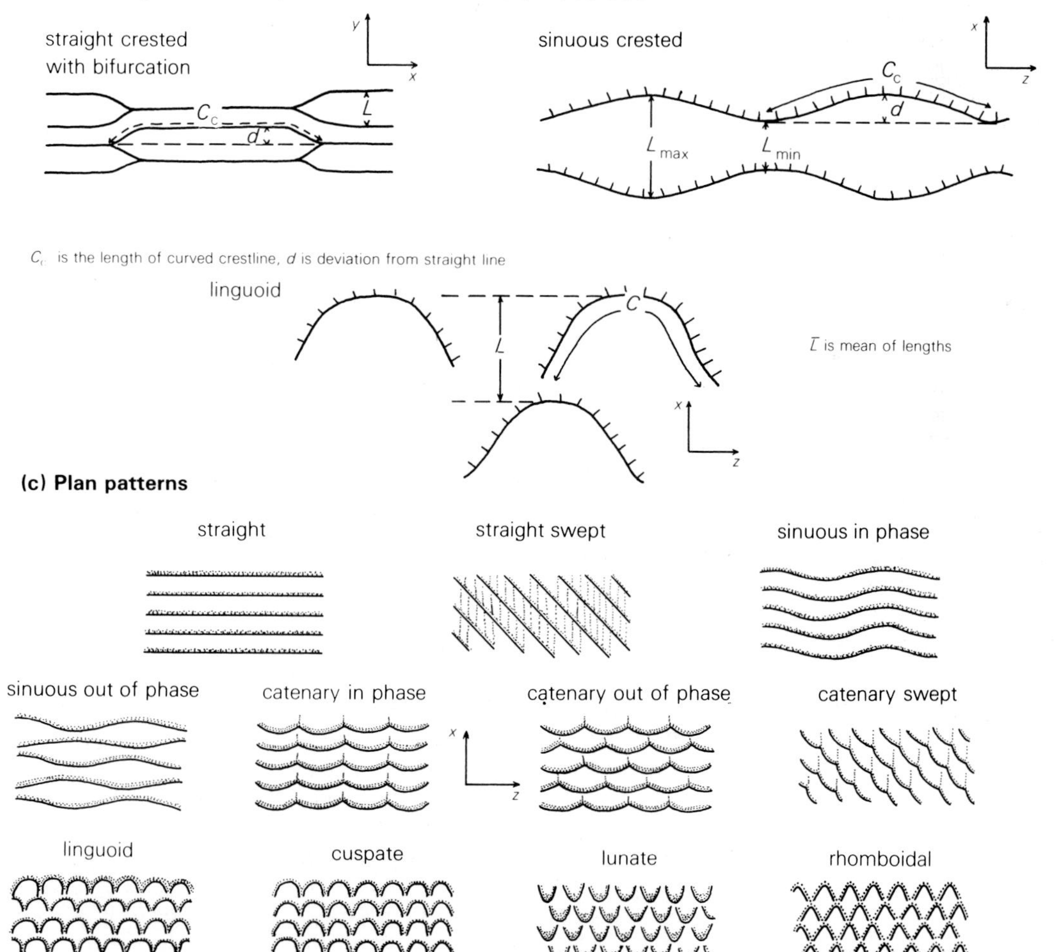

Figure 6.1 Definition diagrams for many of the descriptive terms used in the description of ripples. Most of the terms can also be applied to larger ripple-like bedforms. The reference axes are x parallel to the current, y vertical and z horizontal and perpendicular to the current. (Partly after Allen 1968.)

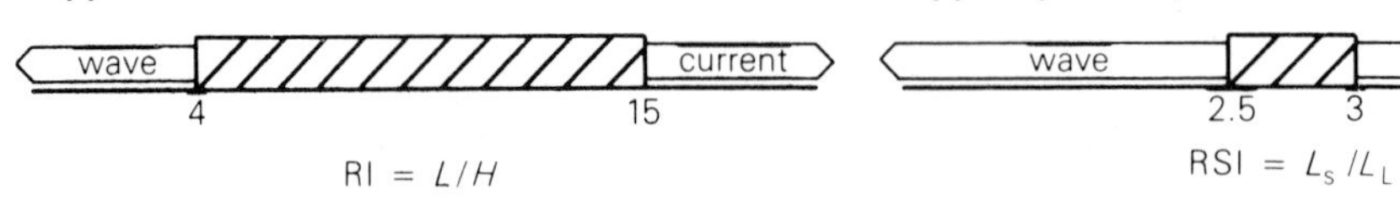

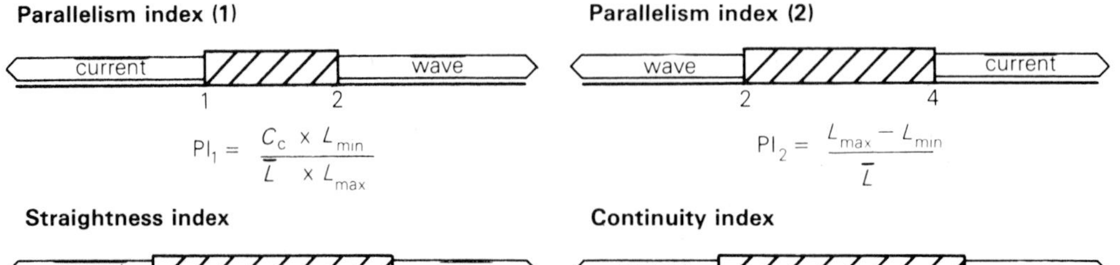

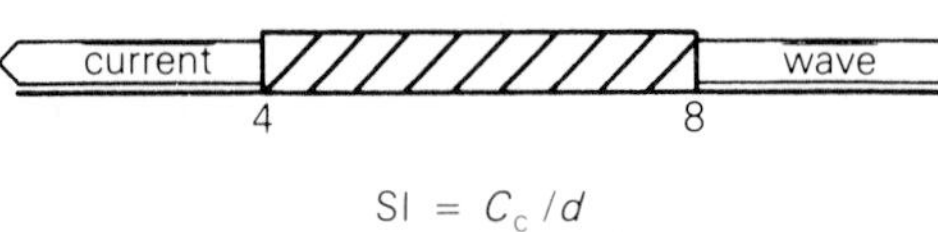

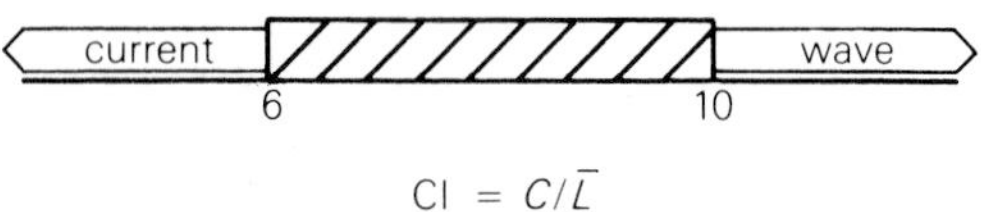

Figure 6.2 Simple ripple indices, showing their ability to discriminate between wave and current activity. The shaded area indicates the range of values over which the indices are overlapping and non-discriminatory.

Figure 6.3 Asymmetrical current ripples with sinuous to linguoid crests on a present-day river bed. Flow was from top left to bottom right. (Knife for scale (arrowed).) Tana River, Finnmark, Norway.

Figure 6.4 Symmetrical wave ripples with continuous crestlines and with some crestline bifurcation. Smaller 'ladder ripples' in the troughs of the main ripple set result from late-stage wave action during emergence. Tana delta, Finnmark, Norway.

Figure 6.5 Symmetrical wave ripples with rather continuous crestlines: (a) rounded crests with concentration of heavy minerals; (b) peaked crests with weakly developed 'ladder ripples' in the troughs. Tana delta, Finnmark, Norway.

Figure 6.6 Asymmetrical ripples with continuous crestlines and with ridges and troughs on their stoss sides. Such ripples might reflect either the co-existence of currents and waves or the asymmetry of the wave orbital velocities. Tana delta, Finnmark, Norway.

very marked repeated rhomboidal shapes which give a fish-scale pattern to the sediment surface. These **rhomboid ripples** are elongate parallel with the current (usually the backwash on a beach) having spacings of the order of a few tens of centimetres and heights of less than 1 cm. They are asymmetrical in profile, being highest at their downstream point. Ancient equivalents are rare.

Symmetrical ripples and those with very straight and continuous crestlines are associated with wave action. They generally lack the scour pits and their crestlines show zig-zag junctions or bifurcation (Fig. 6.4) and may be either smoothly rounded (Fig. 6.5a) or quite sharply peaked in profile (Fig. 6.5b). More complex wave ripples show multiple crests. Others show flattened tops or small steps on their sides, usually produced by shallowing or emergence.

Figure 6.7 Interference pattern in symmetrical wave ripples resulting from at least two, probably coexisting wave sets. Tana River, Finnmark, Norway.

Figure 6.8 Photographs (a–d) show ripples from both present-day sand surfaces and upper bedding surfaces of sandstone. Try to describe the ripples as fully as possible and suggest what processes were responsible for them. In what directions did the currents and/or waves responsible operate?

Try to distinguish between current and wave ripples on the basis of symmetry and crestline continuity, but do not expect complete success. Ripples with straight crestlines but with marked asymmetry may be caused by shoaling waves or by an interaction of waves and currents of similar direction (Fig. 6.6) (see 6.1.5).

More complex patterns, resulting from interference between more than one wave set or between waves and currents with divergent directions, range from slight modification of one dominant ripple type to more complex interference patterns (Fig. 6.7).

In order to assess your descriptive powers and understanding, study Figure 6.8, or better still visit a modern, rippled, sand environment and describe, measure and interpret the ripples you find there.

6.1.4 Internal structure: cross lamination

Where ripples occur on a bedding plane or present-day surface, it is often possible to see an associated internal lamination on surfaces perpendicular to bedding. Recognition of such lamination is of value in interpreting rock sequences. In some sequences of interbedded sand and finer sediment, sand ripples are isolated in finer-grained sediment or are preserved as morphological features on the top surfaces of thicker sand beds. Such units are termed **form sets**, and the relationship between the form and the internal lamination is usually clear (Figs. 6.9 & 10). In many sandstones, however, only internal small-scale trough cross lamination occurs. This comprises units (sets) up to 3–4 cm thick, each made up of inclined laminae (foresets or cross laminae) (Fig. 6.9) which are usually concave upwards with tangential lower contacts and sharp, truncated upper con-

Figure 6.10 A coset of ripple cross-laminated sand, with ripple form sets preserved on the top surface. The cross lamination dips to the right, reflecting the successive positions of the lee faces of the ripples. The sets are inclined to the left (up-current) and 'climb' to give 'ripple-drift cross lamination'.

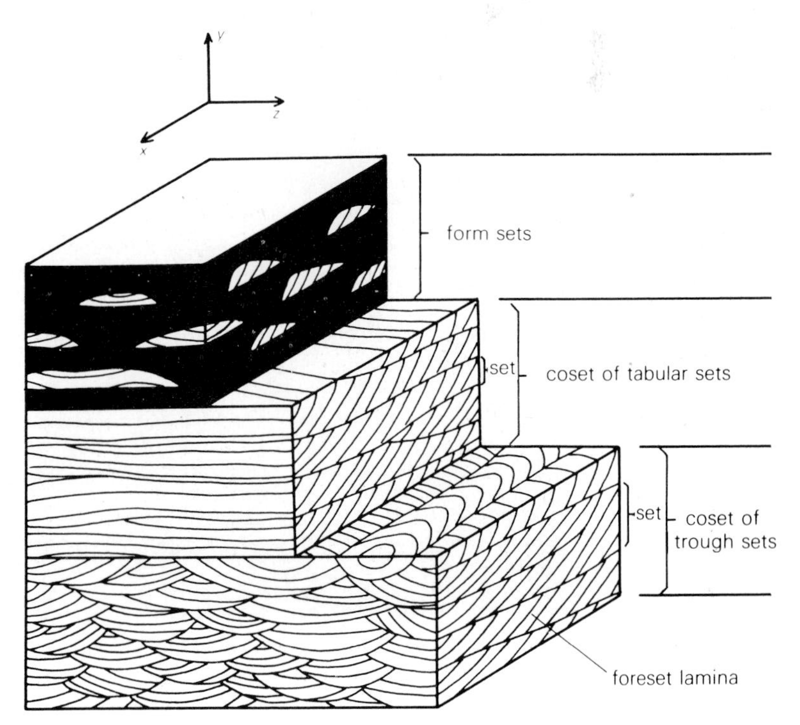

Figure 6.9 Definition diagram for the basic types of cross lamination. The same terms apply at a larger scale to cross bedding (based on Allen 1968).

tacts. Bases of sets are commonly trough- or spoon-shaped, being strongly concave upwards transverse to the mean foreset dip and more gently curved parallel to the dip.

Many exposed, wind-deflated sand surfaces and many bedding planes in ancient, medium- to fine-grained sandstones show a distinctive pattern of curved laminae dipping into the bed in parallel zones (Fig. 6.11). The laminae are concave down dip. The zones are commonly up to 8 cm wide and 20–30 cm long, although sometimes longer. This pattern is termed **rib and furrow** and it is a plan view of the trough cross lamination produced during migration of current ripples (see below). Less commonly, straighter cross laminae intersect bedding planes and may dip in opposed directions. This pattern of opposed cross lamination is generated by certain types of wave ripple. It is often accompanied by an interdigitating of laminae at the ripple crest and by the draping of some laminae over the crest (Fig. 6.12). Cross lamination can *only* be understood by a full appreciation of its three-dimensional nature as different orientations of vertical section (i.e. of exposure) give rather different patterns of lamination.

The organisation of laminae is best understood by reference to block diagrams (Fig. 6.9). Test your understanding by trying to predict the patterns of lamination on horizontal and randomly oriented vertical surfaces through these block diagrams. Two important varieties of cross lamination warrant separate treatment.

Climbing ripple cross lamination (ripple drift). In most cross-laminated sediment, boundaries between sets are erosive and roughly horizontal, but in others the boundaries are inclined and not always erosive. This is **ripple drift** or **climbing ripple cross lamination**. Set boundaries dip in the opposite direction to the dip of the cross laminae and at varying angles (Fig. 6.13). Some set boundaries are not erosive and they preserve laminae deposited on the stoss sides of the ripples.

Always record the inclination of set boundaries when they are erosive. When stoss-side laminae are preserved, try to record the inclination of a line through successive positions of the same ripple crest. Both these measurements record the angle of 'climb' of the ripple sets.

Figure 6.11 Upper bedding surface of sandstone showing 'rib and furrow', the horizontal expression of trough cross lamination. Compare this structure with the more idealised view shown on top of the lowest block in Figure 6.9, and hence determine current direction. Central Clare Group, Upper Carboniferous, Co. Clare, Eire.

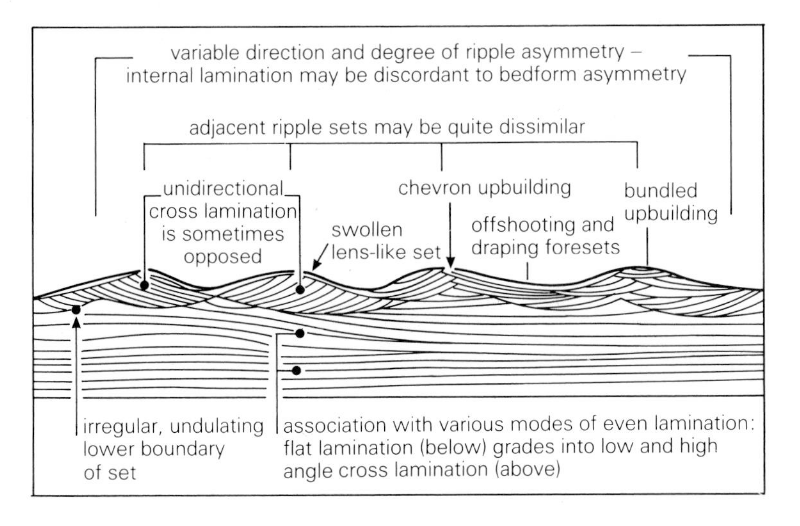

Figure 6.12 Some of the features which help in the diagnosis of wave-ripple cross lamination (after de Raaf *et al.* 1977).

Flaser and lenticular bedding. In some units of ripple cross-laminated sand, the pattern is broken up by interlaminations and lenses of finer-grained sediment (silt and mud). Where fine-grained sediment dominates, sand ripples may occur as isolated ripple form sets (**lenticular bedding:** Fig. 6.14c). In contrast, in **flaser bedding**, the muddy sediment occurs as thin and often discontinuous laminae which drape ripple forms or are confined to ripple troughs (Fig. 6.14b).

6.1.5 *Processes of ripple formation and deposition by water*

Water movement over a sand bed, as unidirectional currents, oscillatory waves or a combination of both may give rise to ripples.

Unidirectional water currents. When the velocity of water flowing over a sand bed exceeds a certain critical value, grains begin to move (see 3.5). With widespread movement of grains finer than about 0.6 mm in diameter, asymmetrical ripples begin to form almost immediately (Fig. 6.15). The earliest ripples are usually rather straight and continuously crested but with gradually increased velocity the ripples change to a more three-dimensional pattern resulting in linguoid forms. Ridges and hollows parallel to flow become more common and more closely spaced on the stoss sides and lee-side scour pits become more clearly defined. In plan, therefore, the shape of the ripples provides a rough qualitative guide to flow velocity (Fig. 6.16), although water depth also plays a part in the case of shallow flows.

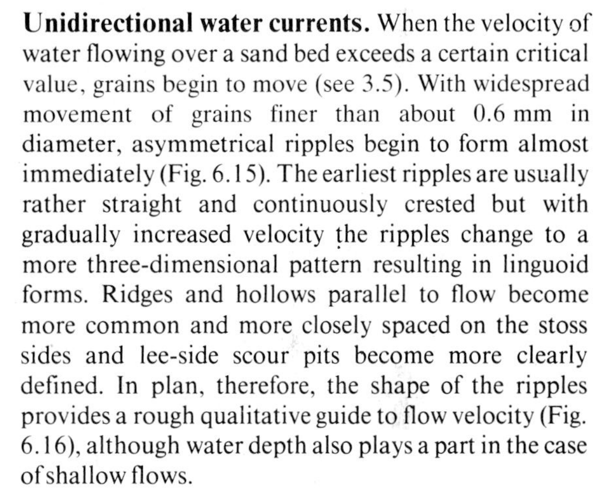

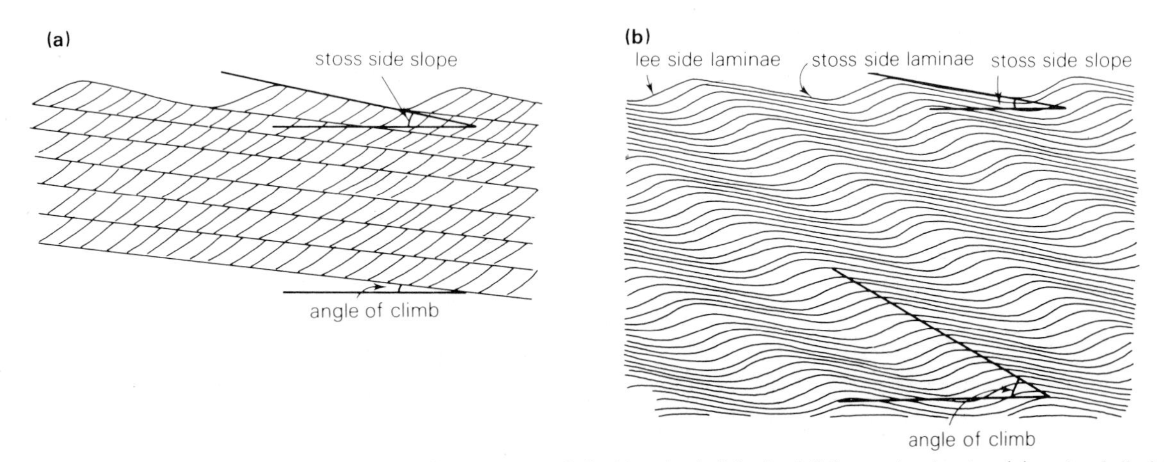

Figure 6.13 Schematic illustrations of different types of climbing ripple ('ripple-drift') cross lamination: (a) angle of climb is less than the angle of the stoss-side slope giving erosion between sets; (b) angle of climb is steeper than the stoss-side slope giving preservation of stoss-side laminae.

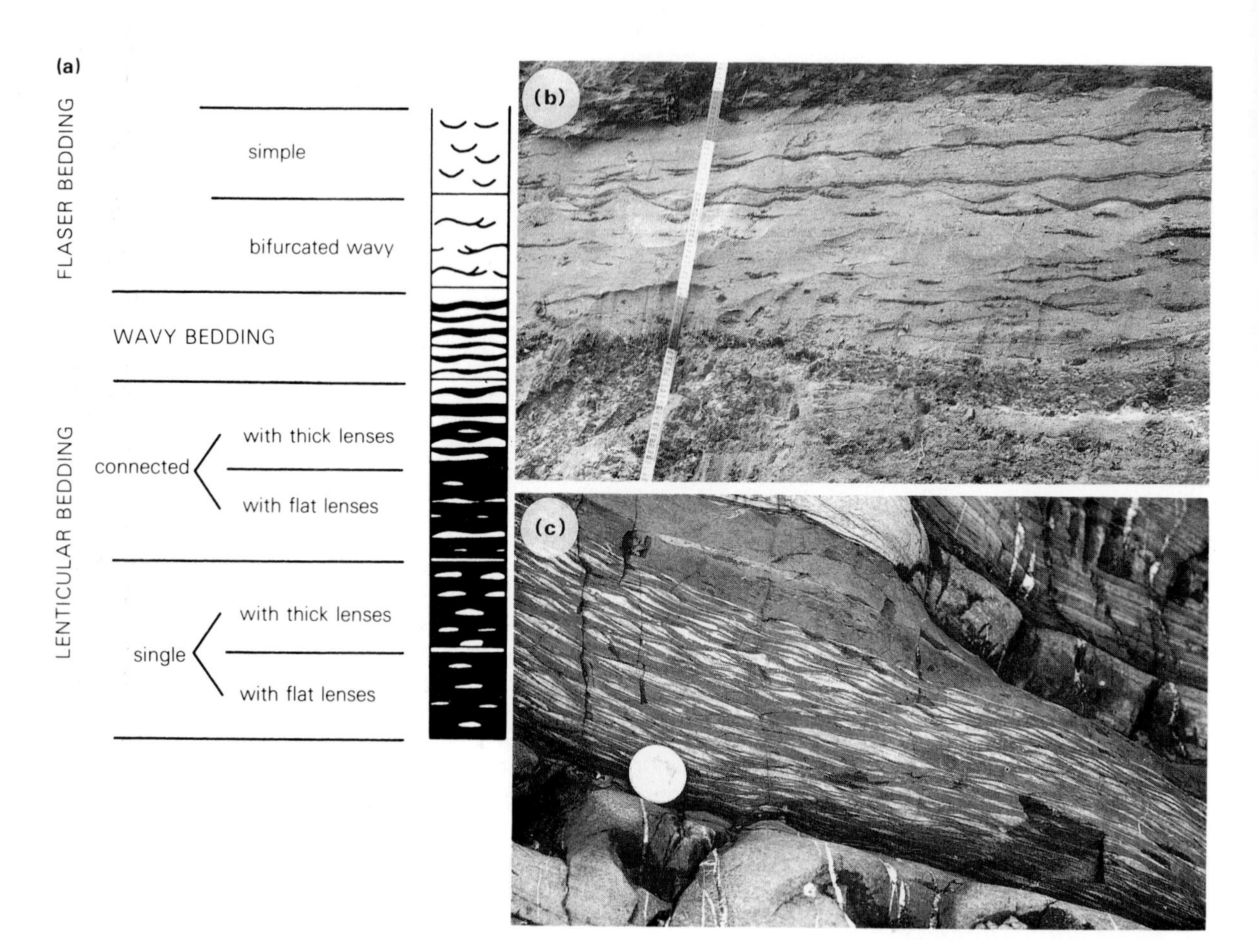

Figure 6.14 (a) The variety of cross lamination resulting from mixed lithologies of sand and mud (after Reineck and Singh 1973). Photographs illustrate: (b) flaser bedding with mud drapes predominantly in the ripple troughs (Haringvliet excavation, Netherlands); (c) lenticular bedding with rather peaked symmetrical ripple form sets. (Northam Formation, Upper Carboniferous, North Devon.)

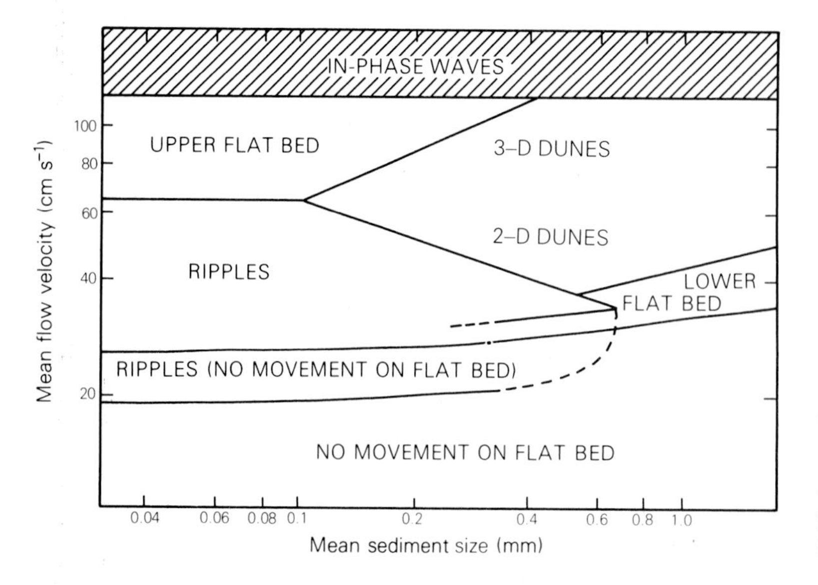

Figure 6.15 The stability fields of different bedforms in relation to flow velocity and grain size for water depth of 20 cm (note that the plot is on a log–log scale). For other depths, the positions of the lines are shifted and a three-dimensional plot is needed to illustrate fully the bedform distribution. (After Harms *et al.* 1975.)

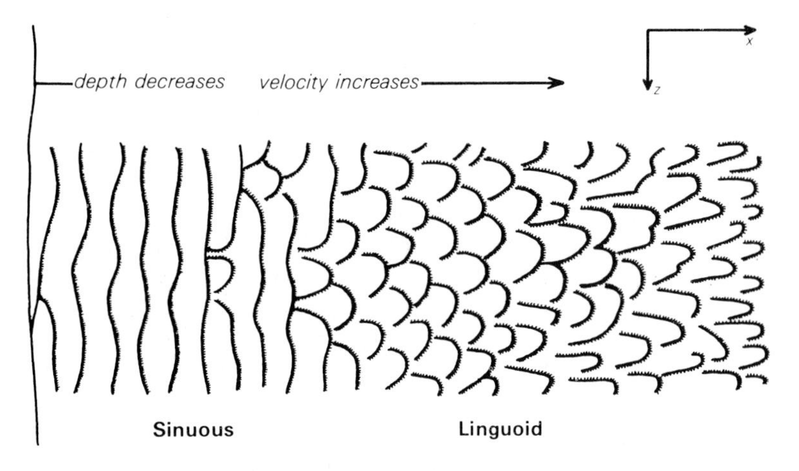

Figure 6.16 The shapes of asymmetrical current ripples, formed without wave influence, related to water depth and velocity. These relationships are only understood in a semi-quantitative way at present. (After Allen 1968.)

Although there is a slight increase in wavelength with increasing flow velocity, the main control on ripple size is the grain size of the sediment, coarser sand giving larger ripples. Current ripples are most readily envisaged as forming in shallow water, but they are also produced in deep water due to the action of ocean bottom currents. Turbidity currents (see 3.7.3) also give rise to ripples and cross lamination. During deceleration of such a current, sand and silt falling from suspension may be reworked on the bed into ripples.

The variety of ripple shape probably relates to the structure of water turbulence close to the bed. Straighter-crested ripples with crestlines transverse to the flow have a rotating eddy in the lee of the ripple, with its axis of rotation parallel to the ripple crestline (Fig. 6.17). With increasing flow velocity, eddies with axial components parallel to flow become more important, producing a more three-dimensional ripple shape. Ridges and hollows on ripple stoss sides also result from eddies with axes of rotation parallel to flow.

Rhomboid ripples, which are not so extensively studied, appear to form under very shallow conditions close to the boundary between ripples and upper flat beds. The ripple crests appear to be associated with small-scale hydraulic jumps (see 3.2.6).

Study and try to map the movement of sand grains over ripples in a small stream or laboratory channel. Identify zones of separation and reattachment. Notice how the flow reattaches to give an area of scour down stream of each ripple. With straight-crested ripples the reattachment is in a more or less continuous zone, but with linguoid ripples reattachment is concentrated in scour pits from where grains move centrifugally. Some of the sand swept from the scour pit moves up stream to mix with sand being deposited in the lee of the upstream ripple. This mixing gives the lee face of the ripple a tangential base. Most of the sand swept from a scour pit moves down stream to supply the next ripple down stream. Grains moving relatively slowly at the crestline stop abruptly and accumulate high on the lee face, oversteepening its slope. As the angle of slip is exceeded, failure occurs and **grainflow** takes place (see 3.7.2). Grains moving more rapidly at the crestline are thrown further out onto the lee side by a process of **grainfall**. The grains' trajectories are influenced by the strong eddying in the separation zone. Together these two processes generate the cross laminae which record the migration of the lee face of the ripple. The downstream movement of linguoid ripples with scour pit – lee face couplets generates trough cross lamination (Figs 6.9 & 11). Straighter-crested ripples produce cross-laminated sets of less pronounced trough shape.

Cosets of ripple cross lamination result from the migration of ripples combined with a net supply of sediment to the bed. With no addition of sediment, ripples migrate down stream but they will only be preserved when movement ceases, and then only as form sets. With a high rate of sediment supply, the bed will grow vertically as ripples migrate producing climbing ripple cross lamination. The angle of climb reflects the balance between the rates of upward bed growth and ripple migration. When the angle of climb exceeds the slope of the ripple stoss side, stoss-side laminae are preserved (Figs 6.13 & 18).

Surface wave processes. All symmetrical ripples, and many asymmetrical ones with straight and continuous crestlines, result from surface wave activity, sometimes operating in conjunction with a current. The morphology and lamination are closely related to the pattern of water movement close to the bed. Look closely at gentle wave action on a beach or try simple experiments in a laboratory wave tank. Crystals of

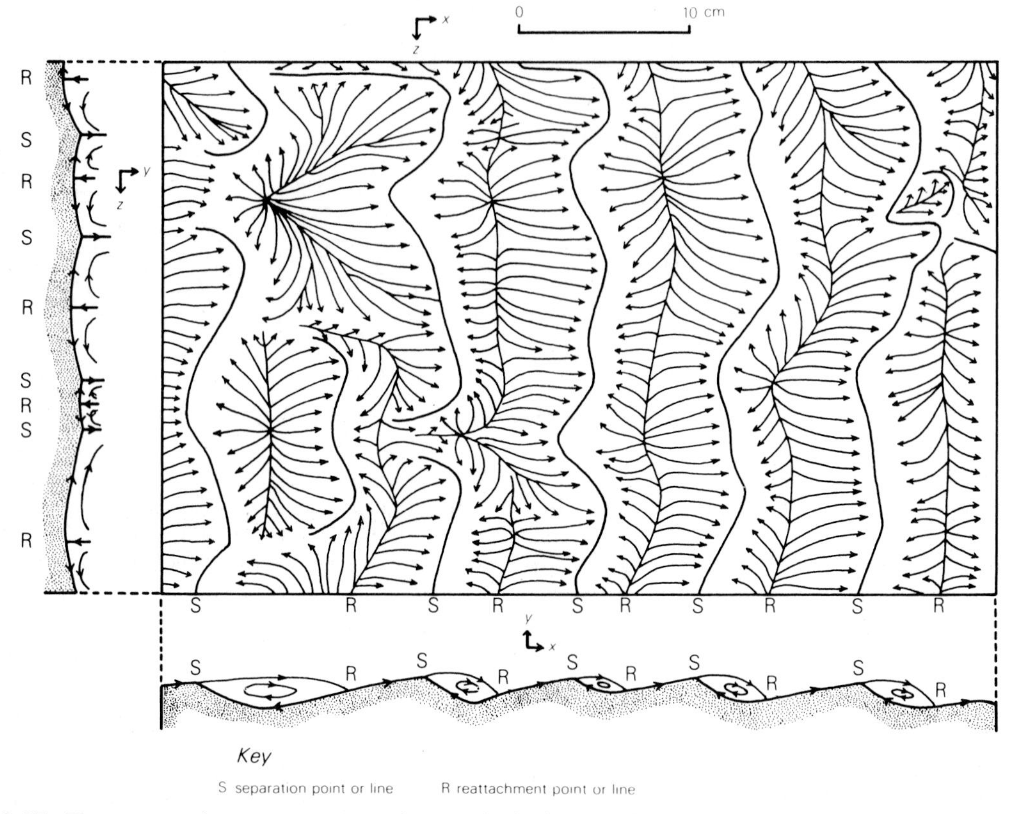

Figure 6.17 The pattern of water movement close to the bed over a field of asymmetrical current ripples. The plan shows flow directions at the bed, the heavier lines representing ripple crestlines. These directions will be similar to the directions of grain movement. Flow separates at the ripple crest and reattaches on the stoss side of the next ripple down stream. Note how the attachment is focused down stream of concave sectors of crestline. (After Allen 1970.)

potassium permanganate on the bed of a wave tank will give a dye stream which shows the pattern of water movement close to the bed. There are three main types of wave behaviour which have a bearing on sediment response: **free gravity waves, forced waves** and **breaking waves.**

Free gravity waves move beyond the area where they were generated by wind action, and the pattern of movement for any water particle is an almost closed loop (see 3.3). Close to the bed wave-orbitals become horizontally flattened first as ellipses and eventually linear with a to-and-fro movement. This oscillatory motion generates straight-crested ripples with their crestlines parallel to the wavefront.

The first ripples to form are **rolling grain ripples**, which are of low relief and reflect movement just above critical erosion conditions. With stronger waves, ripples generate eddies and **vortex ripples** develop (Figs 6.4, 5, 6 & 19).

The size, spacing and symmetry of wave-generated ripples appear to be controlled by four principal factors which define boundary conditions for wave-generated bedforms: the maximum wave-orbital velocity at the bed, the asymmetry of orbital velocities at the bed, the mean grain size and the wave period. The last two factors mainly influence ripple size. Wave ripples occur when maximum orbital velocities fall between those that give no movement and those which give a flat bed (Fig. 6.20). The asymmetry of the orbital velocity determines the boundary between symmetrical and asymmetrical ripples, greater velocity asymmetry giving more asymmetrical ripples. In shallow offshore areas, where waves are shoaling, a zonation of ripple types can sometimes be recognised (Fig. 6.21).

Most waves result from the drag of wind on the water surface, but such processes are complicated and have little direct bearing on sediment response. In shallow water, however, the sediment surface may be strongly

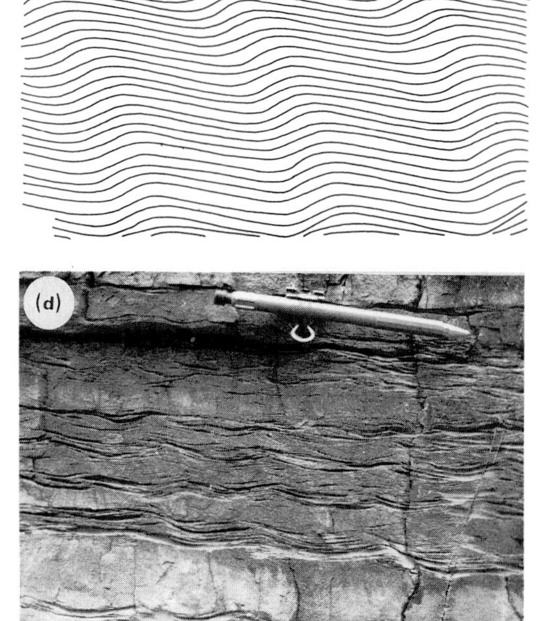

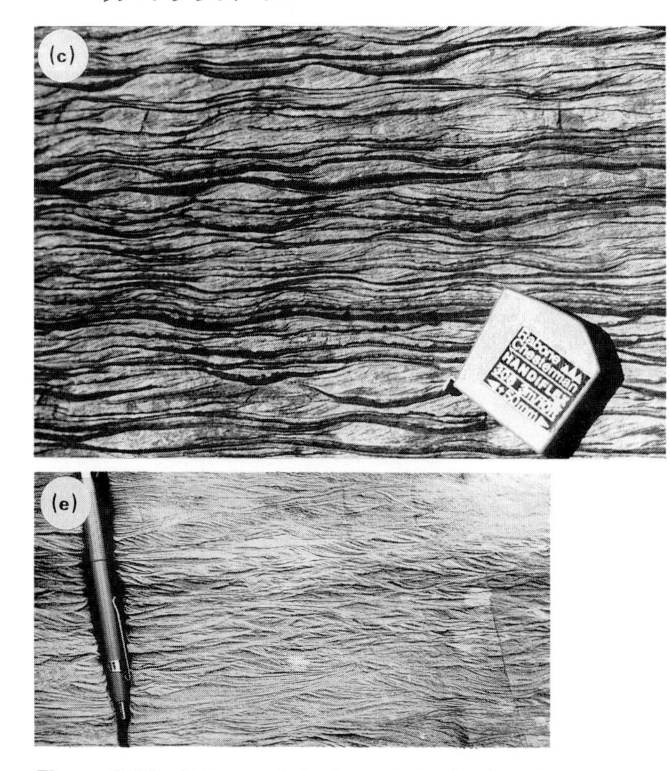

Figure 6.18 Patterns of ripple cross lamination shown schematically and in rocks. Suggest what has happened in terms of depositional process in each case. (Photograph (c) courtesy G. Kelling.)

influenced by waves being actively driven by the wind (**forced waves**). Their pattern of water movement is more complex than that of free gravity waves, involving a combination of orbital motion and a unidirectional component. The resulting ripples are asymmetrical and may be difficult to distinguish from the products of shoaling waves (Fig. 6.6).

Under **breaking waves**, flow is extremely confused. The surge and backwash of the swash zone will generate ripples only if the waves are gentle. Under more active conditions rhomboid ripples or a flat bed develops (see 6.1.2, 6.4.4 & Fig. 6.20).

Shapes of symmetrical ripples are largely a function of water depth. Round-crested forms occur in rather deep water while strongly peaked ripples are more common in very shallow, near-emergent conditions.

Interference effects. In many settings, waves and currents or more than one different wave set may coexist and interact. With wave–current interaction, the ripple

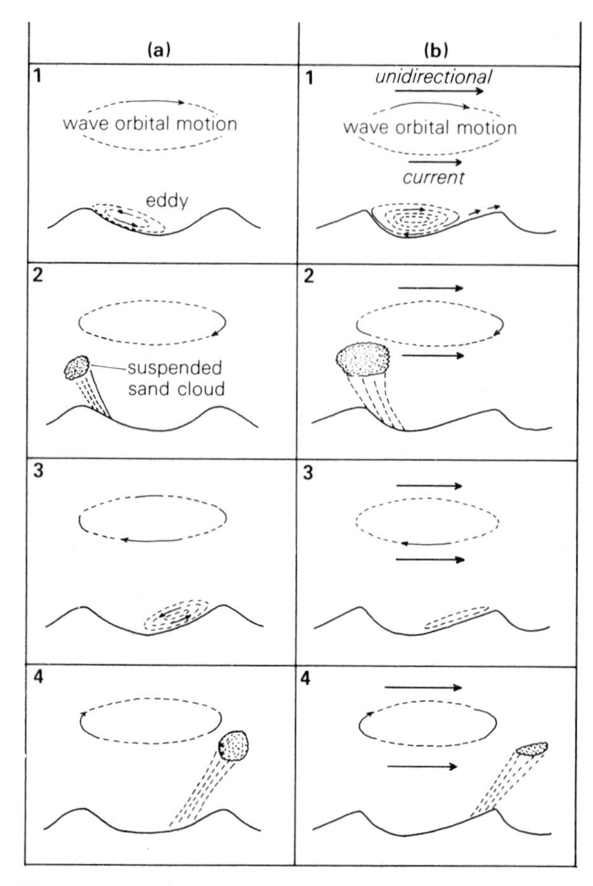

Figure 6.19 The patterns of water and sediment movement over vortex wave ripples: (a) where the wave orbital velocities are similar ripples are symmetrical; and (b) where there is an asymmetry in the orbital currents ripples are asymmetrical (after Inman and Bowen, 1963).

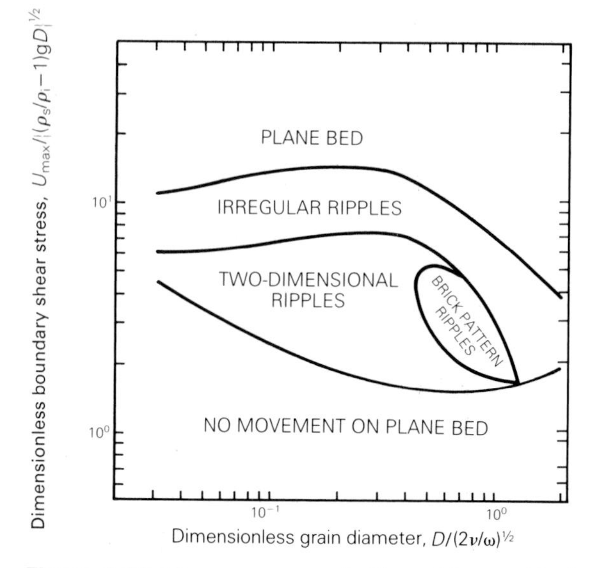

Figure 6.20 The occurrence of different types of bedform as a result of waves acting in a straight channel, under different conditions of wave strength and sediment grain size. The ordinate is a dimensionless expression of boundary shear stress (a measure of wave strength) and the abscissa is a dimensionless measure of grain size. U_{max} is maximum water orbital velocity close to the bed, D is particle diameter, ρ_S and ρ_L are solid and fluid densities, ν is kinematic viscosity of the fluid and ω is the angular frequency of the waves (after Kaneko 1980).

pattern produced depends on the relative strengths and directions of the two processes. If they act in similar directions, although not necessarily with the same sense of motion, straight-crested ripples result. These are difficult to distinguish from those produced by shoaling or forced waves. Waves straighten the crests of what might otherwise have been sinuously crested current ripples. When wave and current directions diverge, interference patterns develop (e.g. Fig. 6.7). However, not all interference patterns imply that the various processes operated at the same time. Separation of different processes in time is particularly common on tidal flats.

When wave motion is superimposed on currents, the water velocity close to the bed is instantaneously increased. Critical erosion velocity may then be exceeded and a bed may become rippled under a current whose time-averaged velocity is subcritical.

6.1.6 *Wind ripples*

Three types of small-scale ripple are common on present-day wind-blown surfaces although they are less common in the rock record.

Impact ripples. Impact ripples have low relief and they form from the coarser-grained fraction of the sand upon which they develop. They have a high ripple form index (Fig. 6.2) and rather straight and continuous crestlines transverse to the wind direction and upon which the coarsest grains are concentrated. These ripples are slightly asymmetrical in profile. Their lee faces are inclined at low angles, below the angle of rest (Fig. 6.22).

With impact ripples, the controlling factor on their form and development is the behaviour of sand grains as high momentum, impacting particles. Under wind shear, grains move principally by saltation, the length of trajectory being directly proportional to wind speed.

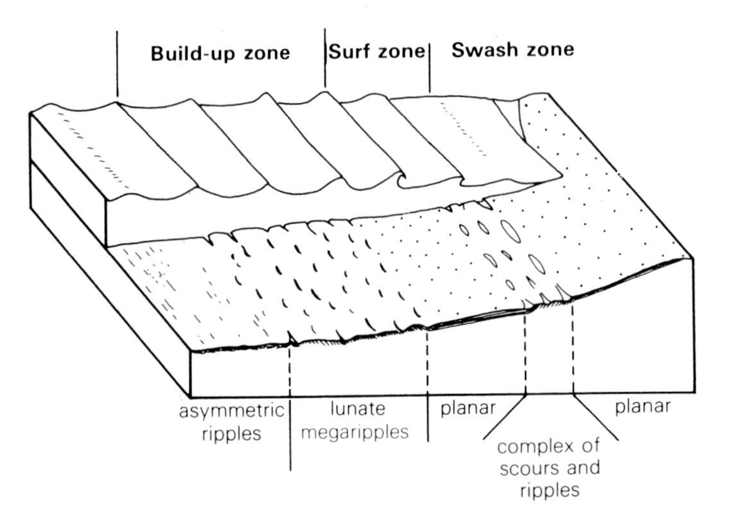

Figure 6.21 Zonation of wave-generated bedforms offshore from the beach on the high-energy coast of Oregon. Shorelines with different energy regimes have different patterns of zonation (after Clifton *et al.* 1971).

Once established, ripples present different angles to impacting grains on their upwind and downwind sides. Grains hitting the upwind side are likely to bounce or set other grains in motion, whereas the lower angle of incidence of grains hitting the downwind side favours their trapping (Fig. 6.23). Impacting grains cause other grains on the bed to move short distances by **creep**, and this process can account for a considerable proportion of total transport. Ripple spacing compares closely with trajectory length and therefore varies with wind speed.

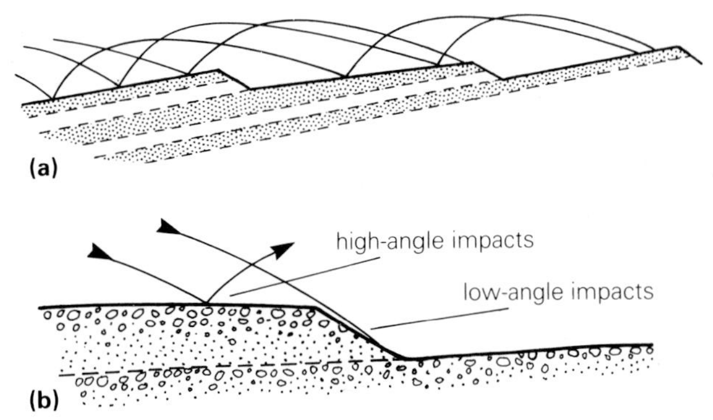

Figure 6.23 (a) Wind impact ripples, showing how the ripple spacing relates in a general way to saltation path length and how the migration of ripples produces sub-parallel lamination. (b) The impact angle of saltating sand grains differs between stoss and lee sides. High-angle impacts promote creep of coarser grains on the bed, and sorting gives rise to lamination with inverse grading.

Figure 6.22 Wind impact ripples with rather continuous crestlines and a slight asymmetry. Wind from top left to bottom right.

Larger grains move by creep and become concentrated towards the ripple crest. Air temperature, which influences viscosity, will influence the grain size of ripples, coarser sand ripples being more common in cold settings.

Aerodynamic ripples. Aerodynamic ripples are low relief features with a high ripple-form index but commonly with a more three-dimensional form than impact ripples.

Adhesion ripples. When dry sand is blown across a wet sediment surface, some grains stick to the surface on impact. Capillary rise of moisture helps to trap further grains by constantly wetting the new surface and an irregular, blistered but sometimes ripple-like surface develops (Fig. 6.24). Steeper sides tend to be up wind. Adhesion ripples have a low preservation potential since, on drying out, they collapse and are further reworked by the wind.

6.1.7 *Uses of ripples and cross lamination*

Ripple marks and cross lamination have three main uses.

'Way-up'. Ripple marks and cross lamination are amongst the most reliable indicators of 'way-up', although there are three ways in which they may be mistaken for structures whose 'way-up' significance is ambiguous or opposite.

TECTONIC RIPPLES
In strongly deformed rocks, bedding-cleavage intersection sometimes causes a pattern of small-scale undulations on bedding surfaces which can look remarkably like ripples. Careful study of the cleavage and joint patterns may resolve the problem, but one should always be cautious when apparent ripple crests are closely parallel to fold axes or cleavage traces. Try to identify cross lamination associated with the 'ripples' before asserting their sedimentary origin.

CASTS OF RIPPLE MARKS ON LOWER BEDDING SURFACES
In interbedded sequences with preserved ripple morphologies, a lower bedding surface may sometimes preserve the underlying rippled surface as a cast. An examination of the internal structure of the beds either side of the rippled surface should show cross lamination *beneath* the rippled surface.

TRANSVERSE SCOURS (see 4.2.2)
Superficially these can resemble current ripples, but they have no cross lamination and may be associated with other types of sole mark.

Conditions of deposition. Ripples indicate deposition by currents and waves strong enough to exceed the critical erosion velocity but not strong enough to form dunes, sandwaves or a flat bed. Ripple symmetry and crestline shape enable estimates of the relative strengths of currents and waves to be made. Hydrodynamic interpretations of preserved bedforms or internal structures are usually based on comparison with equilibrium conditions. All sedimentation, by its very nature, demands non-equilibrium conditions with an excess of sediment supply due to either waning or expanding flow. Climbing ripple cross lamination can help to indicate sediment supply as the angle of climb relates directly to sedimentation rate.

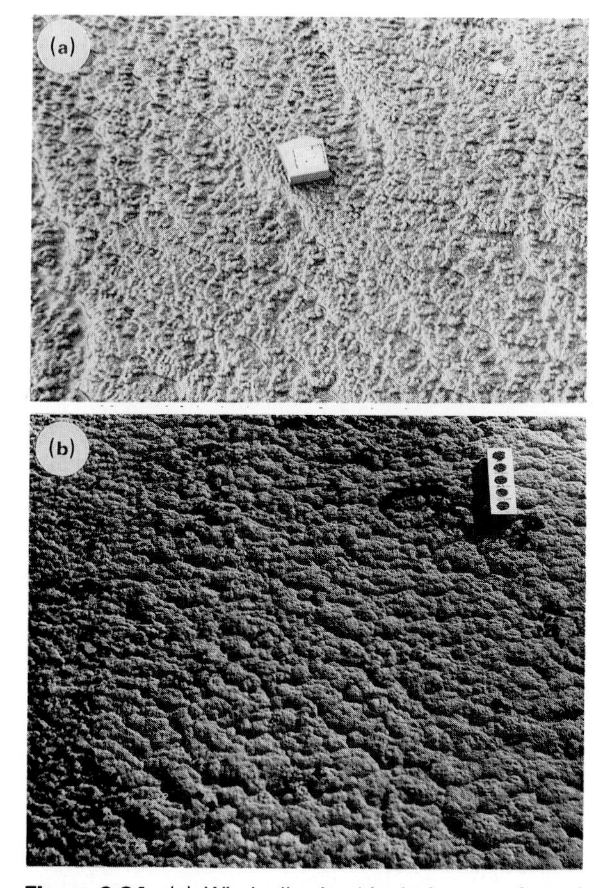

Figure 6.24 (a) Wind adhesion 'ripples' on a surface of wet sand over which dry sand has been blown. Note how the steeper sides of the 'ripples' face into the wind. Tana delta, Finnmark, Norway. (b) Upper bedding surface of sandstone showing an irregular, small-scale topography interpreted as wind adhesion 'ripples'. Independence Fjord Group, Proterozoic, N. Greenland.

Palaeocurrent and palaeowave direction. Both ripple morphology and cross lamination may indicate directions of waves and currents. Ripples respond quickly to local or short-term changes in flow direction, so they may record directions divergent from the overall palaeoslope or the high stage flow.

Ideally, try to measure ripples or cross lamination on bedding planes rather than rely on cross lamination in vertical section. Remember that it is difficult to judge anything but a component of direction in vertical section. Measure ripple crest direction (for wave ripples), general

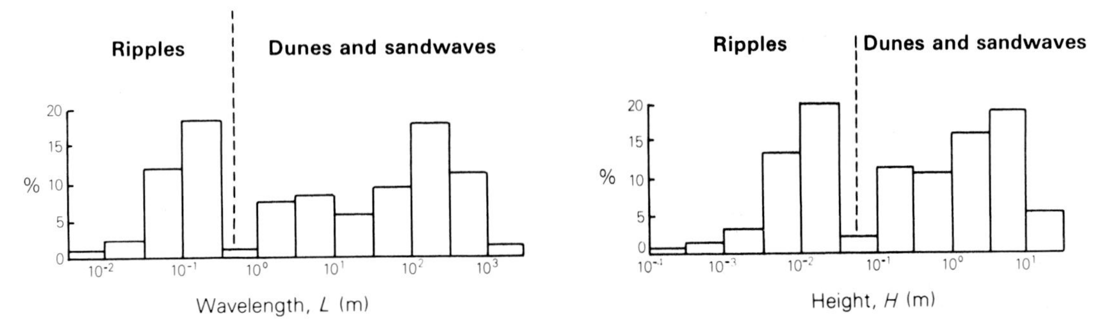

Figure 6.25 Histograms of wavelength and height of ripple-like bedforms from various present-day environments. A conspicuous gap separates ripples from larger forms. The population of larger forms probably includes representatives of both dunes and sandwaves. (After Allen 1968).

ripple trend (for current ripples), or the axes of troughs on bedding planes showing rib and furrow. For symmetrical wave ripples, it may not be possible to judge the sense of wave movement.

6.2 Aqueous dunes, sandwaves, bars and cross bedding

6.2.1 *Introduction*

Many areas of sandy river beds, tidal flats and channels and of sandy sea floor swept by tidal currents show bedforms many times larger than current ripples. These larger forms are separated from ripples by a distinct jump in both height and spacing even though many of the proportions and shape factors are often comparable (Fig. 6.25). It is relatively uncommon to find current-generated bedforms in the height range 3–10 cm and the spacing range 30 cm – 1 m. The larger forms commonly have current ripples superimposed upon their stoss sides. When such superimposition is seen on a tidal flat or a river bed, it is quite likely that it results, at least in part, from continued sediment movement during the falling river stage or the ebb tide, when the large forms were no longer active. However, in controlled experiments, superimposition also occurs under equilibrium conditions.

In the geological record, only fairly small-scale examples of these larger bedforms occur as bedding surface features (form sets) and their presence is most often reconstructed from the patterns of cross bedding to which they give rise.

6.2.2 *Material*

Large-scale bedforms and the cross bedding which characterises their internal structure most commonly occur in sands of medium sand and coarser grain size. They may also occur in gravelly sands and fine gravels of any composition. Cross bedding is, in addition, found in many coarser conglomerates, where its origin may be due to other morphological features. Certain types of large-scale cross bedding in sandstones are also the result of large morphological features, not directly related to bedforms.

6.2.3 *Size, shape and classification of large-scale bedforms*

Large sandy bedforms, ranging in size upwards from 1 m in wavelength have in the past been described as dunes, sandwaves, megaripples, large-scale ripple marks and various types of bar. Lack of agreement on classification and lack of consistency in applying classifications has led to some confusion. Here we try to set out the features which seem to us most important in describing these forms. Most of these features will be visible at low tide or low river stage, but similar criteria can be applied to the description of sub-aqueous features as revealed by echo-sounder.

The first question to ask about large sandy forms is whether or not they form a repetitive pattern on the bed. Do they have a regular spacing and a more or less uniform height and is their plan form consistent? If so, the height and wavelength should be recorded and the plan form described using terminology similar to that for asymmetric ripples (cf. Fig. 6.1).

The second question to ask is whether there is only one scale of bedform present (simple forms), or whether there is superimposition of more than one scale (compound forms). In the case of compound forms, try to judge if the smaller forms are confined to the stoss side of the large form or if they occur on both stoss and lee sides. Also, record the dimensions of both scales of structure and try to see how the orientation of the smaller structures relates to the larger ones.

Figure 6.26 An area of typical 3-dimensional dunes on a present-day river bed with superimposed current ripples. The sharp base of the slip face in the foreground probably resulted from falling-stage flow. During rapid dune migration under strong flows, the slip faces were probably tangentially based. Tana River, Finnmark, Norway.

With superimposition, the larger bedforms may be repetitive and independent of the morphology of the channel. In other cases, the larger forms may be related to bends in the channel, to sinuosity of the flow within the channel or to splitting and rejoining of the flow. This last judgement may, in the case of large rivers or estuaries, be rather difficult to make. Climbing a nearby hill or the study of aerial photographs or of detailed topographic maps will often be very informative. In the very largest systems, there may be several orders of superimposition.

Simple, repetitive, strongly asymmetric forms, whose dimensions are independent of the width of the channel, are best referred to as **dunes**. These may be strongly three-dimensional with sinuous crest-lines and well developed scour pits on their lee sides (Fig. 6.26) or they may be straight-crested without scour pits (Fig. 6.27b). Dunes commonly have small-scale current ripples on

Figure 6.27 Large bedforms on present-day sand surfaces. (a) These large linguoid forms are of the order of 200 m long and up to 2 m high. They are probably best regarded as a type of low-relief dune, although the classification of such forms has been problematical and terms such as sandwaves and linguoid bars are sometimes seen. Tana River, Finnmark, Norway. (b) Straight-crested dunes on a tidal flat with superimposed current ripples. These dunes lack scour pits in their lee side area. The small terraces on the lee sides reflect wave modification during emergence. The ripples immediately in front of the lee side are oriented oblique to the dunes and reflect later-stage run-off of the ebb tide. Loughor Estuary, West Glamorgan, Wales.

their stoss-sides. These commonly face down stream, towards the dune crest. Immediately down stream of a dune slip face they may face up stream. In the scour pits of three-dimensional dunes, the ripples may fan out from the centre of the pit.

In some areas subject to strong tidal currents, usually subtidal, large and apparently simple bedforms occur. These are up to several metres high and hundreds of metres in wavelength. They are most often seen on the records of echo-sounding or side-scan sonar surveys and are usually referred to as **sandwaves**. They are commonly asymmetric and have rather straight and continuous crestlines up to many hundreds of metres in length.

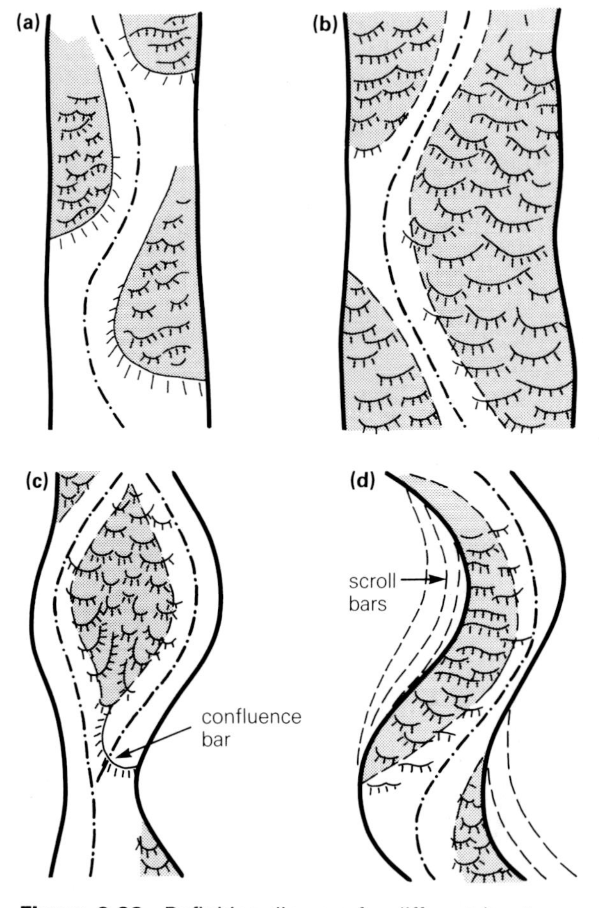

Figure 6.28 Definition diagram for different bar types on a sandy river bed. Note how the dimensions of the bars relate to the channel width. Surfaces of bars have superimposed repetitive bedforms (dunes) whose size is independent of channel width. (a) Alternate bars with their own large-scale slip faces. (b) Side bars without discrete slip faces. (c) Mid-channel bars and possible associated confluence bars. (d) Point bars related to channel curvature; migration of channel gives rise to scroll bars.

Sandwaves are oriented normal to the direction of tidal flow. The asymmetry, which may be very obvious on foreshortened echo traces, is often, in reality, quite slight though a whole spectrum exists from near symmetric forms to strongly asymmetric ones. With strong asymmetry, the steep side may be a slip face, but with progressively reduced asymmetry the angle of the steeper side declines.

With compound bedforms, which apparently are scaled independently of the width of the channel, a term such as **complex** or **compound sandwave** is probably appropriate, although the emergent top of such an area may be referred to as a **sand flat**.

Where a bedform is related in scale to the width of the channel, the general term **bar** is appropriate. This can be suitably qualified depending on its relationship to channel or thalweg curvature, to whether or not it is simple or compound, and whether or not it has its own discrete slip face (Fig. 6.28).

6.2.4 Modification by emergence

Large-scale bedforms, exposed on a river bed at low water stage, commonly show features produced as they emerged. These occur at a variety of scales, and the extent of their development reflects the rate at which emergence took place. With slow emergence, there is more time for modification to occur, slip faces and crestlines of bedforms become rounded off and lobes of sand may extend up stream from the crestline as a result of washover by waves (Fig. 6.29). The same action may also reduce the slope of the slipface, concentrate heavy

Figure 6.29 Waves breaking over the upper part of the lee side of a sandwave as it emerges during the falling-water stage. The waves rework the face to a lower slope and may give washover lobes on the stoss side of the bedform.

minerals and remould current ripples on the stoss side into wave or interference ripples. During falling water stage, the flow may be split by emergent bedforms. Current ripples may be reoriented to reflect this flow and sand lobes may develop at the confluences of these flow threads.

6.2.5 *Internal structures of dunes; medium-scale cross bedding*

Excavation of trenches into dunes reveals patterns of inclined bedding similar to cross lamination, but on a larger scale. This is called **cross bedding**, although the terms 'current bedding' and 'false bedding' are sometimes found in older literature. Although set thickness is usually greater than 10 cm, the terminology of cross lamination (Fig. 6.9) can also be applied. Straight-crested dunes generate tabular sets of wide lateral extent, while three-dimensional dunes give trough-shaped sets (Figs 6.30 & 31). Both types are common in the rock record, particularly in medium- and coarse-grained sandstones.

Trough sets seldom have a maximum thickness greater than 1.5 m and are up to a few metres wide and a few tens of metres long. Most commonly, they are around 30 cm thick, 1–2 m wide and 5–10 m long. Foresets of trough sets are invariably concave upwards with tangential lower contacts. In plan view, trough cross bedding displays a larger version of 'rib and furrow' (Fig. 6.32, cf. Fig. 6.11). Sections perpendicular to the current direction are sometimes described as showing **festoon cross bedding** (Fig. 6.31a).

Tabular sets have a much wider range of size, though sets less than 1 m thick are most common. Sets around 1 m thick commonly extend laterally for tens of metres, often beyond the limits of exposure. Where several sets are stacked in a coset, they may be separated by thin layers of ripple cross lamination. Isolated, single, tabular sets, up to tens of metres thick, are probably not the product of dunes and are described later (6.2.9).

The foresets of tabular sets are either asymptotic (tangentially based) or planar (angular-based) (Fig. 6.33). In plan, they are straight or gently curved.

Within cosets of cross bedding it is usual for the bounding surfaces between sets to be more or less horizontal, although in small outcrops, it may even be difficult to judge the orientation of the depositional horizontal. In some more extensive exposures, however, it is apparent

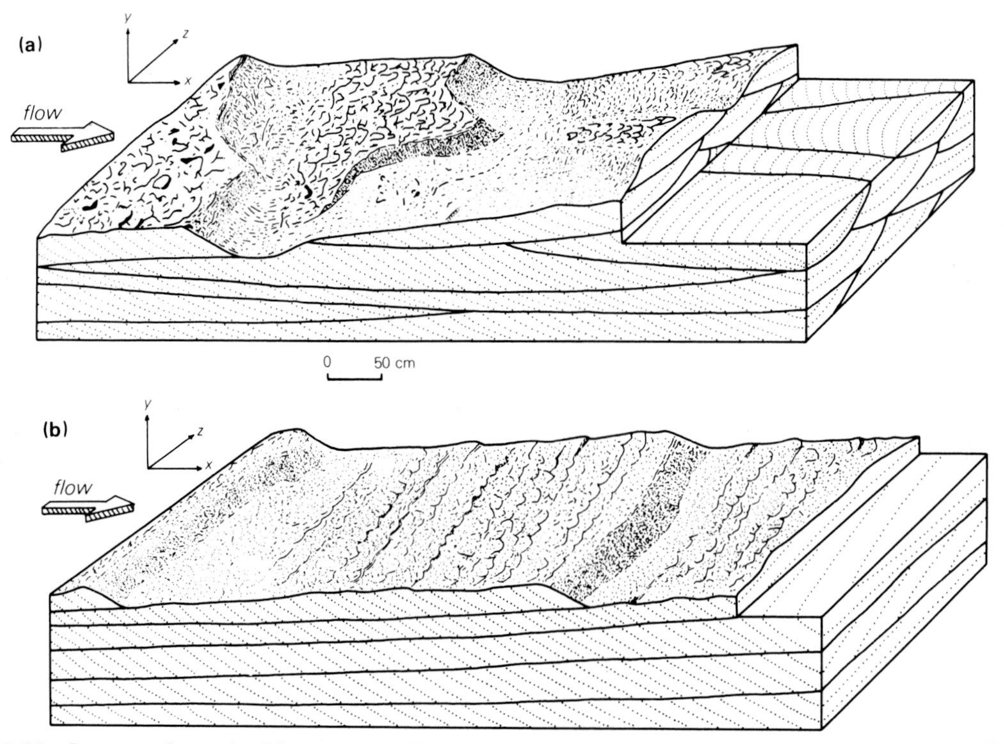

Figure 6.30 Patterns of cross bedding generated by the migration of: (a) dunes (trough cross bedding; see Fig. 6.31a) and (b): sandwaves (tabular cross bedding; see Fig. 6.31c). (After Harms *et al.* 1975.)

Figure 6.31 Patterns of cross bedding in sandstones: (a) trough cross bedding nearly perpendicular to current direction with a slight component to the right (Kinderscout Grit, Upper Carboniferous, Derbyshire); (b) trough cross bedding seen nearly parallel to the current direction (Upper Carboniferous, Nova Scotia, Canada); (c) tabular cross bedding roughly parallel to the current direction. The foresets are planar and angular-based (Roaches Grit, Upper Carboniferous, Staffordshire).

Figure 6.32 Bedding-plane view of trough cross bedding in sandstone showing a larger-scale view of 'rib and furrow' (see Fig. 6.11 and the upper surface of Fig. 6.30a). The concave curvature of the foreset traces indicates flow away from the viewer. Dinantian, East Lothian, Scotland. (Photograph courtesy of G. Kelling.)

that bounding surfaces between sets are themselves inclined, defining larger-scale dipping units. If this is suspected, it can become very important to determine the magnitude and direction of this inclination in relation to the dip direction of the cross bedded foresets. All types of relationship are possible of which climbing (upstream accretion), descending (downstream accretion) and along slope (lateral accretion) are end members.

In some cosets of tabular cross bedding, the directions of dip of foresets in adjacent sets are opposed. Some examples show alternation of direction from set to set, while in others only a few sets show the opposed directions (Fig. 6.34). Such **herring bone cross bedding** is important in interpretation, but care must be taken to distinguish it from festoon cross bedding where sets *appear* to dip in opposite directions (Fig. 6.31a & b).

Some approximately tabular sets are unusual in showing sigmoidal foresets. In these cases, the convex-upwards foreset laminae at the top of the set may pass laterally up-dip into parallel lamination which occurs as a 'topset' unit (Fig. 6.35). Traced down dip, such sets sometimes show a gradual reduction in thickness and in foreset dip.

6.2.6 Discontinuities and modifications in cross bedding

Cross beds are not always simple, particularly in tabular sets. Small-scale ripple cross lamination may occur *within*

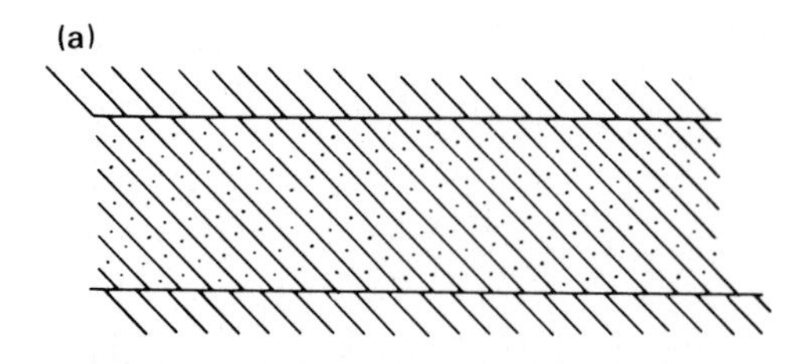

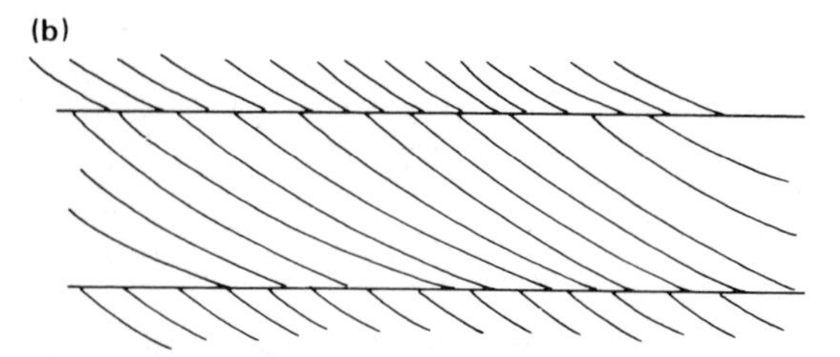

Figure 6.33 Idealised sections in tabular cross bedding, parallel to flow, showing the difference between (a) angular planar foresets and (b) asymptotic or tangential foresets.

Figure 6.34 'Herringbone' cross bedding in sandstone. The sets are tabular in shape and the two directions of cross bedding are roughly opposed to each other. Such cross bedding is most likely in tidal regimes. Lower Greensand, Lower Cretaceous, Bedfordshire.

Figure 6.35 A tabular set of cross bedding in which the foresets extend upstream into topset laminae, giving an overall sigmoidal appearance. The structure probably reflects flows close to the dune-plane bed transition acting under conditions of high net deposition. Scale, centre bottom, is 5 cm wide. St Bees Sandstone, Permian, Cumbria.

the foresets, particularly in the lowest parts of tangentially based foresets, the toeset region. Such cross lamination is commonly directed up the slope of the larger-scale foresets and is termed **counter current cross lamination** (Fig. 6.36). Discontinuities within foresets are sometimes seen in sections parallel to foreset dip. These erosion surfaces (**reactivation surfaces**), are generally less steeply inclined than the foresets on either side and they occur singly or as multiple features within a set (Fig. 6.37). With trough cross bedding or with downstream-inclined bounding surfaces, it may sometimes be difficult to distinguish bounding surfaces of sets from reactivation surfaces.

In cross bedding from some tidal environments, clay drapes occur on the foresets. In some examples these occur in pairs, separating thicker and thinner sand foreset increments. In exceptional cases the spacing of the paired drapes increases and then decreases systematically when traced along the set, to define **foreset bundles**. The number of foreset increments per bundle broadly coincides with the tides of a lunar month.

6.2.7 *Process of formation of dunes, sandwaves, bars and cross bedding*

Dunes and sandwaves are both responses of a sand bed to currents more powerful to those which generate ripples. The differences in dune morphology are the result of differences in flow strength within the dune field, the three-dimensional forms reflecting deeper, more powerful flows. The pattern of eddying around the dunes is closely related to the bedform morphology, with convergence of the separated flow in scour pits characterising the three-dimensional types. In some low-relief

Figure 6.36 Sets of cross bedding seen roughly parallel to the current direction with counter-current ripples developed at the base of the foresets. Counter-current ripples are usually the result of separated flow in the lee of the larger bedform, but in a tidal environment they may relate to the subordinate tidal flow. Subrecent, River Rhine, Netherlands. (Photograph courtesy of R. Boersma.)

Figure 6.37 Discontinuities in the foresets of tabular cross-bedded sets. These 'reactivation surfaces' result from falling-stage modification of the bedforms or from subordinate tidal flow followed by a further period of dominant flow. Tana River, Finnmark, Norway.

dunes (long spacing and small height) the pattern of flow separation and reattachment is confined to the immediate lee side area. In these cases, the bedform heights are much closer to the flow depth and there is often a quite shallow flow over the crestline.

The separated flow in the lee of a dune gives rise to a backflow component which helps produce a tangential slip face. The strongly focused eddying in front of concave sectors of slip face leads to the development of scour pits and to the associated ripple fans. With straight-crested dunes, the strength of flow over the crest can strongly influence lee-side profile and hence foreset shape. With weak flows, grain flow (avalanching) dominates on the slip face, giving angular foresets. With stronger flows, flow separation and grain fall become more important, leading to tangential foresets, sometimes with countercurrent ripples (Fig. 6.38). Reactivation surfaces (Fig. 6.37) in river bedforms result from reworking during emergence (see Fig. 6.39), but the subordinate tide in both intertidal and subtidal settings is also able to produce similar effects.

Tidal sandwaves have morphologies which reflect the imbalance between the two opposing tidal flows. Very slight imbalance is capable of generating a marked asymmetry, but in such cases migration rates are slow. With greater imbalance, migration rates are likely to increase and asymmetry is likely to be more marked. The imbalance is also likely to be reflected in the internal cross bedding. Highly asymmetric bedforms are likely to show relatively simple cross bedding, reflecting the dominant tide, the activity of the subordinate tide being recorded only in reactivation surfaces in the upper parts of sets. More symmetrical bedforms will have more complex cross bedding, showing a greater occurrence of reactivation and of sets of reversed cross bedding, leading to herringbone cross bedding.

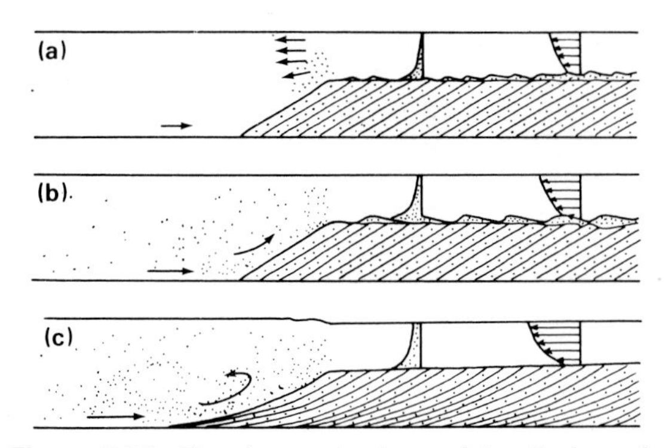

Figure 6.38 The changes in shape of the slip face of a small laboratory delta due to progressive increase in the velocity of flow over it, from (a) to (c) (after Jopling 1965).

Mud drapes are the product of fallout of suspended load at tidal slack water, and when paired they record the slacks on either side of the subordinate tidal flow. The thicker sand layer is the product of the dominant tide, and its strength and duration may change through the lunar month to give bundles of systematic change.

Sigmoidal foresets with parallel topsets record the condition of high vertical bed accretion at the same time as the bedform was migrating forwards. The parallel lamination indicates upper flow regime conditions (see 6.4) on the top of the bedform. The whole assemblage may record a transition between the dunes and upper stage plane bed (see Fig. 6.15).

Patterns of ascending, descending and laterally accreting sets of cross bedding reflect deposition on major bedforms (bars) through the migration of dunes. Point bars, medial bars and side bars could all be sites of lateral accretion. Ascending sets (upstream accretion) are most likely to occur on the upstream sides of medial bars or side bars, while descending cross beds are likely at the downstream ends of those bars. In addition, descending cross beds could reflect changing flow stage (see below and Fig. 6.40). A particular style of lateral accretion structure is dealt with in Section 6.2.10.

Bars, which form at the scale of the channel, probably result from the action of large-scale patterns of water circulation in the channel flow. For example, the flow of water around a channel bend gives rise to a large-scale spiral vortex with surface flow directed towards the

Figure 6.39 Changes in morphology and internal structure due to changing water stage over a sandwave: (a) currents parallel with the side of a linguoid sandwave at lower stage may be able to deposit sand by lateral accretion; (b) wave action during emergence can reduce the angle of the lee-side and truncate steeper foresets (see Figs 6.28 & 37). (After Collinson 1970.)

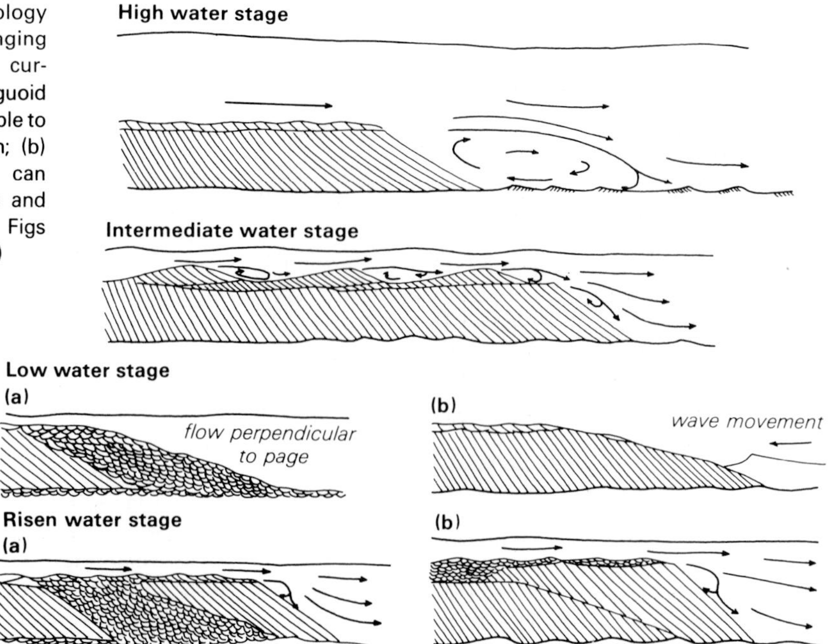

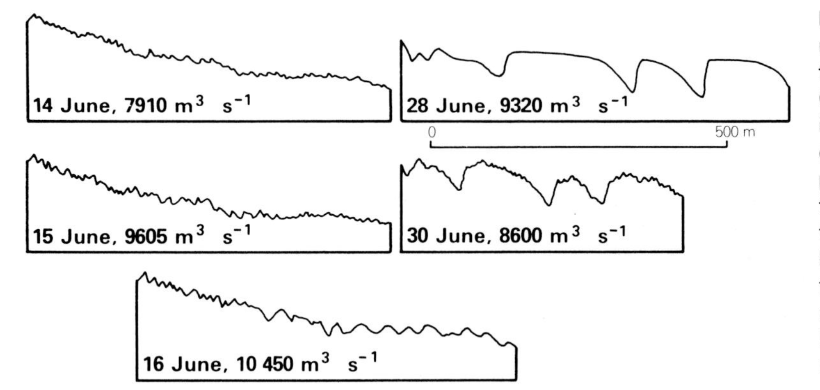

Figure 6.40 Echo-sounding runs made at different discharges of the flood on the Fraser River, British Columbia. The large bedforms increase in size during the rising discharge and continue to increase beyond the flood peak. During falling discharge, smaller forms develop on the backs of the large forms. Both these effects show how the bed response lags behind the prevailing flow due to the large volumes of sediment that must be reworked to modify large bedforms. Discharge is shown in $m^3 s^{-1}$. (After Pretious & Blench 1951.)

outer bank and near-bed flow directed towards the inner bank. This leads to movement of bedload sediment towards the inner bank and to the deposition of a point bar by lateral accretion. Similar processes are probably active along the flanks of medial bars. Converging and diverging flow at the upstream and downstream ends of medial bars also lead to deposition. Large alternate bars along the sides of straight channels are not well documented or understood, but probably involve large-scale flow separation, possibly combined with the development of large-scale spiral eddies.

6.2.8 *Controls on bedform size*

The sizes of dunes, unlike those of current ripples, are related to flow depth and are largely independent of grain size. Deep flows generate higher and longer dunes. An approximate value of 1 : 6 has been suggested for the ratio of dune height to flow depth. This figure should be treated with caution as it results from a two-dimensional analysis when, in fact, many dunes are strongly three-dimensional. Certain very low-relief dunes in sandy rivers show a much higher ratio (*c*. 1 : 2) and the relationships are far from clear. The controls on tidal sandwaves are not well established.

One consequence of this general relationship is that dunes of different sizes may become superimposed as a result of changing flow conditions (Fig. 6.40). When conditions change rapidly, large bedforms cannot change quickly enough to maintain equilibrium. The growth of smaller dunes on larger ones during falling stage could lead to descending cross bedding at the downstream end of the larger forms.

6.2.9 *Isolated large-scale sets of cross bedding*

Very large sets of cross bedding occur in both aeolian and water-lain sediments. Aeolian examples are discussed elsewhere (6.3) and care should be taken to establish this basic distinction. In water-lain sands, large, single tabular sets, several or even tens of metres thick are commonly overlain by a coset of smaller sets showing a similar current direction. Such sets can be very extensive laterally: some sets 20–30 m thick may be traced for several kilometres parallel to the dip direction (Fig. 6.41). These are not readily explained by the migration of repetitive bedforms and at least two alternatives must be considered.

One is by the advance of a delta formed by a stream which carried abundant bedload into quieter water. Such deltas commonly have steep slopes and avalanching causes the delta to advance and create a single, cross-bedded set. Such deltas can easily be modelled in a laboratory tank (cf. Fig. 6.38) and were first recognised around lake margins by G. K. Gilbert in 1885. Examples of Gilbert-type deltas are rare in the rock record except in Pleistocene and Holocene gravels. Their initial description, however, caused confusion as some geologists came to regard all cross bedding as diagnostic of deltaic sedimentation. While deltas dominated by coarse-grained sediment may show cross bedding in their slope deposits, most large marine deltas are dominated by suspended load whose deposition leads to very low-angle delta slopes.

The other possible origin of large-scale, isolated sets is by the advance of large bars either attached to alternating sides of very large, deep channels (Fig. 6.42) or in mid-channel or channel confluence settings. Bars of this sort occur on a small scale in modern streams but a scaling up of these analogues, to sizes beyond anything known at the present day, is necessary to explain large-scale examples in rocks (Fig. 6.32b).

6.2.10 *Epsilon cross bedding*

This type of isolated, single set differs from ordinary cross bedding in several important respects. It consists of a tabular unit, usually between 1 and 5 m thick, in

Figure 6.41 Isolated sets of very large-scale cross bedding: (a) river terrace section in recent deltaic sediments deposited by a sandy river in the head of a protected fjord (Tana River, Finnmark, Norway); (b) large-scale set overlain by a coset of medium-scale cross-bedded sandstone. This large set was probably laid down by a large bar migrating in a deltaic channel (see Fig. 6.42 for model) (Kinderscout Grit, Upper Carboniferous, Derbyshire).

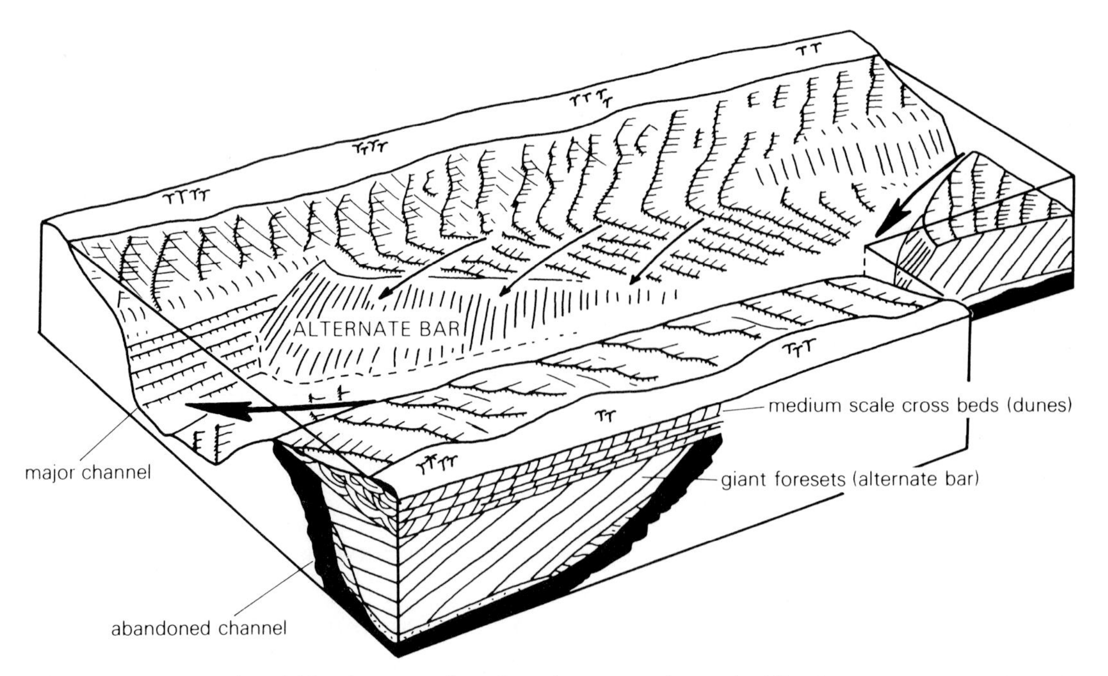

Figure 6.42 A suggested model for the generation of very large sets of cross bedding by the migration of alternate bars in deep channels. This model was devised to explain large cross-bedded sets in the Namurian rocks of northern England (see Fig. 6.41b). (After McCabe 1977.)

which inclined beds dip at angles considerably less than the angle of rest, commonly 5° to 15°; the thicker the unit, the lower the inclination. The inclined beds, which extend over the full thickness of the unit, may be sigmoidal in vertical section and can be defined by differences in grain size between beds (Fig. 6.43). In detail, the inclined beds usually contain smaller-scale internal structures, such as cross lamination and small-scale cross bedding, which indicate flow subparallel to the strike of the inclined beds. The lower surface of the set is erosional, often with a concentration of pebbles or intraformational clasts. There may be an overall upwards-fining of grain size through the set to the extent that the upper part of the unit has interbedded inclined layers of sand and silt. In rare examples where extensive upper bedding planes are exposed, the inclined beds may be seen to be strongly curved through several tens of degrees in plan view (Fig. 6.44). Sets often end laterally at erosion surfaces that dip steeply towards the inclined beds but are separated from them by a unit of siltstone or of disturbed bedding.

This type of cross bedding is an organised assemblage of lithological changes and structures. The similarity to the lateral accretion seen in more uniform cross bedded sandstone is obvious (see 6.2.7). The critical point in its interpretation is that the smaller-scale structures indicate flow subparallel to the strike of the inclined surfaces. From this, one can infer that an inclined depositional surface migrated laterally. The basal erosion surface and the curved plan view of the dipping beds combine with this inference to suggest a curved channel side, probably in a meandering stream. The inclined beds record successive positions of a laterally migrating point-bar surface or, more unusually, lateral migration of the flank of a medial bar. More extended discussions of the processes involved are given in books dealing with sedimentary environments (see reading lists of Chapters 1 and 10).

6.2.11 The uses of sandwaves, dunes and cross bedding

There are three main uses of dunes, sandwaves and cross bedding. They can be used as indicators of 'way-up', conditions of deposition, and palaeocurrent direction.

'Way-up'. When dunes are preserved on upper bedding surfaces or as form sets, they give a fairly positive

Figure 6.43 A unit of epsilon cross bedding. The base of the set is a roughly horizontal erosion surface. The surfaces dipping to the left represent successive positions of the depositional bank of a channel as it migrated laterally. Smaller structures within the inclined sand units indicate flow parallel to the strike of the inclined units. Cloughton Formation, Middle Jurassic, Yorkshire.

Figure 6.44 Upper bedding surface of a sandstone unit made up of several laterally adjacent sets of epsilon cross bedding. The beds within each set are inclined in the direction of convex curvature, suggesting that the epsilon cross bedding is due to lateral accretion on the point bar of a meandering channel. Scalby Formation, Middle Jurassic, Yorkshire.

indication of 'way-up'. Cross bedding is usually an even better indicator. In particular, the sharp cut-off of foresets at the top of many sets contrasts with their tangential bases, particularly in trough sets.

Conditions of deposition. Dunes form under particular conditions of water depth, flow velocity and grain size (Fig. 6.15). It should therefore be possible to put limits on flow conditions based on the forms seen on present-day sand beds and on the cross bedding preserved in rocks. The shape of the foresets in tabular sets can indicate relative current strengths, and changes in foreset shape along a single tabular set could indicate fluctuation of current strength through time. These changes may be associated with reactivation surfaces if depth fluctuation was large. Complex and intensive reactivation, opposed directions and clay drapes can all point towards tidal influence.

Although dune height relates, albeit roughly, to flow depth, there are problems in using the thicknesses of trough sets as indicators of flow depth. It is difficult to know how much of a dune's height is preserved in a trough set. A systematic upwards reduction in set thickness through a coset may, however, suggest a shallowing flow.

Direction of palaeocurrents. Cross bedding is one of the most widely used palaeocurrent indicators. As large bedforms usually respond to a dominant flow and are not easily remoulded by low-stage flows they tend to give a good indication of the palaeoslope.

With tabular sets, the most valuable measurement is the direction of dip of the foresets (foreset azimuth), but you should also record the magnitude of dip as well, particularly if the succession is tectonically tilted. To measure cross bedding in vertical sections, it is necessary to see faces with more than one orientation (e.g. Fig. 6.31). The apparent dip on a single face only shows a component of the true dip. A bedding surface view of the foresets will always give the most accurate measurement of foreset azimuth.

With trough sets these problems are compounded by the curved nature of the structure. It is necessary to measure the direction of trough axes on bedding planes if accuracy is required (e.g. Fig. 6.32). With experience, however, it is possible to judge the orientation of trough axes from vertical exposures to an accuracy of ± 15° which is adequate for many purposes.

When cross bedding occurs between bounding surfaces which are themselves inclined, the relative orientation can provide evidence of the nature of accretion on larger bedforms. It is very important to recognise epsilon cross bedding and to distinguish it from normal cross bedding. An uncritical measurement of dip direction could suggest a palaeocurrent 90° divergent from the true trend.

6.3 Aeolian dunes and cross bedding

6.3.1 Introduction

Small aeolian dunes of coastal belts or inland sand 'seas' (ergs) are comparable in size with aqueous dunes or sandwaves but range up to significantly larger dimensions. Aeolian dunes, however, migrate across larger structures called **draas** which have no aqueous counterparts. Small aeolian ripples and horizontal beds are superimposed on both dunes and draas. The suggestion that ripples, dunes and draas form a hierarchy of equilibrium bedforms provides a basis for classification and description. However, because of the low density and viscosity of air there is a high chance of aeolian processes frequently passing from equilibrium to gross disequilibrium, both in terms of the energy required to form structures and the direction of flow. No thick aeolian deposits are forming today, and most large bedforms are not in equilibrium with the local wind regime. Coastal dunes are comparatively little known.

In the rock record, preservation of aeolian bedforms as relief features is rare, and the former existence of dunes is deduced largely from internal structures. Be sceptical of books which give simple criteria for the recognition of dune types; records of aeolian processes, structures and environments are amongst the most difficult to identify and explain in detail.

6.3.2 Material

Aeolian dunes occur only in sand, rarely extending into granule-sized gravels. The sand is almost invariably composed of quartz, chert or lithic grains of metaquartzite, but coastal dunes of carbonate are known and dunes of gypsum occur adjacent to inland evaporite lakes. Friable, cleavable sand grains, e.g. of feldspar, mica or silt-clay aggregates, are virtually absent in aeolian sands except close to their source. Dunes of dry, sand-size clay–silt aggregates are commonly seen, especially as parabolic dunes and lunettes, forming down wind of dried-out lakes (Fig. 6.45d). On wetting, such dunes become solid masses but retain their cross bedded structure.

6.3.3 Size, shape and classification of large-scale aeolian bedforms

Formerly, aeolian dunes above 1 m in wavelength and 10 cm in height were grouped together as 'dunes' and were classified into many types (Fig. 6.45). Recently, however, space satellites have photographed vast areas of ergs and coastal dune belts. Hence classifications based on ground observation, low-level aerial photography, internal structures and the measurement of wind regimes and sand flows are now supplemented by patterns recognised on remote sensing images and related to regional climatological data on wind regimes.

Observations of areal frequency, width, wavelength, height and, arguably, grain size reveal a hierarchy of bedforms in which similar structures coexist at different sizes and spacings (Table 6.1). Possibly, therefore, the structures are equilibrium bedforms comparable with those in water. At least three distinct scales of structure are noted (Table 6.1): (i) aerodynamic and impact ripples (see 6.1.6), (ii) dunes and (iii) draas (also known as megadunes). Large structures consisting of dunes migrating on draas are sometimes described as compound or complex 'dunes'.

All transverse bedforms possess gentle upwind **stoss** slopes and steeper downwind **lee** slopes. The latter usually comprise a **slip face** (i.e. a foreset slope), down the upper part of which sand avalanches close to the angle of rest. Not all dunes and draas, however, possess such slip faces and some may be **slipfaceless** (i.e. generated at or degraded to lower angles). Slipfaceless draas often have dunes with slip faces migrating across them. Both dunes and draas invariably have ripples migrating across many parts of their stoss and lee slopes.

Movement and growth rates of bedforms are related to the volumes of sand involved, so that draas may take as long as 10 000 years to develop and equilibrate, while ripples may respond almost instantaneously to changes in wind direction and strength.

The shape of the structures, their orientation relative to the resultant of the **effective winds** (i.e. the winds that produce significant sand transport), and the spacing of transverse and longitudinal forms allows the component features of dunes and draas to be isolated and the main patterns to be identified. The classification of many observed combinations of transverse, longitudinal and oblique components is more difficult, as are attempts to relate them to dune and draa types (Fig. 6.46) and the main controlling variables (sand supply, vegetation cover, wind strength and wind regime).

The commonest pattern of dunes and draas is a network, known in Africa as **aklé**, in which there are transverse, longitudinal and oblique components. These sinuous, transverse ridges display alternating linguoid and barchanoid (i.e. concave down wind) sectors which are either in- or out-of-phase in relation to those in an adjacent ridge (Fig. 6.45b). Elsewhere there may be straight-crested dunes (or draas) transverse to the wind (**transverse dunes**) and, close to them, **dome-shaped dunes** with many minor slip faces and rounded flanks inclined at low angles.

Parabolic dunes form U-shapes closing down wind in areas of increasing vegetation. Where there is less sand, the emphasis of transverse, longitudinal and oblique components changes as aklé forms give way to true barchan and longitudinal dunes. **Barchans** are crescent shaped, open down wind, and have corridors of sand-free ground all around; barchan draas sometimes have barchanoid dunes superimposed on their stoss slopes and lateral horns (cf. Fig. 6.45c).

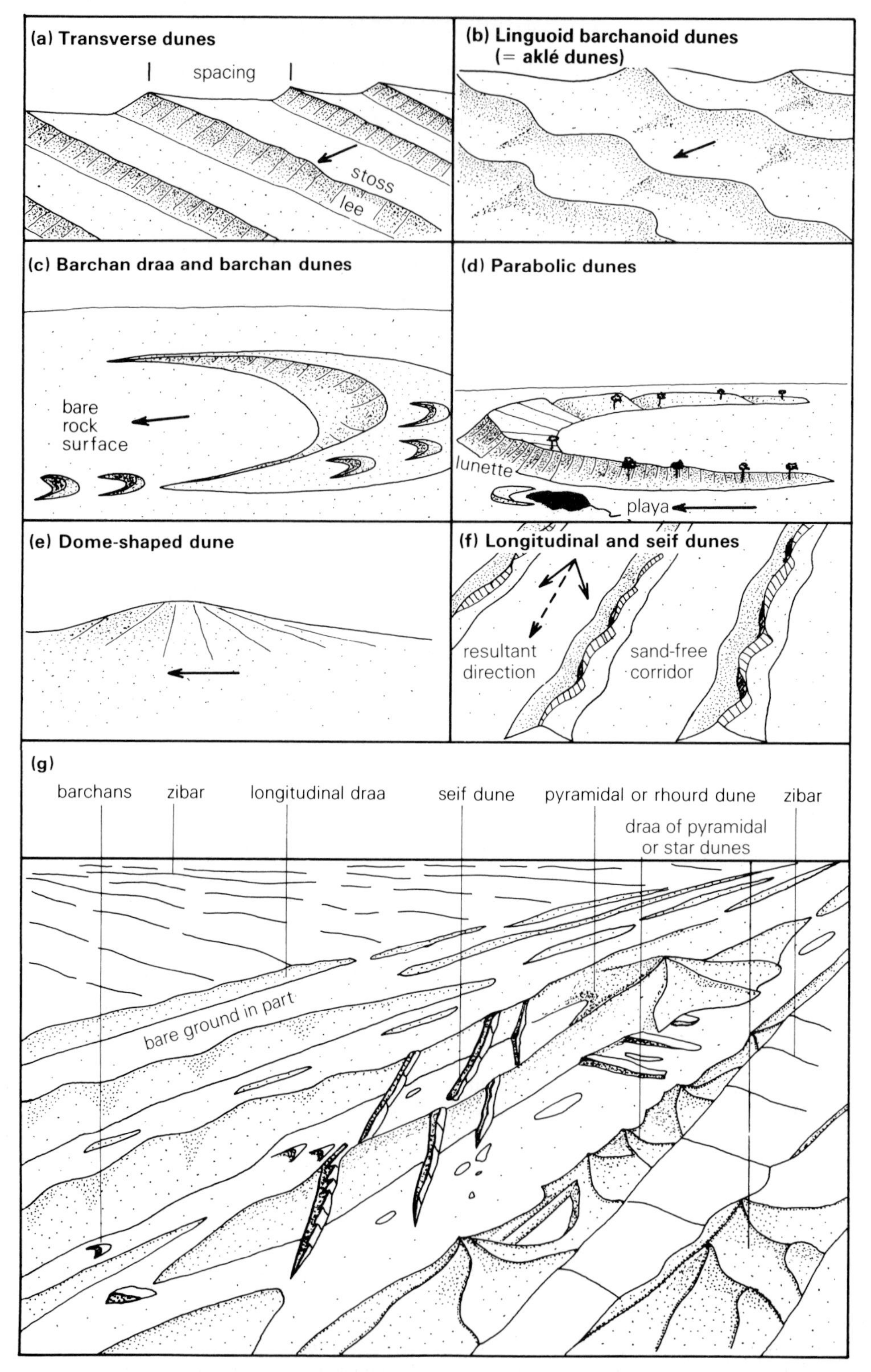

Figure 6.45 (a–f) Three dimensional forms of some common dune types. The arrows mark the dominant directions of the effective winds and, in case (f), the dotted arrow is the resultant effective direction. (g) three-dimensional view of draas, dunes and other structures, e.g. zibar. Note that the relationships shown are diagrammatic, and such varied close associations are not likely to be found in nature. (After Cook & Warren 1973.)

Table 6.1 The classification of aeolian bedforms (after Wilson 1972, p. 197).

Order	*Wavelength*	*Height*	*Orientation to wind*	*Possible origin*	*Suggested name*
1st	300–5500 m	20–450m*	longitudinal or transverse	primary aerodynamic instability	draas* (= megadunes)
2nd	3–600 m	0.1–100 m	longitudinal or transverse	primary aerodynamic instability	dunes
3rd	15–250 cm	0.2–5 cm	longitudinal or transverse	primary aerodynamic instability	aerodynamic ripples
4th	0.5–2,000 cm	0.05–100 cm	transverse	impact mechanism	impact ripples
	1–3,000 cm	0.05–100 cm	longitudinal	secondary horizontal spiral vortices	secondary ripple sinuosity

* Draas can have their own slip faces or may have their lee sides covered in dunes.

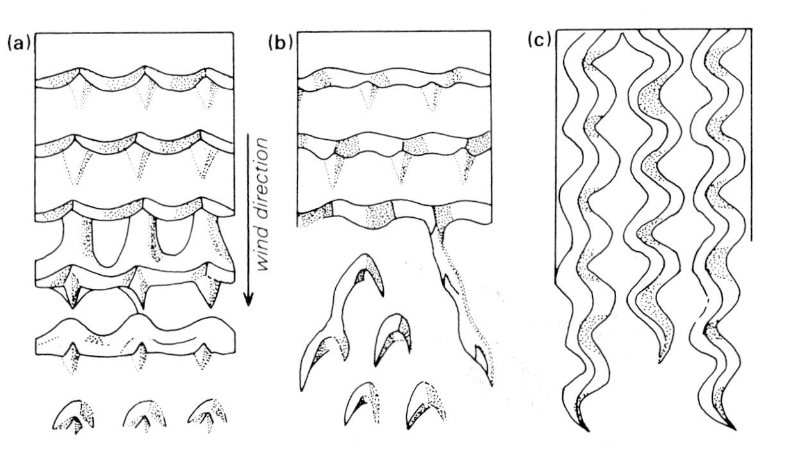

Figure 6.46 Bedform patterns formed by combinations of longitudinal and transverse elements: (a) grid iron, largely transverse in-phase; (b) fish scale, largely transverse out-of-phase; (c) braided, largely longitudinal out-of-phase patterns. A progressive decrease in sand cover is shown towards the bottom of the diagram. Some kinds of barchan and seif dunes form fish-scale patterns where there is incomplete sand cover. (After Wilson 1972.)

Longitudinal dunes and **draas** vary between narrow and relatively straight (the **sand ridges**) and sinuous (the **seif** type). Spacing between dunes is twice their mean width. The crests of such dunes exhibit a regular sinuosity with slip faces on alternative flanks. Upwind ends of ridges are rounded, and along their length Y-shaped forks show 30–50° angles to the flow and open up wind. At their downstream ends the ridges are pointed. Most seifs are parallel to the resultant vector of the effective winds. Longitudinal draas sometimes have smaller seifs aligned diagonally across their slipfaceless flanks at 30–50° (Fig. 6.45g). **Zibar**, low ridges of coarse-grained, hard-packed sand without slip faces and aligned transverse to the wind, sometimes occur in the corridors between seifs and as independent patches.

Prediction of the pattern of cross bedding in dunes and in simple draas (i.e. those with their own slip face) is relatively easy (Fig. 6.47a & b). For transverse, aklé, barchan, parabolic and, to some extent, dome-shaped dunes, the volumetrically most abundant and steepest laminae will show rather consistent dip azimuths, related to the effective wind. Variance will increase from transverse to dome-shaped dunes. If deposition were to occur from time to time on stoss surfaces, then rare, less steep laminae with azimuths at 180° from the mode could occur. The predicted pattern for longitudinal, especially seif, dunes is of two modes about 120° apart. For complex or compound draas and for bedforms which are out of equilibrium, more complex patterns will occur.

Cross bedding data derived from ancient sequences (both from outcrop and borehole dipmeter) can be interpreted through comparison with these predictions

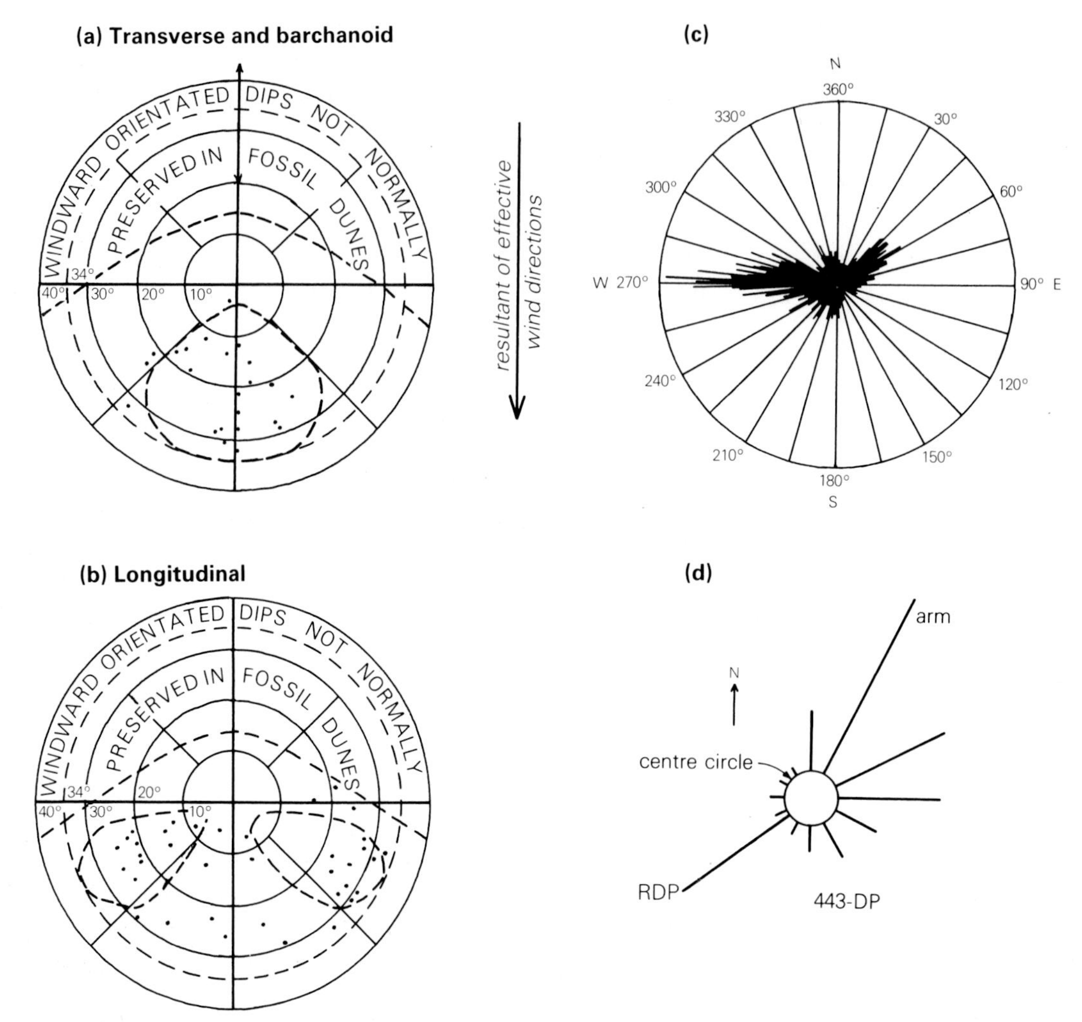

Figure 6.47 (a) Palaeocurrent pattern predicted for barchan, barchanoid and transverse dunes, plotted as poles to foreset laminae, on an upper hemisphere stereographic projection. (b) The same prediction for seif dunes (after Glennie 1970). (c) A typical wind rose. Wind observations taken at 15 min time intervals over a period of time. The lengths of arms are proportional to the time the wind blew from a given direction. Winds from the west and north-east were dominant. (d) A typical sand rose. Components are as follows: *arm* (vector unit totals) proportional in length to potential amount of sand drift *towards* the centre. *DP* (drift potential in vector units): measure of relative sand-moving capability of the wind, derived from reduction of surface-wind data through a weighting equation. *RDP* (resultant drift potential and direction, in vector units): net trend of sand drift. [(c & d) after McKee 1979.]

only with great caution, especially where complex or compound forms are suspected. Before proceeding further, try to analyse and describe the bedforms in Figure 6.48.

As most of us are unable to observe large-scale aeolian processes at first hand, we must gain experience from aerial photographs and the field study of dunes in coastal belts. Small-scale processes can be studied in simple laboratory wind tunnels. The angle of rest of dry, damp and wet sand of different grain sizes, shapes and densities can be determined experimentally (see 3.4). Few will observe much of the internal structure of modern draas, and even the internal structures of dunes will rarely be seen in relation to surface form. In coastal areas, however, horizontal blow-out surfaces may provide plan views of internal structures.

In describing aeolian dunes, measure width, wavelength, height, geometry of crests (Fig. 6.1), angle of

Figure 6.48 (a) Aerial photograph of part of a large sandy desert. Describe and analyse the bedforms and other morphological elements in relation to the possible effective wind pattern. What are the relationships of the different scales of bedform? Erg Oriental, Algeria. (b) Aerial photograph of stabilised seif dunes with wide inter-dune areas. What is the likely pattern of effective winds? Dunes are around 100 m wide. Strzelecki Desert, South Australia.

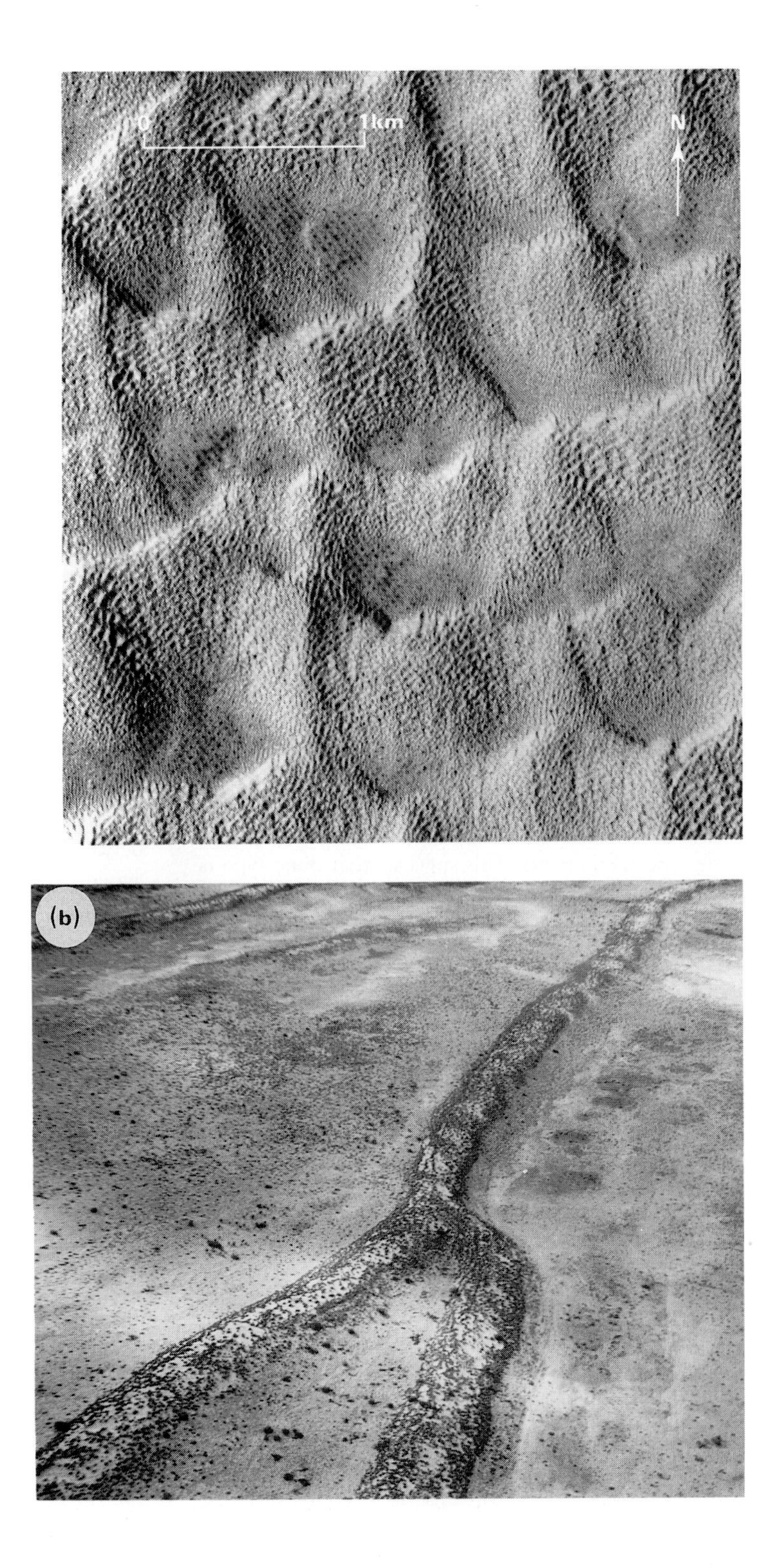

inclination of the top, middle and bottom of the stoss and lee slopes, mean grain size and distribution of grain sizes, the coherence and porosity of the sand, and the nature, distribution and direction of superimposed smaller structures. Take lacquer peels of structures from both vertical and horizontal excavated surfaces (see Appendix B). Smoke pots may help to discern the local airflow over active dunes. Look for the presence or absence of a lee-side eddy, a cross wind and a reattachment point of the flow. The use of stakes may help to monitor the rate of advance of parts of a dune. It may be possible to measure sand flow into and off the dune and to plot a sand rose diagram (Fig. 6.47d). Monitoring of wind direction and speed by a hand-held anemometer may be helpful locally. Regional climatological data should be plotted so that the weighting of the effective wind direction is properly represented (Fig. 6.47).

Dune form has been related to wind variability, sand transportability and spread-out equivalent sand thickness (*EST*). It is possible to determine drift potential, *DP*, the relative amount of sand potentially moved by the wind in unit time, weighted for velocity. Sand drift roses can be calculated for any station (Fig. 6.47d) from which the magnitude of the vector resultant drift potential, *RDP*, is derived. High values of *RDP/DP* indicate low directional variability. *RDP/DP* values, plotted against *EST*, yield a thoroughly discriminant plot for different dune types (Fig. 6.49). This suggests that dune type is controlled by wind **regime** as well as by the availability of sand. The degree of vegetation cover may complicate the situation further.

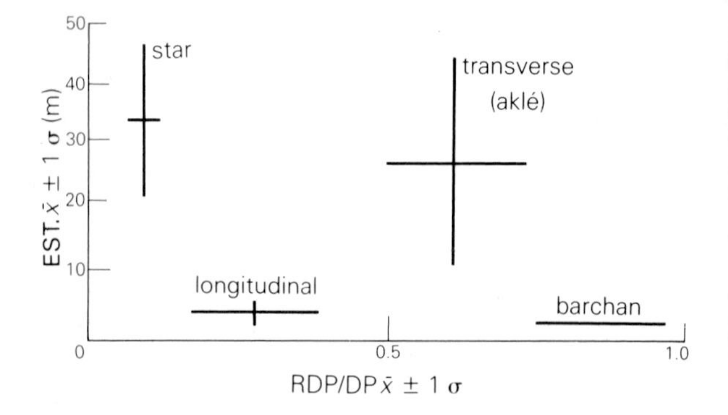

Figure 6.49 Two variables may be used to separate four basic dune types at a highly significant statistical level ($P = 0.001$). These variables are (i) a measure of variability of wind direction (*RDP/DP*) and (ii) the equivalent sand thickness (*EST*). The standard deviation (σ) is very low for barchans and does not show as a vertical bar on the plot (after Wasson & Hyde 1983).

6.3.4 *Internal structures of modern and ancient aeolian sands*

When trying to describe and interpret any suspected ancient aeolian sandstone it is important to realise at the outset the very sparse data which are available for either inland ergs or coastal dunefields today.

Very few vertical sections through present-day dunes and only small parts of draa have been excavated. Horizontal surfaces produced by natural agencies or by engineering are more frequent, more easily studied and add considerably to an appreciation of three-dimensional structure, but they are still few.

Our knowledge of the internal structure of large present-day transverse, barchanoid, dome-shaped and parabolic dunes rests upon single excavations made in easily stabilised gypsum sand dunes in New Mexico. It is naive to expect that the organisation of each of these dunes is unique to, or even representative of, each dune type. Other studies of large present-day longitudinal, seif, barchan, star and reversing dunes and draas composed of quartz sand are limited by the logistics of stabilising and excavating them. Vertical cross sections are seldom more than 2 m – less than skin deep. On the other hand, studies of much smaller dunes of various types, less than 2 m high, on Padre Island, Texas, have greatly advanced understanding of the relationship between small-scale internal structures and processes of aeolian deposition.

When trying to understand the structure and genesis of ancient aeolian strata bear in mind, therefore, that existing studies give only a crude guide to the type of dunes present, not their size and complexity. The larger the fossil dune or draa structure, the less likely it is, through erosion, that anything more than a small part, usually the lower lee slopes, will be available for study (see the lower part of Fig. 6.51).

LARGE-SCALE INTERNAL STRUCTURES

Before generalising about the internal structure of large dunes, we suggest that you draw up a table similar to Table 6.2 and fill in the columns based on your own observations of Figures 6.50–53.

Figures 6.50–52 present the results of excavations through present-day gypsum-sand dunes, dominantly transverse to the wind. While generalisation is somewhat dangerous the following features are apparent:

(a) Medium- and large-scale cross bedding is dominant.
(b) The cross bedding is of several types, the relative proportions of which vary amongst the examples. Tabular planar cross beds appear to be more common than wedge planar and trough types. Bundles of convex upward laminae are present in several dune types.

(c) Stacked sets are separated by **bounding surfaces** which are mostly horizontal or inclined at low angles down wind. Other low-angle bounding surfaces steepen to 20–28° down wind and truncate more steeply dipping foresets.

(d) Trough cross beds often occur as solitary sets or thin cosets in the upper parts of dunes.

(e) Upwards, sets of cross beds become thinner and foreset laminae flatter. Thin units of flat bedding often occur on the stoss sides and crestal areas of dunes.

By contrast, Figure 6.53 shows the results of shallow trenching into a seif dune. Planar laminae are inclined roughly parallel to the dune flanks. Bundles of laminae

Table 6.2 Internal dune structures.

Type of stratification / *Type of dune*	*Transverse*	*Barchanoid*	*Dome*	*Seif*
(1) tabular–planar cross-bedded (horizontal or low angle bounding surfaces) set size, frequency & location				
(2) tabular–planar cross-bedded or simple set (moderate to high angle bounding surfaces) set size, frequency and location				
(3) wedge–planar cross-bedded set size, frequency and location				
(4) trough cross-bedded set size, frequency and location				
(5) convex-upwards cross-bedded: set size, frequency and location				
(6) planar bedding: set size, frequency and location				

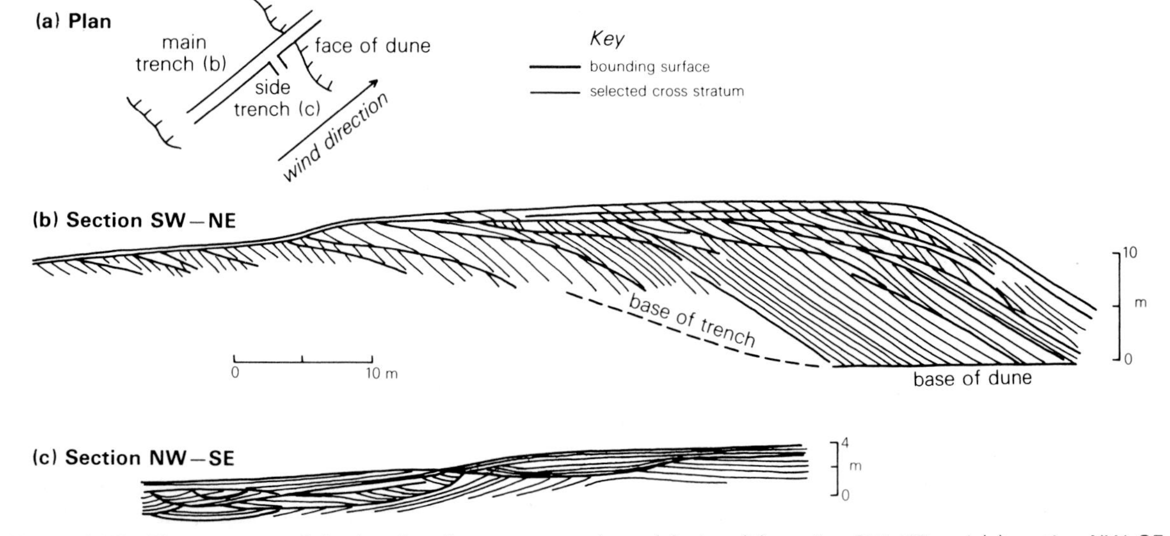

Figure 6.50 The structure of the interior of a transverse dune: (a) plan, (b) section SW–NE and (c) section NW–SE (modified after McKee 1966, McKee 1979).

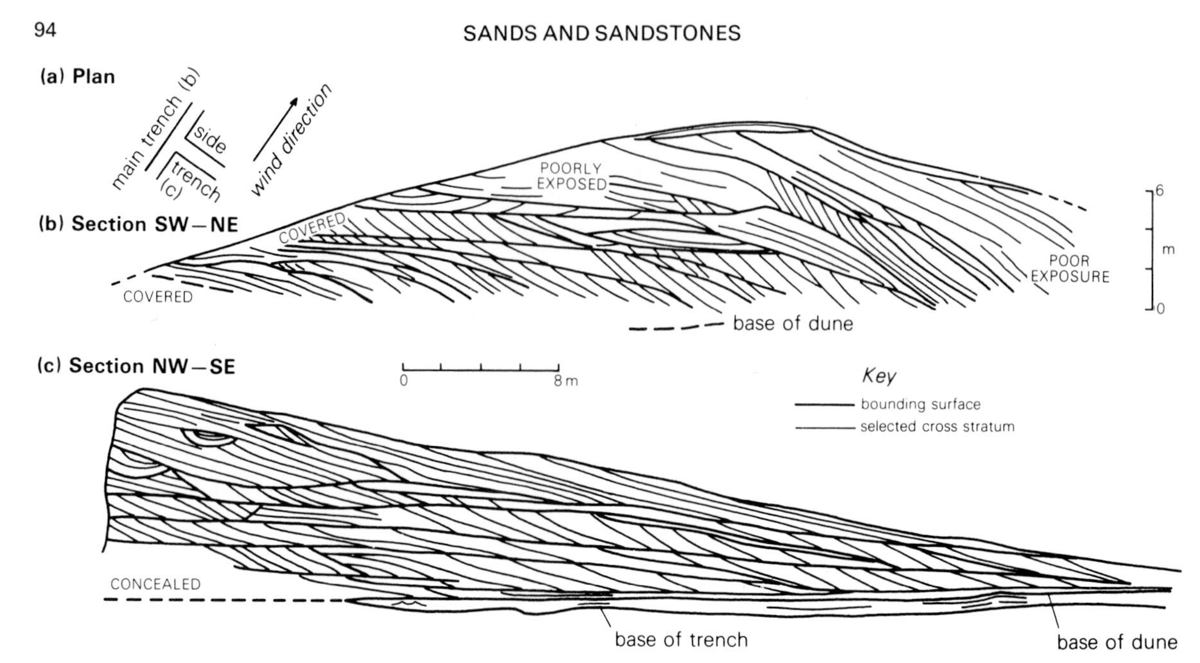

Figure 6.51 The structure of the interior of barchanoid ridge dune: (a) plan, (b) section SW–NE and (c) section NW–SE (modified after McKee 1966, McKee 1979). Of the modern dunes excavated, this particular example shows a complexity of internal lamination which would not have been expected from the external morphology and suggests a complex evolution.

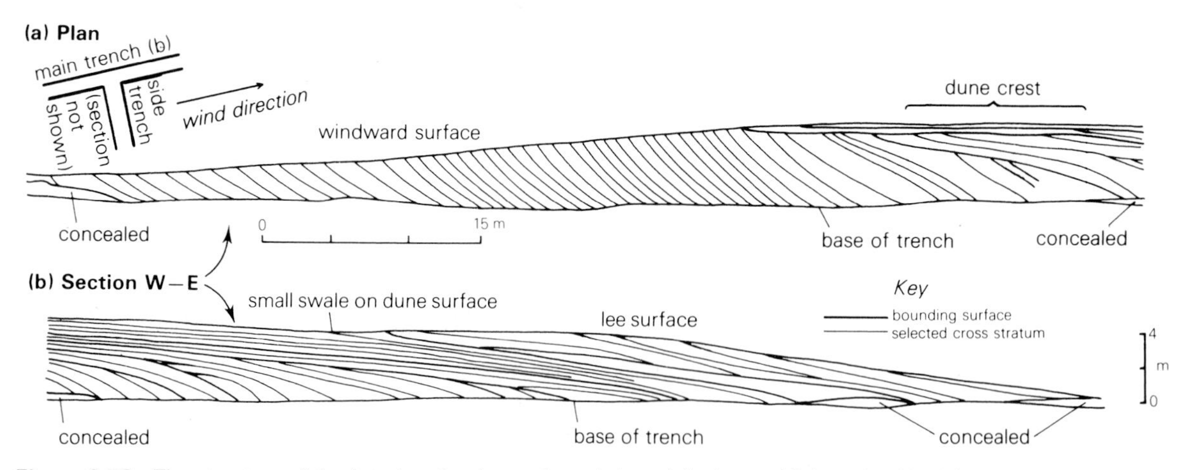

Figure 6.52 The structure of the interior of a dome-shaped dune: (a) plan and (b) section W–E (modified after McKee 1966, McKee 1979).

are separated or truncated by apparently planar inclined bounding surfaces. The results appear to confirm that wedging cross beds dip steeply and obliquely to alternating sides of the crestline (Fig. 6.45f, 6.46c & 6.47b).

The internal structures of draas are little known, but the complexity of the orientation of crests and slip faces of draas does not seem to be reflected by the cross-bed patterns found in one superficially excavated example. The single mode of cross-bed directions observed in the excavations suggests that this draa was formed by a strong effective wind.

This limited information from modern dunes is greatly augmented by studies of ancient aeolian sandstones wherein large-scale cross bedding is the most striking feature, with sets commonly ranging in thickness up to 10 m and occasionally to 35 m. In many exposures such

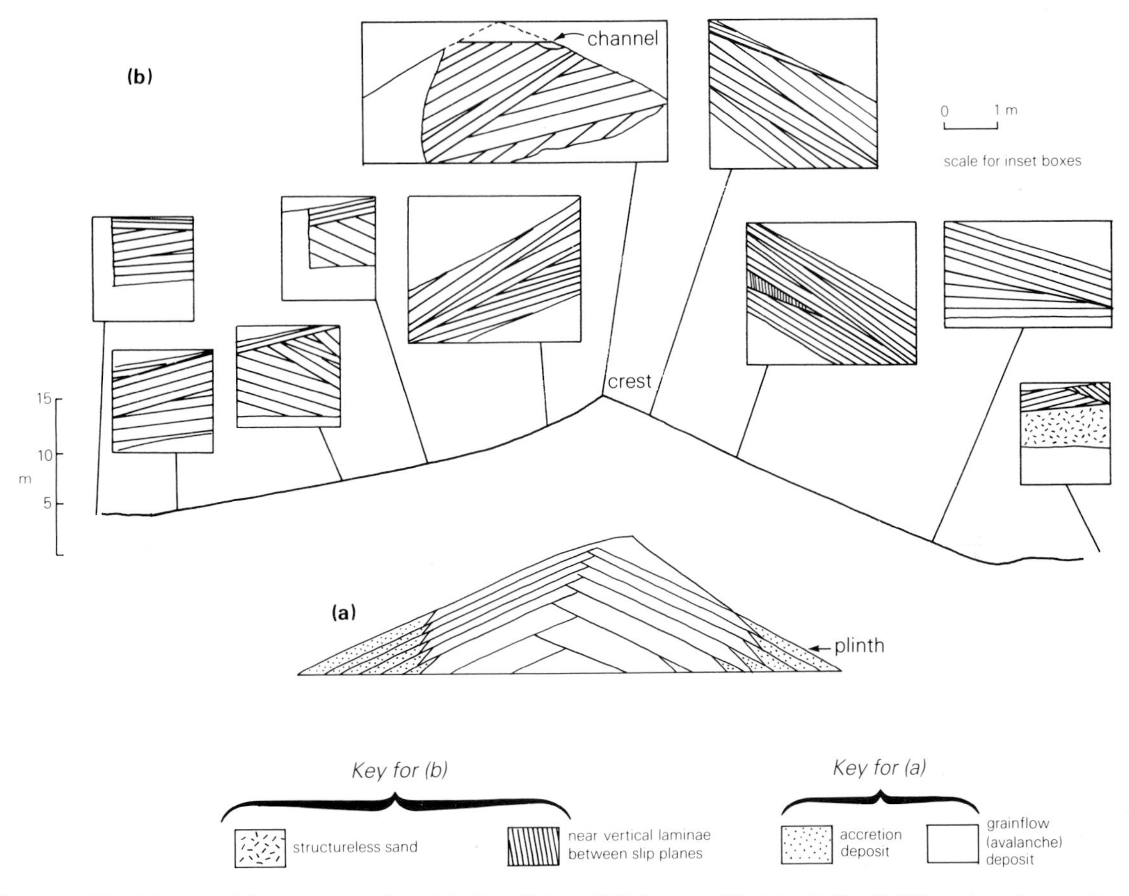

Figure 6.53 (a) Bagnold's hypothetical model of a seif dune (1941) as modified by McKee (1979) to show foreset dips in opposed directions. (b) Direct evidence of a real example, based on shallow excavations of a seif dune in Libya (after McKee & Tibbits 1964). Note that the depiction of avalanche and accretion deposits does not take into account the distribution of small-scale structures predicted for such dunes by Hunter's studies (1977).

cross-bedded sets may appear tabular, but observations of large exposures suggest increasingly that trough sets are the dominant form. Careful analysis of aeolian sandstones has led to the recognition of bounding surfaces at three distinct scales (Fig. 6.54), the smallest-scale surfaces truncate and separate bundles of foreset laminae within the same set. Such **third-order bounding surfaces** are most common in the upper part of sets and commonly have a convex-upwards shape. **Second-order bounding surfaces** separate the sets themselves and are the most commonly recognised type. They mainly define the bases of trough-shaped sets, but in sections parallel to the dominant foreset dip they are either horizontal or may be inclined upward or downward at angles up to 20°. Occasionally they may be associated with thin units of low-angle or horizontally bedded sand or finer-grained sediments. **First-order bounding surfaces** are the most difficult to recognise and are characterised by great lateral extent and very low angles of inclination. It is often difficult to detect whether they diverge significantly from the horizontal. Only where they clearly truncate, or where they are angularly overlain by, second-order surfaces, can their presence be recognised. They, too, may have associated with them units of horizontally bedded sediments. The significance of these various structures is discussed in Section 6.3.5.

Contorted cross bedding may occur in sets of virtually any scale. Massive or structureless sands usually occur near areas of contortion which are commonly confined to single sets, the contortion being truncated by the overlying bounding surface.

SMALL-SCALE INTERNAL STRUCTURES AND TYPES OF LAMINATION

Recent trenching and planing-off of present-day dunes has revealed further small-scale structures and lamination types within sets up to 10 m thick. They have such

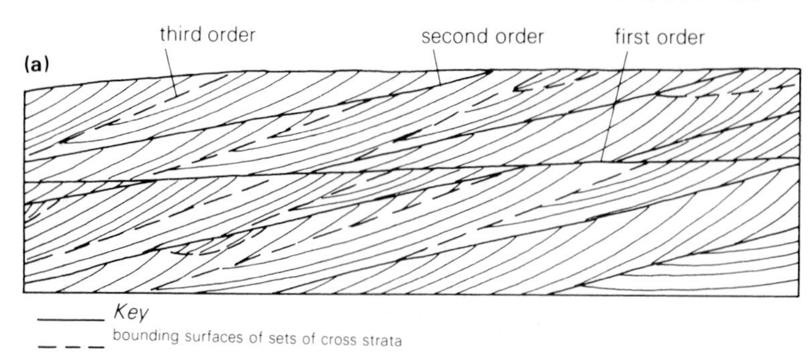

Figure 6.54 Definition diagram for the concept of first-, second- and third-order bounding surfaces within cross bedded sands and sandstones. While this terminology was first proposed for aeolian sandstones, it can also be applied to water-lain sediments (after Brookfield 1977).

distinct relationships to particular processes as to be very significant in identifying and understanding ancient aeolian strata (Figs 6.55 & 56).

(a) Foresets dipping at 18–34°, 2–5 cm thick but thinning upwards, show a tongue- or cone-shaped structure which displays *en echelon*, scalloped or tabular shapes in horizontal section in large dunes, and lenses in small dunes. The strata have sharp, concave-upwards erosional bases, are internally structureless, and show inverse grading normal to the base, as well as lateral grading along their length. At the foreset toe, coarse grains may appear at the base and top of the layer. Heavy minerals line the base of the upper parts of foresets. Packing is very loose and porosity is very high (around 45%).
(b) Foresets dipping at 20–28° with indistinct but parallel internal laminae which apparently lack grading and follow pre-existing topography. Contacts are non-erosional; porosity is about 40%. These units inter-bed with type (a).
(c) Ripple-marked strata, dipping at 0–25° in units 1–15 cm thick with contacts of laminae varying from erosional to non-erosional, sharp to gradational. Inverse grading of laminae is more common than normal grading. Cross lamination or in-phase waviness may be seen. Porosity averages 39%. The strata appear on the stoss side, crest and convex-downwind projections of sinuous dunes, the 'horns' of barchans, the 'noses' of parabolic dunes and on sand drifts to the lee of obstacles. They also appear on the slip faces of dunes where asymmetric ripple marks are aligned parallel to the dip. The strata inter-bed with types (a) and (b), sometimes in rhythmic sequences.
(d) Flat-bed laminae, dipping at 0–15°; parallel, even, with sets of laminae 1–10 cm thick picked out by slight variations of grain type and size. Sharp or gradational non-erosional contacts are present and porosity is less than in types (a), (b) and (c). These laminae occur rarely and always on upwind, or gently sloping, lee surfaces.

Small-scale deformation structures show plastic, brittle and structureless features (see Fig. 6.57).

6.3.5 Processes of formation of dunes, draas and aeolian cross bedding

EXTERNAL FORM

The formation of dunes and draas requires the availability of abundant sand-size grains which are moved as bed-load, whereas silt and clay are removed in suspension and gravel is left as a lag. Observations in modern deserts suggest that for a patch of sand to develop into a dune it has to be at least 4–6 m long. The wind must be sufficiently retarded by the patch's saltation cloud for deposition to occur. Each patch grows to a critical height which depends on grain size and wind strength, whereupon a slip face develops. Dunes and draas are regularly repeated bedforms which record the response of a sand bed to the shearing action of the wind. The bedforms record attempts to reach a dynamic equilibrium in response to consistent fluctuation in the flow pattern.

Historically, ideas on the controls of dune/draa shape have included the variable incidence of vegetation, the structure of organised turbulence in the lower atmosphere, the distribution of thermal convective plumes, the variability of effective winds and the amount of sand available for transport. Recent analysis suggests that the last two factors exert the greatest primary control (Fig. 6.49). Barchans, longitudinal and seif structures are associated with low sand availability, while transverse aklé (barchanoid ridge) and star forms seem to develop where abundant sand is available. Barchans and transverse forms reflect low diversity of effective wind directions, while longitudinal and star dunes reflect high diversity.

In dunes, large-scale units of low-angle, stoss-side

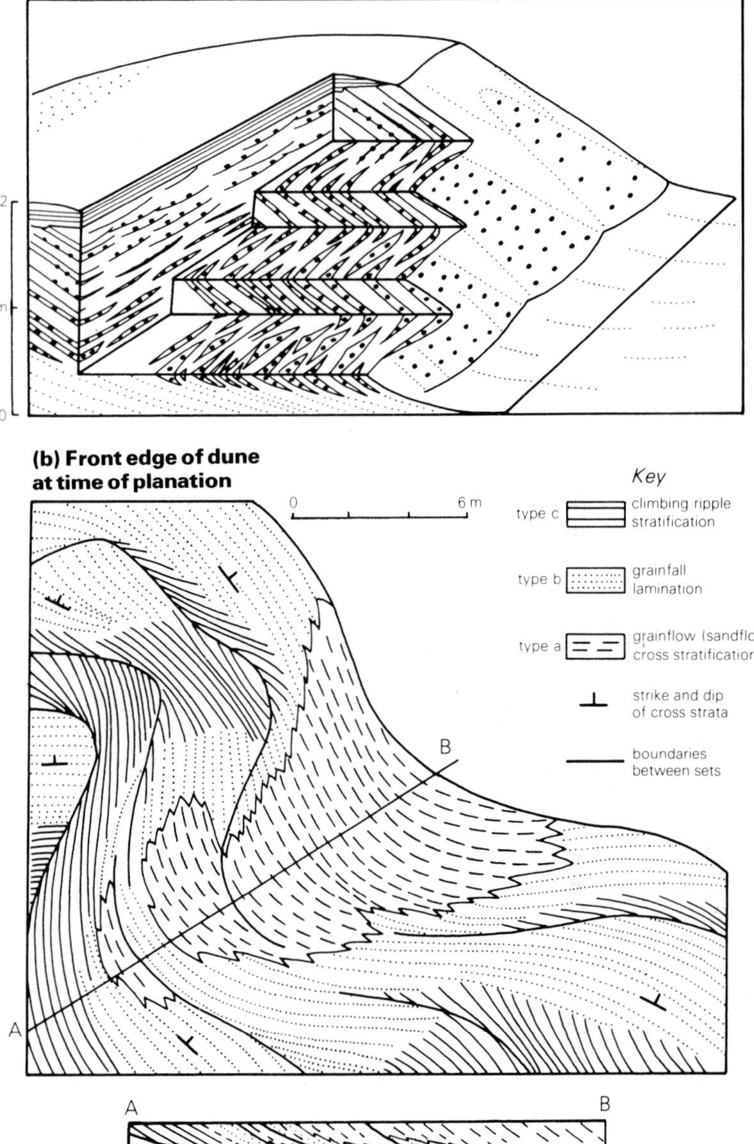

Figure 6.55 (a) Schematic block diagram showing the different small-scale structures of foresets of different type: type (a) simple cone-shaped flow foresets (heavily stippled); type (b) grainfall laminae (unpatterned) on foresets, bottom sets and in accretion deposits; type (c) climbing ripple strata (thinly lined) developed here only in the top set and lee-side accretion deposits but appearing within the foresets of large dunes; type (d), plane bed lamination, would be developed in exposed places on the top of the dune, but is not shown here. (After Hunter 1977.)

(b) Map and cross section of dune foreset cross strata exposed on a planed-off sinuous transverse or barchanoid ridge dune, showing the distribution of small-scale foreset structures of types (a), (b) and (c) described in the text. Simplified from an exposure on Padre Island, Texas. (After Hunter 1977.)

accretion deposits with surface ripples and flat beds grow during saltation. The steep lee-side slip faces form largely from avalanching at angles close to the angle of rest (33–42°) (see 3.7.2). This generalisation will be qualified later when the evidence of small-scale internal structures and the incidence of grain fall are considered.

INTERNAL STRUCTURES

Medium- and large-scale cross beds record the progradation and growth of large bedforms. The slip faces of dunes generate the foresets of the cross beds, and studies of their small-scale structure show that these have a varied origin. High-angle foresets of type (a) (28–40°) with scallop-shaped features (Figs 6.55a & 56) represent avalanche grain flows initiated by slumping and the attainment of the angle of rest. Their inverse grading is due to the dispersive pressures and kinetic sieving generated by colliding grains (see 3.7.2). The largest, roundest grains flow to the toe most rapidly and this accounts for the lateral grading and the vertical inverse grading in those parts. Lower-angle foresets of type (b) (20–28°)

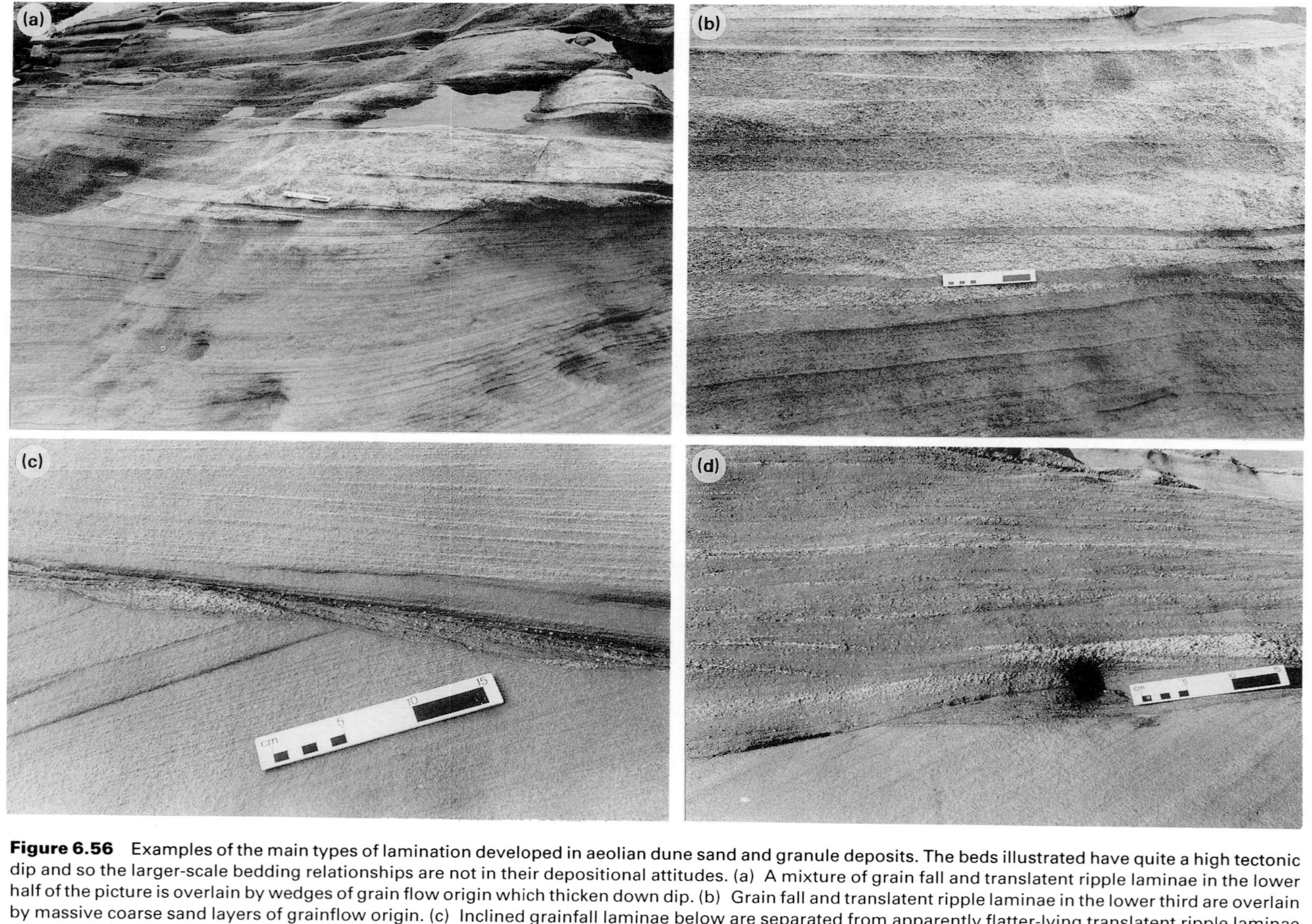

Figure 6.56 Examples of the main types of lamination developed in aeolian dune sand and granule deposits. The beds illustrated have quite a high tectonic dip and so the larger-scale bedding relationships are not in their depositional attitudes. (a) A mixture of grain fall and translatent ripple laminae in the lower half of the picture is overlain by wedges of grain flow origin which thicken down dip. (b) Grain fall and translatent ripple laminae in the lower third are overlain by massive coarse sand layers of grainflow origin. (c) Inclined grainfall laminae below are separated from apparently flatter-lying translatent ripple laminae above by a thin layer comprising ripple form sets developed in coarse sand and granules. (d) Try to describe this one for yourself. All from Corrie Sandstone Permian, Arran, Scotland. (Photographs courtesy of L. B. Clemmensen.)

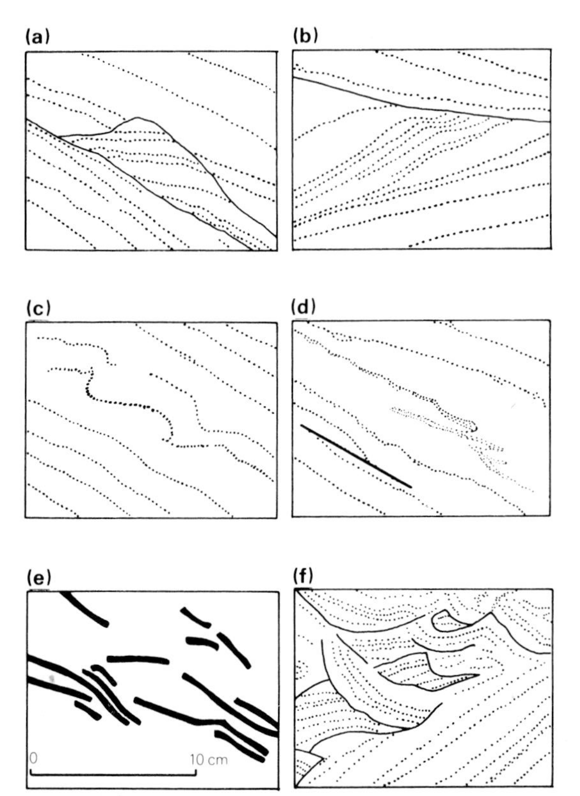

Figure 6.57 Principal types of deformation structures in the foresets of dunes: (a) rotated structures, (b) drag folds, (c) high-angle asymmetrical folds, (d) overturned folds and overthrusts, (e) break-apart structures and (f) brecciated foresets (modified after McKee 1979).

are the product of grain fall in zones of flow separation within the lee-side eddy or on the flanks or apron of the dune where cross winds are not strong enough to form ripples. Distinct lamination is probably due to grain-size variations produced during fluctuations of transporting power. Porosity is high because of a lack of sustained saltation, which generates tight packing. Other low-angle foresets of type (c) in lower parts of dunes are produced by the migration of climbing ripples which may occasionally have evidence of foreset laminae or high index ripple marks (form sets). Generally, however, they generate pseudo-laminae each consisting of the coarser grains at the top of the laminae which have migrated along the crests of the ripples and finer grains at the base which have migrated in the lower part of the ripple (Figs 6.22 & 23). Flat lamination, type (d), arises when saltation which produces ripples is inhibited during gale-force winds. It arises in conditions equivalent to those of the lower part of the upper flow regime in aqueous flows.

These lamination types occur in all sizes of cross bed set, the size of which can be crudely related to the size of the bedform responsible. Only the lower parts of the bedform are normally preserved, though extrapolation of concave-up foresets to the angle of rest may allow the reconstruction of bedform height. The selective erosion of dune sands will lead to the preferential preservation of ripple and grainfall lamination in the lower parts of structures. Where present, the thickness of grainflow layers may offer a qualitative indication of original bedform size.

The third-order bounding surfaces (Fig. 6.54; Section 6.3.4) which are confined within sets are thought to record short-term fluctuations in the strength and direction of the effective winds. The surfaces record erosion during anomalous wind intervals and compare with reactivation surfaces in aqueous cross bedding. The second-order bounding surfaces separating sets record the migration of one dune over another. The degree of curvature of these bounding surfaces, viewed transverse to the palaeowind direction, gives some measure of the three-dimensional shape of the bedform responsible (cf. 6.1.4). Where the bounding surfaces dip upwind, some form of dune climbing must be envisaged; where the surfaces dip downwind, migration of dunes down the lee side of some larger bedform (?draa) is necessary. The first-order bounding surfaces have a less certain origin. One possibility is that they represent the migration surfaces of draas (Fig. 6.58) but it seems also possible that deflation to the water table (Fig. 6.58) or erosion due to a change of climate, may also have operated.

Thin units of interlaminated sands and silts associated with both first- and second-order surfaces are most likely to have formed in inter-dune or inter-draa areas (Fig. 6.58). Details of the inter-dune environment may be deduced from careful examination of the sedimentary structures of these beds.

The origins of deformation in aeolian sands are much debated. Warps, gentle folds, drag folds and flame structures are characteristic of dry sand. Break-apart breccias, rotated slabs and blocks, and high-angle asymmetric folds develop where sand is wet and cohesive. Some deformations are due to tension near the tops of slip faces, others to compression near the bases of foresets. Deformation features due to brittle strain can usually be used to distinguish aeolian from aqueous deposits. Rain or hail storms may leave circular pits; escape of air or water produces injection structures; fronds of plants swinging in the wind produce concentric semi-circular marks; animals leave prints and trails on the flat beds of the stoss face and occasionally on the foresets of the lee when they walk up slope. Walking down the lee provokes avalanches and slumps which destroy prints. In dry sand,

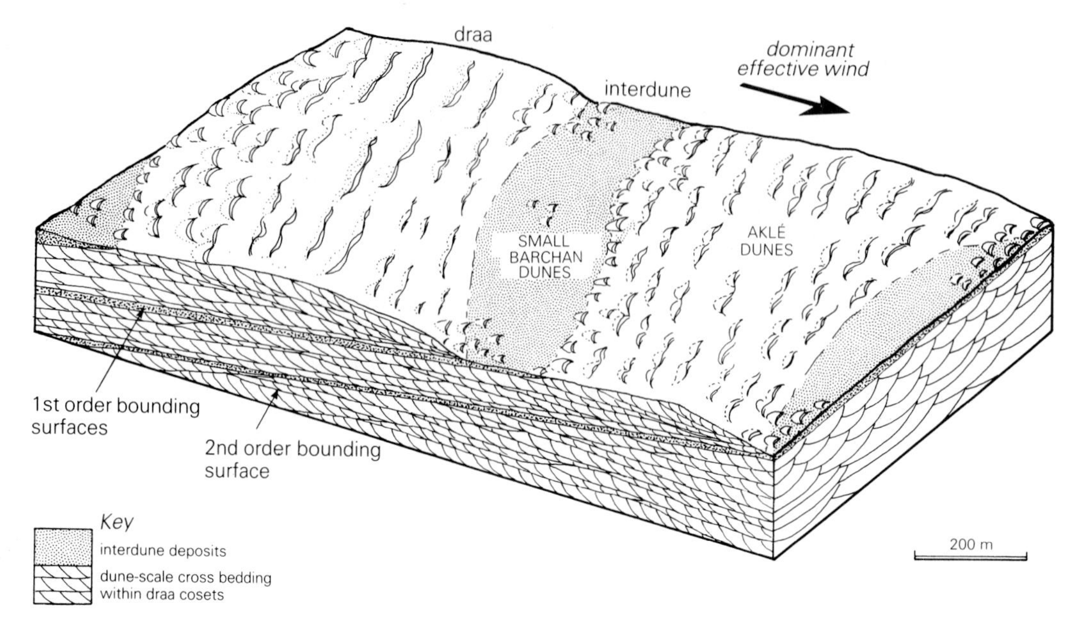

Figure 6.58 Schematic reconstruction of a series of transverse draas with superimposed barchanoid (aklé) dunes. Note how first- and second-order bounding surfaces are generated by the migration of such a system under conditions of net sedimentation. In this example, inter-dune areas separating the draas generate thin units of sediment along the first-order bounding surfaces. Bounding surfaces of both types are shown much steeper than they would appear in nature (modified after Clemmensen & Abrahamsen 1983).

the marks are subsequently preserved by dampening of the sand by mist or dew. Insects burrow in search of safety and a haven from extreme temperatures. Plant roots cut through established laminae, producing a feature known as **dikika**, and lightning strikes develop forked tubular structures of fused sand with a star-shaped cross section, known as **fulgurites**.

6.3.6 *The uses of aeolian dunes, draas and cross bedding*

The greatest use of these structures is in interpreting palaeoenvironments. Aeolian rocks contribute to palaeoenvironmental reconstructions at continental and global scales, for they help to locate major oceanic and meteorological systems. Palaeowind studies are particularly helpful in this regard, but remember that dunes and draas reflect the effective winds, not necessarily weak prevailing and/or seasonal winds. Coarser-grained sand may be moved only by strong winds which may relate to only one season and/or direction. There is no reason why palaeowind directions should have any relationship with palaeoslope.

Aeolian structures could help to establish way-up in highly dipping sequences. Care is needed, however, for straight, angular-based foresets and convex-upwards cross bedding may not be recognised to be the right way up without consideration of the whole sequence and context.

Finally, thick aeolian sequences of porous, permeable, cross bedded sandstone with few impermeable barriers are important as potential reservoirs for oil, gas, water and hydrothermal metalliferous deposits.

6.4 Flat beds and parallel lamination

6.4.1 *Introduction*

Many sandy surfaces in modern aqueous and aeolian settings and many sandstone bedding planes are completely flat. These flat surfaces are related to parallel lamination within the underlying deposit.

Flat-bedding surfaces and parallel lamination occur mainly in sands and sandstones of fine–medium grain size, including those rich in mica. They can, however, occur in material up to very coarse sand size.

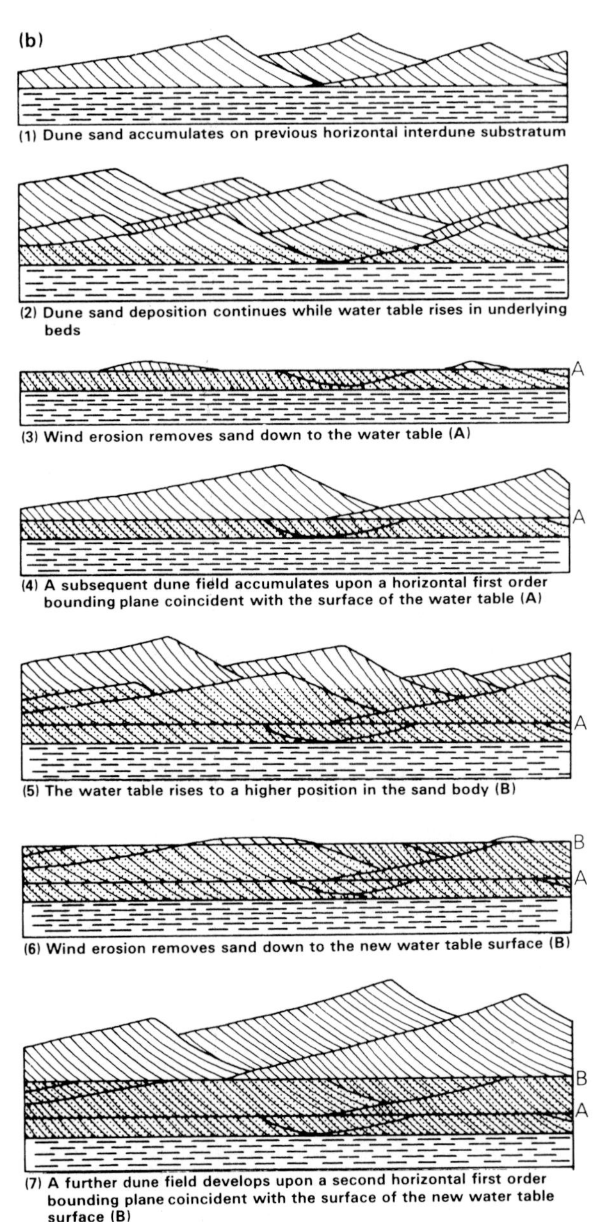

Figure 6.59 Possible origin of multiple parallel first-order bounding planes between medium- and large-scale cross beds. (After Stokes 1968.)

6.4.2 *Flat-bed morphology and primary current lineation*

Go to a beach with very little relief or to a flat area of a recently exposed river bed and look carefully at the small-scale relief. Alternatively look very carefully at the bedding surfaces of parallel-laminated sandstone. With low angle lighting it is common to see a distinct linear pattern superimposed on the flat surfaces.

In very coarse-grained or slightly pebbly sandstone, a lineation is most commonly developed on the surfaces of slightly finer-grained layers, whereas in micaceous sandstone, mineral segregation often gives a colour lineation as well as a relief. This lineation is primary current lineation or parting lineation (Fig. 6.60).

The lineation occurs as a series of closely spaced ridges and hollows. Typical spacing is a few millimetres and relief is of the order of the grain diameter. Individual ridges and hollows persist parallel to the lineation for a few centimetres or even tens of centimetres. There are no systematic differences between opposite ends of ridges and hollows and they cannot be used to determine the sense of current direction.

6.4.3 *Internal structure: parallel lamination*

Excavations into flat sand surfaces and vertical sections in sandstones with flat bedding planes usually show parallel lamination. Laminae are usually only a few grain diameters thick and, in coarse, less well sorted sandstones, they may barely exceed the thickness of the coarsest grains. The laminae are defined by slight grain-size differences or by concentrations of mica.

6.4.4 *Process of formation*

A flat bed is a distinct bedform produced by particular flow conditions. Where the sand is medium- to fine-grained and reasonably free of mica, those conditions

Figure 6.60 Primary current lineation or parting lineation on the bedding surface of parallel-laminated sandstone. The steps between adjacent laminae trend parallel to the lineation. Cloughton Formation, Middle Jurassic, Yorkshire.

are of high flow velocity and shallow water depth, the so called 'upper flow-regime flat bed' mode of transport (Fig. 6.15). Under these conditions ripples and dunes are destroyed and turbulence appears suppressed. The water surface takes on a smooth, glassy appearance and the conditions are similar to those of 'rapid flow' (see 3.2.6). These conditions may also develop during the deposition of sand from turbidity currents, as parallel lamination is a common feature of turbidite sandstone beds.

The lineations, which run parallel to the flow, may relate to streaks of faster- and slower-moving water close to the bed, such as occur in the viscous sub-layer see (3.2.4). As such a sub-layer only occurs with fine-grained sediments, this process may not explain lineation in coarse-grained sands.

Experimental work shows that ripples do not form in sands coarser than about 0.6 mm, but that movement occurs on a so called 'lower flow-regime flat bed' for flows above the critical erosion velocity and below those forming sandwaves (Fig. 6.15). Little is known about the ability of this mode of transport to produce parallel lamination and primary current lineation.

Other experiments suggest that abundant mica inhibits the formation of ripples whose existence depends, amongst other things, on the ability of grains to avalanche down the lee faces of the bedforms. Some high micaceous, parallel-laminated sands and sandstones could reflect deposition from suspension, but concentration of mica into layers implies some sorting on the bed. Lineation in such micaceous sands could well be due to high- and low-velocity streaks in the viscous sub-layer.

Whilst flat-bed mode of transport is most readily envisaged for unidirectional flow, waves can also lead to the development of a flat bed. The flat beds that we see on exposed beaches relate to unidirectional backwash, but high-energy waves beyond the surf zone can also give such a bed.

The formation of lamination, by its very nature, demands some process of grain-size segregation. Fluctuations in flow strength may be involved, but it is also possible that grain segregation on the bed produces layers of moving grains with different size characteristics which, when they stop, deposit laminae. Parallel lamination in aeolian successions is dealt with in Sections 6.3.4 & 6.3.5.

6.4.5 *Uses of flat beds and parallel lamination*

Flat beds and parallel lamination have little use as indicators of way-up, but they do provide useful information on (a) current strength and (b) palaeocurrent or palaeowave direction.

(a) It is fairly certain that flat beds and parallel lamination indicate upper flow-regime conditions, provided that the sand is in the medium to fine range and that it is not very micaceous. With coarser-grained or very micaceous sands, lower flow-regime conditions may have applied. Flat beds due to waves and currents cannot be differentiated by internal characteristics.

(b) To establish a palaeocurrent direction from flat beds it is necessary to see the primary current lineation. Because this is a simple, linear feature, it only indicates the direction and not the sense of movement.

6.5 Undulating smooth surfaces and lamination

6.5.1 *Introduction*

On modern sand surfaces and sandstone bedding planes an uncommon, gentle wave form is associated with parallel but gently undulating lamination.

This structure is probably confined to sands but its rarity prevents a precise delineation of grain-size limits. A similar style of bedding is sometimes associated with pyroclastic ash deposits of subaerial base-surge flows.

6.5.2 *Morphology*

The sand surface, or more commonly the sandstone bedding surface, is smooth with a gentle undulation, either two-dimensional waves or three-dimensional domes and hollows. The amplitude is commonly measured in centimetres, and the wavelength in tens of centimetres, sometimes being of the order of one metre (Fig. 6.61). The bedding surfaces show a primary current lineation (6.4) which in the case of the two-dimensional waves is roughly normal to the wave crests. In vertical section, lamination is thin and roughly parallel to the surface undulation. Slight divergences in otherwise parallel-looking lamination may be related to this structure even though truly undulating surfaces are not seen and a low-angle cross bedding may sometimes be seen below the waves or domes.

6.5.3 *Processes of formation*

The association of primary current lineation with this structure and its general similarity to parallel lamination suggests a related origin. If the velocity of water flowing over an upper flow-regime flat bed is increased,

Figure 6.61 Undulating lamination associated with thin, roughly parallel lamination in sandstone. Primary current lineation occurs on the bedding surfaces. Morœnesø Formation, Proterozoic, N. E. Greenland.

a pattern of ephemeral waveforms develops on both the water and the underlying sediment surfaces.

You can easily see these waves in small, steep streams cutting across sandy beaches or in stormwater flowing down steep gutters. The dimensions of the water surface waves depend on the water depth and flow velocity. The waveform on the sediment surface is in phase with the water surface wave but is of lower amplitude (Fig. 6.62). When the position of the waves is relatively stable, they are known as **standing waves.** However, if you stop to watch these waves in a natural setting, you will probably notice that they often move upstream rather abruptly. In doing so the water surface wave may break and collapse giving a flat water surface from which new waves grow. When the waves move upstream, they are called **antidunes**. They may in rare cases give cross bedding that is inclined upstream. The underlying sediment wave is also destroyed. It is easy to see that the chances of preservation of undulatory lamination formed under such flow conditions are very low.

6.6 Hummocky cross stratification

6.6.1 Introduction

This structure is becoming widely recognised in ancient sandstones, though until quite recently it had probably been dismissed as 'wavy', 'irregular' or 'undulating' bedding or lamination. It is most common in fine- to medium-grained sandstone of shallow-marine or near-shore origin. It occurs both within thicker sandstone units and within sharp-based sandstone beds of interbedded sandstone/mudstone sequences. No unequivocal modern examples have been described.

6.6.2 Morphology

The structure comprises sets of curving lamination with both convex-up (**hummocks**) and concave-up (**swales**) sectors. The laminae seldom dip at more than 12° and sets intersect one another at low angles (Fig. 6.63). Laminae may thicken into swales and thin over hummocks, so that the undulations gradually die out upwards. Heights of undulations seldom exceed 20 cm and wavelengths are of the order of 1 m. In many cases there is no apparent preferred orientation to the inclination of laminae, suggesting a more or less uniform three-dimensional pattern. In some cases however, a preferred orientation is apparent giving a form of low-angle cross bedding. Where the structure occurs in sharp-based sandstone beds, there are commonly signs of erosion in the form of sole marks. Upper contacts are broadly horizontal and often are characterised by wave ripples.

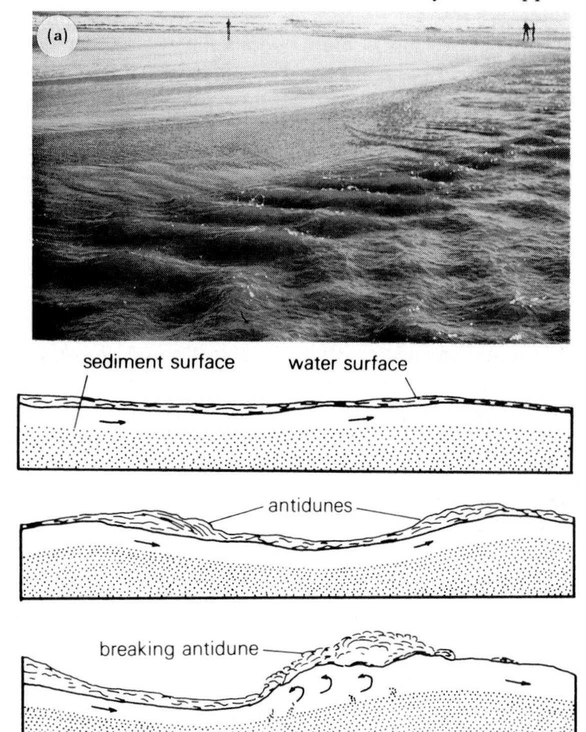

Figure 6.62 (a) Antidunes developed on the water surface of a small stream crossing a beach, Pendine, Dyfed, Wales. (b) Standing waves and antidunes as seen in a laboratory flume experiment. Note how the water and sediment waves are in phase with one another and how the antidunes break in an upstream direction. Each section is about 1 m long. (Based on Kennedy 1961.)

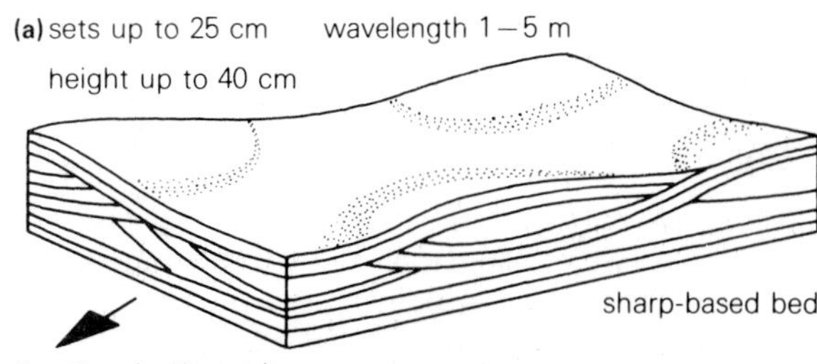

Figure 6.63 (a) Block diagram showing the main features of hummocky cross bedding (after Walker 1979). (b) Sandstone bed with low-angle cross bedding of hummocky origin. Note the wave ripples on the top surface which reflect the late-stage waning of the storm event responsible for deposition of the bed. Carolinefjell Formation, Cretaceous, Spitsbergen.

6.6.3 Process of formation

The occurrence of the structure in a shallow-marine setting (as deduced from associated fossils, trace fossils and overall context) and its clear association with wave action and an episodic style of deposition (as shown by interbedded sequences) has led to its interpretation as the product of strong and complex wave activity, mainly in areas below fair weather wave base. In interbedded sandstones its occurrence suggests a phase of vigorous activity which eventually decayed into recognisable wave oscillation. This is most likely in a storm at the peak of which wave action is most vigorous. Storm bedforms are not well described for obvious reasons, but the inference from these ancient examples is that a complex pattern of erosion and rapid deposition occurs on an irregular undulating surface. Examples showing a recognisable preferred orientation to the inclined layers perhaps indicate a coexistence of strong wave action and a unidirectional current, a form of combined flow. Arguments about how the sand was transported to the site of deposition during storms are unresolved with wind-driven currents, storm surge and turbidity currents all being advocated. Such discussions demand a wider knowledge of the context of particular examples and are beyond our scope here.

6.7 Massive sand beds

6.7.1 Introduction

The structures of sands and sandstones described and discussed so far are all associated with well defined lamination. There are, however, many sands and sandstones which lack recognizable lamination and which are described as **structureless**, **massive** or **unlaminated**.

6.7.2 Description

These beds may occur in sand and sandstone of virtually any grain size or sorting. Some such beds are lenticular, while others are parallel-sided and may be interbedded with finer sediment. As a rule, it is much more difficult to establish the absence of a particular feature (in this case lamination) as opposed to its presence. In looking at an apparently structureless sandstone, perhaps one is simply not seeing the lamination. Is it not weathering out in that particular exposure? Would some more sophisticated technique of observation reveal hidden lamination? Staining, etching and polishing of cut surfaces can indeed reveal previously unnoticed structures. X-radiography of thin slabs cut normal to bedding can be even more effective. Despite these methods, however, there are still beds which lack detectable lamination. In the field it is reasonable to apply terms like massive, unlaminated and structureless, especially where there is a clear contrast between the structureless beds and neighbouring laminated beds which have undergone similar weathering.

6.7.3 Processes of formation

Absence of lamination may reflect conditions of deposition or it may be the result of destruction of original lamination. A primary lack of lamination most commonly results from rapid deposition, most probably through the deceleration of a heavily sediment-laden current. Grains arrive at the bed so rapidly that they are buried before any bedload movement can occur and thus give rise to sorting into laminae. A 'frozen' grainflow or fluidised flow may also appear structureless.

Destruction of lamination can come about through the intense reworking of sediment by organisms living within it and also by physical disruption of waterlogged sediment due to liquefaction and movement. In the case of organic reworking, burrows may be visible in adjacent sediments or may be revealed by x-radiography (see Fig. 9.40). Where lamination has been destroyed by liquefaction, structures due to associated water-escape may be present (see 9.2.2). In aeolian sandstones, remnant blocks of brecciated sand or plastically folded patches occur within structureless sand and suggest a secondary destruction of lamination (Fig. 6.57).

6.8 Graded beds and the Bouma Sequence

6.8.1 *Introduction*

Certain sharp-based sandstone beds, most commonly in interbedded sandstone/mudstone sequences, show an assemblage of grain-size changes and sedimentary structures which, together, are highly diagnostic of depositional process. Such beds occur in a wide range of depositional settings and can involve sandstones ranging in grain size from very coarse and pebbly sand to very fine sand or even silt. The assemblage of features may involve both changes of grain size through the bed and also the presence of a sequence of different styles of lamination (the **Bouma Sequence**).

6.8.2 *Graded beds*

A bed which shows a progressive upwards reduction in grain size from top to bottom is said to be **graded** or **normally graded** (Fig. 6.64). The grain size change can take one of two forms: **content grading** where the *mean* grain size of the sediment reduces upwards; and **coarse-tail grading** where the size of the *coarsest* grains diminishes, the rest of the population remaining roughly constant. It is not always possible to judge this difference in the field, especially in finer-graded sandstones, but one should attempt to discriminate wherever possible. In addition, record the range of size variation through the bed. Rarely in sandstone, grain size may show an upwards increase through the bed (**inverse grading**), but this is more common in conglomerates. Grading is also seen in finer-grained sediments, especially siltstones (see 5.2.5).

Graded beds are, in many cases, structureless or unlaminated, and where lamination is present it occurs in the upper part.

Figure 6.64 Normal grading in a sandstone bed.

6.8.3 *The Bouma Sequence*

In addition to showing structureless and sometimes graded intervals, sharp-based sandstone beds may show parallel lamination or ripple cross lamination. When these are all present, they tend to occur in a particular vertical order (Fig. 6.65) which has been called the **Bouma Sequence** after its discoverer. In its complete development structureless sand which may or may not be graded (A division), is overlain by parallel lamination which may show primary current lineation (B division). This in turn is overlain by ripple cross lamination (C division). The D division, the least often recognised, is also parallel laminated, but more silty and diffusely defined. The final interval, the E division, is fine-grained mudstone or siltstone, which may be difficult to separate from the fine-grained interbeds of the sequence.

This complete sequence is very much an ideal development and, in reality, it is commonly the case that one or more interval is missing. Only the vertical order remains constant. The relative thickness of the intervals also varies. Some beds are dominated by the laminated divisions, while others consist almost entirely of the A division, and may only have a thin capping of interval B or C. Beds which end with interval C commonly preserve ripple morphology on their top surfaces. Beds with a thick C division commonly show climbing ripple cross lamination (ripple drift). In addition to the internal structures of these sharp-based beds, it is quite common for their lower surfaces to carry sole marks. If this is the case, it is often valuable to compare the palaeocurrents from such marks with those derived from primary current lineation and ripple cross lamination in the B and C divisions.

6.8.4 *Processes of formation*

Sharp-based sandstones, in interbedded sandstone/mudstone successions suggest a pattern of episodic deposition, the sands recording high energy events and

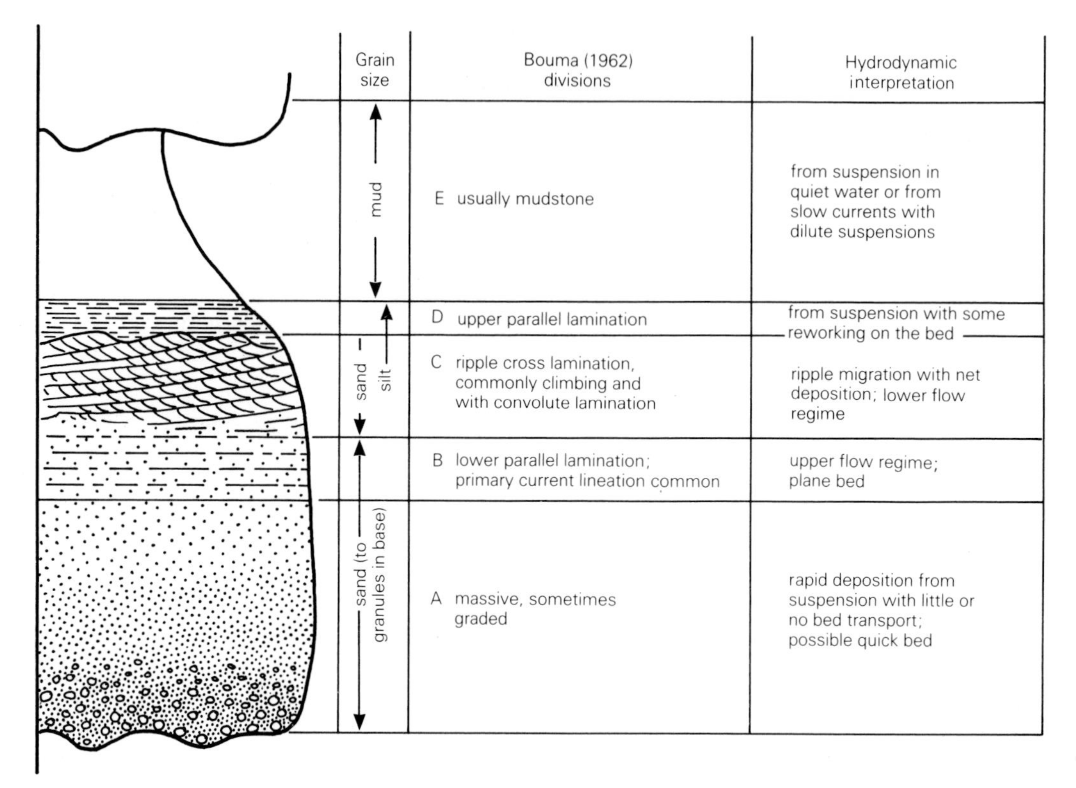

Figure 6.65 The Bouma Sequence of internal sedimentary structures which occur in sandstone beds generated by sudden, decelerating unidirectional currents. In most examples of such beds, one or more of the divisions may be missing.

the mudstones recording longer intervals of deposition from suspension in quiet conditions. As discussed in Section 6.7, massive or laminated sand is attributable to rapid deposition from a heavily sediment laden suspension. Associated with grading, this suggests a decelerating current, with coarsest particles falling to the bed first. Graded beds can be produced very simply in the laboratory by stirring up a suspension of mixed grain sizes in a beaker and allowing it to settle.

The upward passage into laminated sand records a reduction in depositional rate, while the style of lamination records the flow strength when transport and sorting on the bed began. Parallel lamination (B division) records upper flow regime plane bed, while the C division records a weaker current in the lower flow regime. The D and E divisions are mainly the result of direct deposition from suspension. The whole assemblage suggests a decelerating flow with material being deposited from suspension throughout.

The nature of the decelerating current can only be deduced from the context of the sediments. The Bouma Sequence was first described from deep-water sediments where the currents were interpreted as turbidity currents (see 3.7.3), but the sequence is in no way diagnostic of such currents. Similar beds can occur as a result of sudden decelerating flows in many settings.

6.8.5 *Uses of grading and the Bouma Sequence*

As well as containing valuable information about depositional processes, graded beds provide one of our best way-up indicators. The relative rarity of inverse grading in sandstones makes its use almost unambiguous. However, it is best always to check the direction of grading in several beds. The Bouma Sequence is also a useful check on way-up. Both sole marks on the bases of beds and structures in the laminated intervals provide useful palaeocurrent data.

Reference and study list

References marked with an asterisk * are suitable for 16–19 year old advanced level students in schools and colleges in the UK as well as for undergraduates.

Field experience

Present-day processes, bedforms and structures in sands are open to study in most environments. Useful observations can be made in driven snow. The ancient record is equally easy to study, since sandstones are commonly well exposed and structures are easy to observe, measure and record.

Laboratory experience

Most of the appropriate experience that relates particularly to sands has been cited in the equivalent section in Chapter 3.

Allen, J. R. L. 1985. *Experiments in physical sedimentology*. London: Allen & Unwin.

Allen, J. R. L. 1975. Sedimentation to the lee of small underwater sand waves: an experimental study. *J. Geol.* **73**, 95–116.

Ashley, G. M., J. B. Southard and J. C. Boothroyd. 1982. Deposition of climbing ripple beds: a flume simulation. *Sedimentology* **29**, 67–79.

Friend, P. F. 1982. Film: *Current and wave ripples*. 16 mm; colour, sound; *c*. 8 min. Department of Earth Sciences, University of Cambridge, England. [Detailed illustration of processes observed in laboratory flume experiments.]

Fryberger, G. S. and C. J. Schenk 1981. Wind sedimentation tunnel experiments on the origin of eolian strata. *Sedimentology* **28**, 805–21.

Klein, G. de V. 1975. Film: *Intertidal sedimentation in the bay of Fundy*. Urbana, Illinois: University of Illinois, Department of Geology.

*Larsen, F. D. 1968. A tank for demonstrating alluvial processes. *J. Geol. Education* **16**, 53–5.

*Open University 1980. Film: *Shoreline processes – an East Anglian case study*. Colour, sound. Milton Keynes: Open University Enterprises.

Open University 1984. Videotape: *Deserts*. 25 min. Milton Keynes: Open University Enterprises.

Open University 1987. Videotape: *Deep sea sands*. 25 min. Milton Keynes: Open University Enterprises.

Open University 1987. Videotape: *Deltaic sediments*. 25 min. Milton Keynes: Open University Enterprises.

*Open University 1980. Film: *When the river meets the sea – a visit to the Tay Estuary*. Colour, sound. Milton Keynes: Open University Enterprises.

Seppala, M. and Linde 1978. Wind tunnel studies of ripple formation. *Geogr. Ann.* **60A**, 29–42.

*Shelton, J. (ed.) 1967. Film: *The beach: a river of sand*. Encyclopaedia Britannica Corporation (available in UK from Fergus Davidson Associations). [A film originally produced for the Earth Science Curriculum Project and aimed at 14–16 year olds in US schools, but so good that it can be appreciated by all ages in understanding wave, longshore and density-current processes.]

*Tanner, W. F. 1962. Inexpensive models for study of helical flow in streams. *J. Geol Education* **10**, 116–18.

United States Geological Survey 1969. Film: *Flow in alluvial channels*. Colour, sound; *c*. 40 min. Bandelier Productions for USGS. [Flume experiments on bedforms and grain size, developing an appreciation of bed roughness and frictional response for stable channel design. A useful illustration of the generation of bedforms in sand.]

*Walton, E. K. and J. B. Wilson 1969. Film: *Shoreline sediments*. Colour, sound, 29 min. Edinburgh: Audiovisual Aids Unit, Department of Genetics, University of Edinburgh. [The relationship between present-day natural processes and sedimentary structures is outlined at elementary level before being applied to selected examples in the past.]

Yoxall, W. H. 1983. *Dynamic models in earth-science instruction*. Cambridge: Cambridge University Press. [Chapter 10 is on the use of the wind tunnel.]

Aqueous processes and structures in sands

Essential reading

A selection from:

Allen, J. R. L. 1970. *Physical processes of sedimentation*. London: Allen & Unwin. [A good deal of Chapter 2 is to do with structures in sands. Many of the environmental chapters (3–7) discuss features in sands.]

Allen, J. R. L. 1985. *Principles of physical sedimentology*. London: Allen & Unwin.

Boersma, J. R. 1967. Remarkable types of mega cross-stratification in the fluviatile sequence of a subrecent distributary of the Rhine, Amerongen, The Netherlands. *Geol. Mijnb.* **46**, 217–35.

Bouma, A. H. 1962. *Sedimentology of some flysch deposits*. Amsterdam. Elsevier. [The first outline of the 'turbidite' model in which sequences of their internal sedimentary structures (the Bouma Sequence) are analysed.]

Cant, D. J. and R. G. Walker 1978. Fluvial processes and facies sequences in the sandy braided South Saskatchewan River, Canada. *Sedimentology* **25**, 625–48.

Clifton, H. E. 1976. Wave-formed sedimentary structures – a conceptual model. In *Beach and nearshore sedimentation*, R. A. Davies Jr and R. L. Ethington (eds), 126–48. Tulsa: SEPM Sp. Publ. 24.

Coleman, J. M. 1969. Brahmaputra River: channel processes and sedimentation. *Sed. Geol.* **3**, 129–239. [The best description to date of a large sandy braided river and its processes.]

Collinson, J. D. 1970. Bedforms of the Tana River, Norway. *Geogr. Ann* **52A**, 31–56. [A description of bedforms and their modification during changing river stage.]

Dalrymple, R. W. 1984. Morphology and internal structure of sandwaves in the Bay of Fundy. *Sedimentology* **31**, 365–82.

Harms, J. C., J. B. Southard, D. R. Spearing and R. G. Walker 1975. *Depositional environments as interpreted from primary sedimentary structures and stratification sequences*. Short Course Notes no. 2. Dallas: SEPM.

Jopling, A. V. 1965. Hydraulic factors controlling the shape of laminae in laboratory deltas. *J. Sed. Petrol.* **35**, 777–91.

Klein, G. de V. (ed.) 1976. *Holocene tidal sedimentation*. Stroudsburg, Pa: Dowden, Hutchinson & Ross.

Langhorne, D. N. 1982. A study of the dynamics of a marine

sandwave. *Sedimentology* **29**, 571–94.

Leeder, M. R. 1982. *Sedimentology: processes and product*. London; Allen & Unwin. [Several chapters have very useful information on structures in sand.]

McCabe, P. J. 1977. Deep distributary channels and giant bedforms in the Upper Carboniferous of the central Pennines, northern England. *Sedimentology* **24**, 271–90. [Description and interpretation of very large-scale tabular cross bedding.]

McCave, I. N. and A. C. Geiser 1979. Megaripples, ridges and runnels on intertidal flats of the Wash, England. *Sedimentology* **26**, 353–69.

Miall, A. D. (ed.) 1978. *Fluvial sedimentology*. Can. Soc. Petrol. Geol., Mem. 5. [Many useful papers on structures in sand river deposits: see papers by Puigdefabregas & Van Vliet and Nami & Leeder for good descriptions of epsilon cross bedding.]

Middleton, G. V. (ed.) 1965. *Primary sedimentary structures and their hydrodynamic interpretation*. Tulsa: SEPM Sp. Publ. 12. [Greatly concerned with processes and structures in sands.]

Pettijohn, F. J. and P. E. Potter 1964. *Atlas and glossary of primary sedimentary structures*. Berlin: Springer-Verlag. [A compendium of photographs of structures, very many in sandstones.]

Raaf, J. F. M. de, J. R. Boersma and A. Van Gelder 1977. Wave generated structures and sequences from a shallow marine succession, Lower Carboniferous, County Cork, Ireland. *Sedimentology* **24**, 451–83.

Reineck, H. E. and I. B. Singh 1980. *Depositional sedimentary environments*, 2nd edn. Berlin: Springer-Verlag. [Structures in sands or sandstone in nearly every chapter.]

Schmincke, H. U., R. V. Fisher and A. C. Waters 1973. Antidune and chute and pool structures in base-surge deposits of the Laacher See area, Germany. *Sedimentology* **20**, 553–74.

Schreiber, B. C. 1978. Environments of subaqueous gypsum deposition. In *Marine evaporites*, W. E. Dean and B. C. Schreiber, 43–73. Short Course Notes no. 4. Tulsa: SEPM. [Depositional structures in aqueous gypsum sands.]

Swift, D. J. P., D. B. Duane and O. H. Pilkey (eds) 1972. *Shelf sediment transport, process and pattern*. Stroudsburg, Pa: Dowden, Hutchinson and Ross. [A collection of papers dealing with present-day processes: (a) water motion and processes of sediment entrainment; (b) patterns of fine sediment dispersal; and (c) patterns of coarse sediment dispersal.]

Tanner, W. F. 1967. Ripple mark indices and their uses. *Sedimentology* **9**, 89–104.

Walker, R. G. (ed.) 1984. *Facies models*, 2nd edn. Waterloo, Ont.: Geol. Assoc. Canada. [Many chapters are concerned with structures in sands especially 4 (volcaniclastics), 6 (fluvial), 8 (deltas), 9 (barrier islands), 10 (shallow marine), 11 (turbidites) and 17 (subaqueous evaporites).]

Further reading

Allen, J. R. L. 1982. *Sedimentary structures: their character and physical basis*. Amsterdam: Elsevier. [Two volumes in hardback and 1 volume in softback: the most complete and advanced treatment of sedimentary structures available, and likely to remain the definitive source for most aspects for some time.]

Allen, J. R. L. 1963. Asymmetrical ripple marks and the origin of water-lain cosets of cross-strata. *Liverpool & Manchester Geol. J.* **3**, 187–234.

Allen, J. R. L. 1965. The sedimentation and palaeogeography of the Old Red Sandstone of Anglesey, North Wales. *Proc. Yorks. Geol. Soc.* **35**, 139–85.

Allen, P. A. and P. Homewood 1984. Evolution and mechanics of a Miocene tidal sand wave. *Sedimentology* **31**, 63–81.

Buck, S. G. 1985. Sand-flow cross-strata in tidal sands of the Lower Greensand (Early Cretaceous), Southern England. *J. Sed. Petrol.* **55**, 895–906. [See also Hunter (1985).]

Clifton, H. E., R. E. Hunter and R. L. Phillips 1971. Depositional structures and processes in the non-barred, high-energy nearshore. *J. Sed. Petrol.* **41**, 651–70.

Collinson, J. D. 1968. Deltaic sedimentation units in the Upper Carboniferous of northern England. *Sedimentology* **10**, 233–54.

Collinson J. D. and J. Lewin (eds) 1983. *Modern and ancient fluvial systems*. Sp. Publ. Int. Assoc. Sediment. 6. [Useful papers on structures in sand by Allen, Sanderson & Locket and Haszeldine.]

Dalrymple, R. W., R. J. Knight and J. J. Lambaise 1978. Bedforms and their hydraulic stability relationships in a tidal environment, Bay of Fundy, Canada. *Nature* **275**, 100–4.

Davidson-Arnott, R. G. D and B. Greenwood 1976. Facies relationships on a barred coast, Kouchibouguac Bay, New Brunswick, Canada. In *Beach and nearshore sedimentation*, R. A. Davis Jr and R. L. Ethington (eds), 149–68. Tulsa: SEPM Sp. Publ. 24.

Hand, B. M., J. M. Wessel and M. O. Hayes 1969. Antidunes in the Mount Toby Conglomerate (Triassic), Massachusetts. *J. Sed. Petrol.* **39**, 1310–16.

Hunter, R. E. 1985. Subaqueous sand-flow cross-strata. *J Sed. Petrol.* **55**, 886–94. [See also Buck (1985).]

Jones, C. M. 1979. Tabular cross-bedding in Upper Carboniferous fluvial channel sediments in the southern Pennines, England. *Sed. Geol.* **24**, 85–104.

Jopling, A. V. and R. G. Walker 1968. Morphology and origin of ripple-drift cross-lamination, with examples from the Pleistocene of Massachusetts. *J. Sed. Petrol.* **38**, 971–84.

Klein, G. de V. 1977. *Clastic tidal facies*. Champaign, Ill.: Continuing Education Publishing Co.

Leeder, M. R. 1980. On the stability of lower stage plane beds and the absence of current ripples in coarse sand. *J. Geol. Soc. Lond.* **137**, 423–9.

Manz, P. A. 1978. Bedforms produced by fine, cohesionless, granular and flakey sediments under subcritical water flows. *Sedimentology* **25**, 83–103.

Middleton, G. V. and A. H. Bouma (eds) 1973. *Turbidites and deepwater sedimentation*. Short Course Notes. Los Angeles: Pacific Section, SEPM.

Nio, S.-D., R. T. E. Shuttenhelm and Tj. C. E. van Weering (eds) 1981. *Holocene marine sedimentation in the North Sea Basin*. Sp. Publ. Int. Assoc. Sediment. 5. [Papers on structures in sands by Terwindt, Kohsich & Terwindt, Boersma & Terwindt, Elliott & Gardner.]

Reading, H. G. (ed.) 1986. *Sedimentary environments and facies*, 2nd edn. Oxford: Blackwell Scientific. [Chs 3 (alluvial), 4 (lakes), 5 (deserts), 6 (deltas), 7 (shorelines), 9 (shallow seas), 12 (deep seas) and 13 (glacial) have frequent reference to structures in sands.]

Reineck, H. E. and F. Wunderlich 1968. Classification and origin of flaser and lenticular bedding. *Sedimentology* **11**, 99–104.

Rubin, D. M. and R. E. Hunter 1982. Bedform climbing in theory and nature. *Sedimentology* **29**, 121–38. [A geometrical analysis of the generation of cross bedding. Not an instruction manual on how to climb up sand dunes!]

Scholle, P. A. and D. Spearing (eds) 1982. *Sandstone depositional environments*. Tulsa: AAPG. [An authoritative and well illustrated account of depositional environments of sands.]

Swift, D. J. P. 1976. Continental shelf sedimentation. In: *Marine sediment transport and environmental management*. D. J. Stanley and D. J. P. Swift (eds). New York: Wiley.

Walton, E. K. 1967. The sequence of internal structures in turbidites. *Scot. J. Geol* **3**, 306–17.

Aeolian processes and structures in sands

Light reading

Bagnold, R. A. 1935. *Libyan Sands: travel in a dead world*. London: Hodder & Stoughton.

Hedin, S. 1896. A journey through the Takla–Makan Desert, Chinese Turkestan. *Geog. J.* **8**, 264–78.

Thesiger, W. 1964. *Arabian sands*. London: Longman (also in Penguin).

Thomas, B. 1938. *Arabia Deserta: across the empty quarter of Arabia*. London: Readers Union.

Essential reading

A selection from:

*Allen, J. R. L. 1970. *Physical processes of sedimentation*. London: Allen & Unwin. [Ch. 3, especially the descriptive parts.]

Bagnold, R. A. 1941. *The physics of blown sand and desert dunes*. London: Methuen. [The classic book, dealing especially brilliantly with processes and bedforms, but little about internal structures.]

Brookfield, M. E. 1977. The origin of bounding surfaces in ancient aeolian sandstones. *Sedimentology* **24**, 303–32.

Brookfield, M. E. 1984. Eolian sands. In *Facies models*, 2nd edn, R. G. Walker (ed.), 91–104. Toronto: Geoscience Canada.

Cooke, R. J. and A. Warren 1973. *Geomorphology in deserts*. London: Batsford. [Part 4 is devoted to aeolian processes and bedforms, but little on internal structures.]

Chepil, W. S. 1945. Dynamics of wind erosion. *Soil Sci.* **60**, 305–20,397–411, 475–80.

Embleton, C. and J. Thornes (eds) 1979. *Process in geomorphology*. London: Edward Arnold.

Hunter, R. E. 1977. Basic types of stratification in small aeolian dunes. *Sedimentology* **24**, 361–87. [Processes of origin of small-scale features of dunes at Padre Island Texas, the counterparts of which are not usually described or interpreted in fossil analogues.]

McKee, E. D. 1966. Structure of dunes at White Sands National Monument, New Mexico. *Sedimentology* **7**, 1–70. [A monumental field investigation. A good deal on techniques and the internal structure of various types of dunes developed in gypsum sand.]

McKee, E. D. (ed.) 1979. *A study of global sand seas*. Prof. U. S. Geol Surv. Pap. 1052, Washington: US Geol Surv. [Chapter E describes and illustrates a great many structures, mostly types of cross bedding. Other chapters deal with the great variety of aeolian dunes and draas].

Sharp, R. P. 1964. Wind-driven sand in Coachella Valley, California. *Bull. Geol Soc. Am.* **75**, 785–804.

Wilson, I. G. 1972. Aeolian bedform – their development and origins. *Sedimentology* **19**, 173–210.

Further reading

Bigarella, J. J., R. D. Becker and G. M. Duarte 1969. Coastal dune structures from Parana (Brazil). *Mar. Geol.* **7**, 5–55.

Brookfield, M. E. and T. S. Ahlbrandt 1983. *Eolian sediments and processes*. Amsterdam: Elsevier . [A symposium volume with most of the papers devoted to aspects of depositional structures in wind-blown sands and sandstones.]

Clemmensen, L. B. 1986. Storm-generated eolian sand shadows and their sedimentary structures, Vejers Strand, Denmark. *J.Sed. Petrol.* **56**, 520–7. [Careful observation of a commonly encountered structure.]

Clemmensen, L. B. and K. Abrahamsen 1983. Aeolian stratification and facies association in desert sediments, Arran Basin (Permian), Scotland. *Sedimentology* **30**, 311–39.

Cooper, W. S. 1958. Coastal sand dunes of Oregon and Washington. *Mem. Geol. Soc. Am.* **72**.

Folk, R. L. 1971. Longitudinal dunes of the northwestern edge of the Simpson Desert, Northern Territory, Australia 1 – Geomorphology and grain size relationships. *Sedimentology* **16**, 5–54.

Freeman, W. E. and G. S. Visher 1975. Stratigraphic analysis of the Navajo Sandstone. *J. Sed. Petrol.* **45**, 651–68. [A controversial interpretation which is discussed in a later issue of the same journal 1977, **47**, by Picard, 475–83; Folk, 483–4; Steidtmann, 484–9; and Rusyla, 489–91.]

Glennie, K. S. 1970. *Desert sedimentary environments*. Amsterdam: Elsevier. [See chapter on aeolian structures, sequences and predicted palaeocurrent patterns.]

Gradzinski, R., J. Gagol and A. Slackza 1979. The Tumlin Sandstone (Holy Cross Mts, Central Poland): Lower Triassic deposits of aeolian dunes and interdune areas. *Acta Geol. Pol.* **29**, 151–75.

Greeley, R. and J. D. Iversen 1985. *Wind as a geological process*. Cambridge: Cambridge University Press.

Land, L. S. 1964. Eolian cross-bedding in the beach dune environment, Sapelo Island, Georgia. *J. Sed. Petrol.* **34**, 389–94.

McKee, E. D. and G. C. Tibbitts 1964. Primary structures of a seif dune and associated deposits in Libya. *J. Sed. Petrol.* **34**, 5–17.

McKee, E. D., J. R. Douglass and S. Rittenhouse 1971. Deformation of lee-side laminae in eolian dunes. *Bull. Geol. Soc. Am.* **82**, 359–78.

Rubin, D. M. and R. E. Hunter 1985. Why deposits of longitudinal dunes are rarely recognised in the geological record. *Sedimentology* **32**, 147–57.

Sharp, R. P. 1963. Wind ripples. *J. Geol.* **71**, 617–36.

Stokes, S. L. 1968. Multiple parallel-truncation bedding planes – a feature of wind deposited sandstone formations. *J. Sed. Petrol.* **38**, 510–15.

Talbot, M. R. 1985. Major bounding surfaces in aeolian sandstones – a climatic model. *Sedimentology* **32**, 257–65.

Thompson, D. B. 1969. Dome-shaped aeolian dunes in the Frodsham Member of the so-called 'Keuper' Sandstone (Scythian–?Anisian: Triassic) at Frodsham, Cheshire (England). *Sed. Geol.* **3**, 263–89. [Photographic mosaics analysed to reveal sedimentary structures. See also discussion by Brookfield (1977).]

Tsaor, H. 1982. Internal structure and surface geometry of longitudinal (seif) dunes. *J. Sed. Petrol.* **52**, 823–31.

Walker, T. R. and J. C. Harms 1972. Eolian origin of flagstone beds. Lyons Sandstone (Permian), type area, Boulder County Colorado. *Mountain Geol.* **8**, 279–88. [Hard to locate in libraries, but a good description and interpretation of aeolian cross bedding and associated features in ancient rocks.]

Wasson, R. J. and R. Hyde 1983a. Factors determining desert dune type. *Nature* **304**, 337–9.

Wasson, R. J. and R. Hyde 1983b. A test of granulometric control on desert dune geometry. *Earth Sci. Proc. Landforms* **8**, 301–12.

Wilson, I. G. 1973. Ergs. *Sed. Geol.* **10**, 77–106.

Extra references relating to figure captions

Inman, D. L. and A. J. Bowen 1963. Flume experiments on sand transport by waves and currents. *Proc., 8th Conf. Coastal Engng.* 137–50. Am. Soc. Civ. Engrs.

Kaneko, A. 1980. The wavelength of oscillation sand ripples. *Rep. Res. Inst. Appl. Mech. Kyushu Univ.* **28**, 57–71.

Kennedy, J. F. 1961. *Stationary waves and antidunes in alluvial channels*. Report KH-R-Z, W. M. Keck Laboratory of Hydraulics and Water Resources. Pasadena: California Institute of Technology.

Pretious, E. S. and T. Blench 1951. *Final report on special observations of bed movement in the lower Fraser River at Ladner Beach during 1950 freshet*. Vancouver: Natl Res. Council Canada.

Wilson, I. G. 1970. *The external morphology of wind-laid sand deposits*. Ph.D. thesis. University of Reading.

7 Depositional structures in gravels, conglomerates and breccias

7.1 Introduction

The depositional processes and structures of the rudites have been studied relatively little until recently. As our knowledge of processes becomes more refined, so the features to be observed, measured and recorded become clearer. A practical reason for this is the difficulty of making direct observations of such sediments while they are actually being entrained, transported and deposited in rivers, alluvial fans, reef talus slopes, storm beaches, submarine canyons and on volcanic slopes. Direct-recording instruments, including the human body, tend to be severely damaged by the motion of large clasts. The installation of sediment traps in stream beds can give useful information on transport *rates* during floods but tells little else of the *style* of transport and deposition. Some workers have attempted to overcome these problems by studying the day-by-day results of diurnal rise and fall of discharge on bedforms in proglacial outwash areas. Processes are deduced from the products, revealed both both the surface morphology and the internal structures as observed in trenches. Such methods are only applicable in accessible sub-aerially exposed settings. Laboratory experiments on gravels are increasingly attempted, but have been restricted by the need to build large and costly flumes or wave tanks. Even then, the scales of flows and structures are much smaller than the real phenomena. The study of conglomerates is yet another area of geology in which detailed observation and interpretation of ancient deposits can aid the better understanding of present-day processes, particularly in deep-water settings. The careful analyses which have allowed these advances have involved the recording of bed contacts, bed thicknesses, the presence of framework or matrix support and the sizes of the larger clasts.

7.2 Problems of classification

7.2.1 Defining rudites

There is no universal agreement on the percentage of clasts above 2 mm (-1ϕ) which must be present in a deposit before it is classified as a rudite (Fig. 7.1). Where there is a mixture of mud, sand and gravel, we recommend that the rock should contain more than 30% of clasts larger than 2 mm before the terms **gravel, conglomerate** and **breccia** are used. In the field or in the laboratory, first try to estimate the percentage of gravel, sand and mud present, and refer the sediment roughly and provisionally to the classes shown in Figure 7.1. Judgments about whether a very clayey conglomerate is a till (i.e. of glacial origin) or an agglomerate or lapilli-ash (i.e. of volcanic origin) must await detailed description of its composition and the shape of its clasts, etc. and consideration of its overall context.

7.2.2 Defining a sedimentary 'structure' in rudites

The term sedimentary 'structure' is here interpreted broadly to include several mass properties which include textural features: (a) features based on composition; (b) features such as shape, roundness and surface morphology of the constituent clasts; (c) stratification and cross stratification; (d) features based on grain-size distribution, sorting and clast-support systems; (e) features based on fabric, packing and porosity; and (f) the presence and type of graded bedding. The first two properties may be recorded in any preliminary survey and they provide useful pointers concerning the provenance (i.e. the source regions) and the transportation history of the clasts. The last five properties demand particular attention if the aim is to understand processes and environments of deposition.

7.2.3 Composition and classification of rudites

One of the first properties to record in the field is the composition of the larger clasts, which allows useful, preliminary conjectures concerning their possible provenance and their processes of origin, for example whether the rocks may be pyroclastic. Further observations relate to whether the materials originate from within the basin of deposition and are penecontemporaneously eroded, i.e. are intraformational (e.g. reef talus), or whether they come from a source area of older rocks, i.e. are extraformational or exotic.

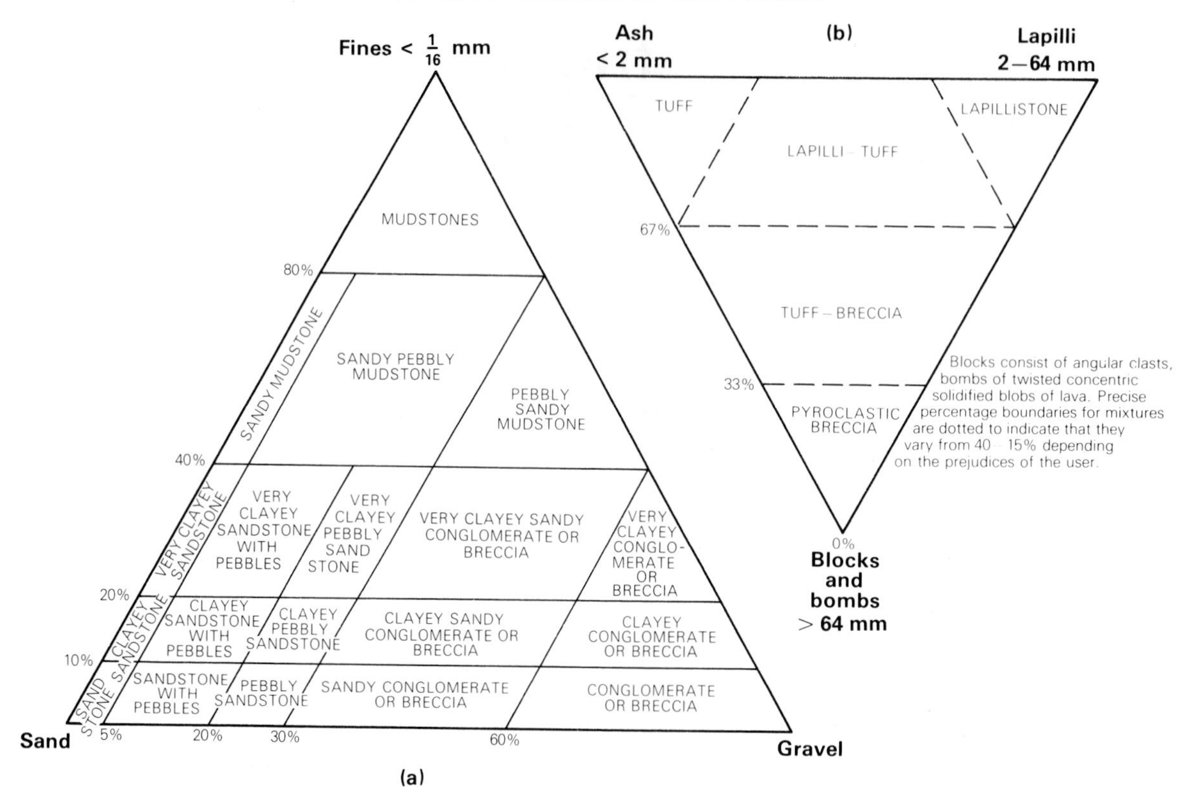

Figure 7.1 The nomenclature of rudites: (a) consolidated rudites defined by the proportions of the grain sizes of their component clasts (modified after Piper & Rogers 1980); (b) pyroclastic rocks defined by the proportions of the grain sizes of their component clasts.

7.2.4 Misidentification of the structures of rudites

Some rocks provisionally classified as rudites will, with further analysis and experience, be regarded as 'pseudoconglomerates' or 'pseudobreccias'. These may originate from processes of *in situ* diagenetic growth (e.g. concretions; see 9.3.1), tectonic disruption (e.g. fault breccias), or simply weathering (e.g. dolerite blocks, in process of spheroidal weathering, surrounded by altered clayey matrix). Likewise, features thought to be clasts will later prove to be trace fossils (e.g. rounded burrow fills, see 9.4).

7.3 Morphology and general settings of gravel deposition

7.3.1 Introduction

The large grain size of gravels prevents the formation of many of the bedforms and structures that are common in sands. For example, ripples will not form in clasts whose diameters approach the size of the ripples themselves. Large dunes may, however, be formed in coarse sand with a subsidiary component of gravel. Subaqueous dunes formed entirely of fine gravel are known in some rivers. The morphological features developed in gravel are relatively large and they may have distinctive packing fabrics developed within or superimposed upon them (see 7.4.4). Such fabrics commonly form the basis of interpretation of ancient gravels and conglomerates.

Unlike sands, there is no all-embracing scheme for gravels wherein the occurrence of particular bedforms can be related to specific flow conditions and grain sizes. But increasingly, these coarse deposits are interpreted as the products of two major sets of processes, namely those associated with bed load transport and those associated with sediment gravity flows. Despite this, no systematic description or classification is attempted here. Instead we outline some gross structures, and environments of gravel deposition.

7.3.2 *River channels*

Streams with gravel beds commonly show highly variable patterns of bars and channels. In braided rivers, bars (elongated parallel to flow) split the stream on several scales. Deposition takes place on the flat bar tops, where pebbles lodge in an imbricate packing (see 7.4.4), and at the down-stream ends, where avalanching gives rise to cross beds. Bars vary in relief from lobate sheets, only one clast thick, to forms several metres high. In more sinuous streams, gravel accumulates on lateral or point bars. In large rivers, bars are commonly composites of smaller features and the resultant gravel bodies are complex, tabular sheets. Channel fills are more restricted in extent and may appear lensoid. Deeper channel areas often display beds covered with the coarsest gravel, but this may be masked by finer sediment laid down during falling river stage when the bed is observed subaerially exposed at low stage.

7.3.3 *Fans and steep slopes*

Episodic floods on alluvial fans give rise to localised tongues and lobes of coarse sediment deposited by mud-flows or as sieve deposits. In mudflows, the presence of fine-grained sediment gives the flow a high viscosity and density (see 3.3). In the resulting deposits, large clasts commonly 'float' in a finer matrix. Gravel lobes are produced by a sieving process when water transporting the gravel sinks into a permeable substrate and leaves clast-supported gravel on the surface.

On very steep slopes, loose blocks of rock fall down and form screes. Such material tends to be angular and slopes of up to 35°, much steeper than the normal angle of rest, may develop. Such deposits have a low preservation potential in sub-aerial settings.

7.3.4 *Fan deltas*

Where a gravel-rich alluvial fan builds into a body of water such as a lake or the sea, a local wedge-shaped, gravel-rich fan delta may be formed. Gravel is not readily redistributed by waves and currents, and a steep delta slope forms at or close to the angle of rest. Some of the gravel may be transported into deeper water by processes of sediment gravity flow (see 3.7).

7.3.5 *Shorelines*

Debris at the foot of sea cliffs is quickly broken up and transported away, enabling wave attack to proceed further. Gravel accumulation is most important on beaches, particularly storm beaches. These gravel ridges are broadly linear but they sometimes migrate to produce sheet-like forms. Beach gravels are commonly very well sorted and the clasts are well rounded. The clasts show subparallel inclination if they are flattened (i.e. imbrication: see 7.4.4).

7.3.6 *Deep-water fans*

At the foot of major submarine slopes, turbidity currents and other sediment gravity flows may lead to the emplacement of gravels, particularly close to the mouths of submarine canyons, both as lobes and as channel fills. Knowledge of this setting comes mainly from facies analysis of ancient sedimentary rocks.

7.3.7 *Reefs and associated settings*

On steep reef fronts, pounded by ocean waves, irregular carbonate clasts, sometimes of very large size and composed of reefal material, fall and accumulate. These steeply inclined wedges (up to 40°) comprise interlocking frameworks with high initial porosities (up to 40%). Such deposits have high preservation potential and are well studied in the ancient record.

7.3.8 *Glacial and associated settings*

Glaciers carry sediment of all sizes and deposit it in a variety of settings, with and without the aid of water. Some debris is dumped directly from melting ice as **till**; other coarse-grained material falls from floating icebergs and shelf ice to give isolated **dropstones**. Material deposited both sub-aerially and sub-aqueously from melting ice is commonly subjected to further movement and deposition by mudflows. Melt water flowing within, above, below or lateral to a glacier may sort, transport and deposit sand and gravel bodies, such as **kames, eskers** and **outwash plains**. These gravels are better sorted than those dumped directly from ice. The complex variety of morphological features associated with glacial and glaciofluvial activity is covered in several more specialised textbooks (see *Further reading*).

7.3.9 *Deposits associated with volcanoes*

Coarse debris ejected during volcanic eruptions accumulates as widespread sheets and as more restricted bodies. **Autoclastic** processes generate clasts through mechanical breakdown and gaseous explosion during movement of magma and lava; **hyaloclastic** processes occur when lava is quenched and shattered by entry into water, water-saturated sediment or ice; and **pyroclastic** debris is produced by explosion of magma and country rock. The last mentioned form sub-aqueous and sub-aerial **pyroclastic fall** deposits, which include many drop-stones which deform underlying strata. **Pyroclastic flow** deposits are transported *en masse*, some of them as **base-surge** density currents. Many primary pyroclastic fragments become mixed during transport with 'normal' siliciclastic and carbonate deposits, and are resedimented as **epiclastic** conglomerates.

7.4 Structures and other descriptive features: their mode of formation

It is useful to note the composition, colour, shape and surfaçe features of the clasts in rudites, but the descriptions of other features usually repay the greatest attention when it comes to understanding the processes and environments of deposition.

7.4.1 Shape (roundness and sphericity) and surface features of the clasts

With experience, sphericity and roundness can be judged visually, and the clasts classified fairly accurately in the field. Such experience is best acquired in the laboratory by measuring and estimating sphericity and roundness parameters and by referring clasts to the categories of shape devised by Zingg. Classes of roundness may be indicated by descriptive terms or the numerical scales of Powers or Pettijohn (see *Reference list*). Experience of measuring and handling individual clasts of various shapes is important in appreciating their three-dimensional properties, as clasts are commonly seen only in two dimensions in many rock exposures (see 7.4.4).

Some shapes suggest processes of origin that allow one to make conjectures about the environment or transport history of the sediment which can then be tested by other means. Flat-sided clasts with two, three or four facets, the smooth surfaces of which may be pitted, fluted and polished, are known as **ventifacts**, i.e. wind-fashioned, sand-blasted objects. In most cases ventifacts will be rolled and reworked into later deposits, but occasionally they may be found *in situ*. Similar, but unpolished, unfluted pebbles can be produced by wet-blasting. Tough, flat-iron pebbles bearing parallel or subparallel scratches and snubbed edges record glacial abrasion and attrition prior to deposition; softer material may be scratched in several environments. Concentrations of well worn, matt-surfaced, disc-shaped pebbles generally become more common as one goes from rivers to low- then high-energy beaches. This is due to increased rates of abrasion by sliding. Highly rounded spheroidal pebbles arise through constant abrasion, attrition and reworking, sphericity falling and roundness increasing from fluvial to low- then high-energy tidal environments. The indices of sphericity and roundness are a measure of the maturity of the sediment. Note, however, that it is more difficult, given constant conditions, to round sand or small pebbles than it is to round large clasts. Percussion rings due to high-velocity collisions are apparent on the surfaces of some clasts. These should not be confused with pressure solution pits mentioned later.

7.4.2 Sorting, grain-size distribution, porosity and clast-support characteristics

Disaggregation and sieving of gravels, and the determination of their size distribution, are rarely attempted even in the laboratory. When ancient conglomerates are well cemented, the task is impossible. Nevertheless, it is important to estimate and describe these characteristics in an approximate way (see Fig. 7.2).

First ask yourself whether particular beds are well sorted or poorly sorted, and record whether there are

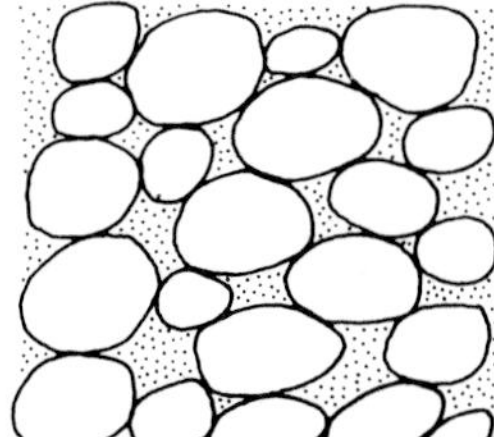

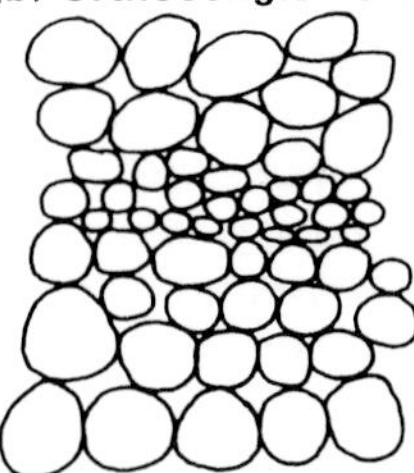

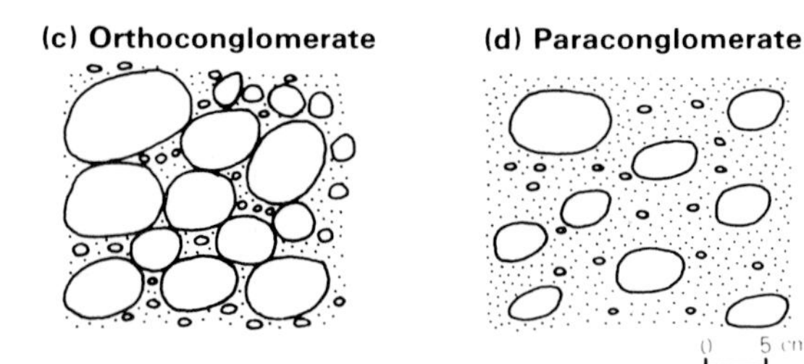

Figure 7.2 Descriptive features of sorting and size distribution in rudites. There is a spectrum in nature between types (a) and (d): (a) Bimodal; clast-supported framework; well sorted matrix. (orthoconglomerate); (b) clast-supported, open-work framework (orthoconglomerate); (c) Polymodal; clast-supported framework; poorly sorted matrix (orthoconglomerate); (d) Polymodal; matrix supported; lacking a framework (paraconglomerate). (After Walker in Harms *et al.* 1975.)

clear grain-size modes. Very commonly there is a mode in the pebble–cobble size and another in the sand size; hence the rock is **bimodal**, the term **clasts** being applied to the coarse mode and **matrix** to the fine. Some deposits are **unimodal** and well sorted, lacking a well defined matrix and hence having very high porosity (>40%) and permeability. Other beds are unimodal and poorly sorted, and others are **polymodal**.

Secondly, determine whether the deposit is clast- or matrix-supported. Are larger clasts in contact with each other, forming a framework, or are they dispersed in a matrix? This is not always easy to judge in a two-dimensional outcrop without disaggregating the rock. A search for indentations of one clast by another (i.e. pressure pittings) might help. Matrix-free, clast-supported, framework conglomerates commonly have a high porosity and permeability, and the term **openwork** is used to describe them. Alternatively, if the clasts are dispersed and supported by the matrix, try to establish whether the supporting matrix is of sand (e.g. in a sandy conglomerate), whence the rock may be very porous and permeable, or of mud (e.g. in a till or where the pebbles are glacial or volcanic dropstones within marine or lacustrine muds), whence the rock may have variable but commonly low porosity and permeability.

The distinctions made in the last two paragraphs provide the basis of a classification of rudites into the **orthconglomerates** and the **paraconglomerates**. In nature, conglomerates appear to range from orthoconglomerates with well sorted, clast-supported, openwork or matrix-filled frameworks, to paraconglomerates which are poorly sorted, matrix-supported and with very variable size distributions.

Following these considerations, further questions can be asked which help to generate ideas about transport and depositional processes:

(a) If the bed is clast-supported and has a matrix, was the finer-grained interstitial material deposited along with the larger clasts or did it fill the framework at a later time?
(b) If the bed is clast-supported but retains an openwork structure, did the current winnow away sediment and/or maintain it in suspension while rolling and sorting larger clasts on the bed?
(c) What does the bed tell us about the abundance of fine sediment during transport and deposition?

Although the possible threshold velocities for the erosion and transport of each mode can be read off the Hjulström-Sundborg graph (Fig. 3.15), bear in mind that these data relate mainly to rivers with low suspended sediment concentrations and floored by relatively uniform material. Figure 7.3 relates the sizes of clasts transported by rolling to the size of sand suspended by the same flow. When the bed has a pebble framework which is filled with sand, however, these graphical relationships do not suggest whether the infilling was largely contemporaneous with the deposition of the gravel, as when part of a heavy suspended load is trapped in the quieter interstices of the gravel bed, or if it largely postdated the deposition of the gravel, as when a river previously in flood, and carrying relatively little sand in suspension, resumes a more leisurely flow with sand transported as bed load. In many cases it is difficult to judge both the timing and the process of emplacement of the matrix. Wind-blown

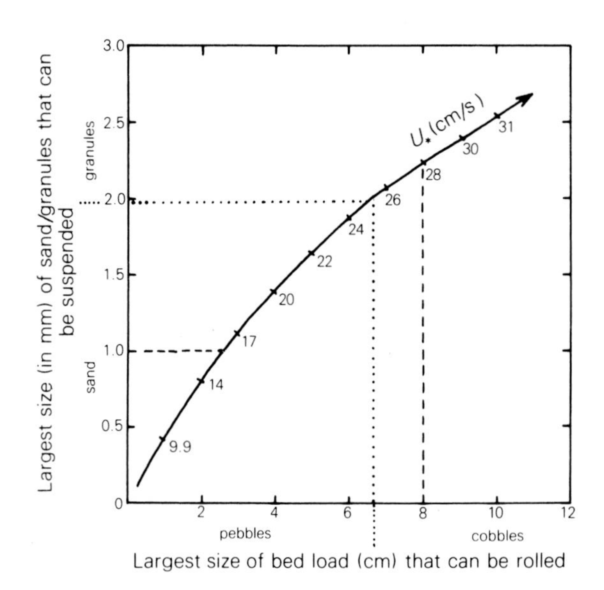

Figure 7.3 Graphical representation of size of clasts transported by rolling, compared with the size of sand grains suspended by the same flow. Numbers on the curve refer to critical values of shear velocity U_* for bed rolling (according to Shield's criterion $\sqrt{(\tau_0/\rho_0)}$ where τ_0 = boundary shear stress and ρ_0 = density) and transport of suspended load determined by the following value of the ratio of the shear velocity to the settling velocity: $U_*/V_0 > 1{\cdot}0$ (where $V_0 >=$ the settling velocity). The dotted lines show that the coexistence in an orthoconglomerate or paraconglomerate of pebbles of 6·3 mm and sand of 1·95 mm relates to slight velocity fluctuations around a mean value of $U_* = 25$ cm/s. The dashed lines show that the coexistence of clasts in a bimodal matrix-filled framework orthoconglomerate composed of cobbles of 8 cm and sand of 1 mm would probably result from the deposition from bed load of cobbles at high velocity (c. 28 cm/s) while sand of 1 mm was still in vigorous suspension. With diminishing velocity, the 1 mm sand would fall into the framework and probably fill it. (After Walker 1975.)

sand, for example, could fill a water-lain but sub-aerially emergent gravel framework without leaving much evidence of the nature of the process.

If the rudite is matrix-supported, consider whether the large clasts were transported along with the finer-grained matrix as a high-viscosity flow or whether they were dropped into already deposited finer-grained sediment. The occurrence of disturbed laminae in the matrix may help to decide the case. If the clasts are transported with the matrix, for example in a debris flow, we still have no internal indication of the environment of the flow. Alluvial fans, deep-water turbidite fans and channels, the surfaces of glaciers and the sides of volcanoes are all possible settings. If the clasts were dropped into a deposit, we do not always know whether this took place from a melting iceberg, from a floating and rotting tree root, or as the result of a volcanic explosion. In all cases, this could have taken place in lacustrine or marine settings. Clasts of pumice can float for long distances before sinking and not greatly disturb the underlying sediment when they settle. Only information on the context of the rudites derived from above and below any particular bed will help resolve such problems of interpretation.

7.4.3 Grading within beds

The main styles of grading described in Section 6.8 for sands apply equally well in gravels and conglomerates (Fig. 7.4). Content or distribution grading and coarse-tail grading are both common in normally graded rudites. Inverse grading is much more common in rudites than in sands. This can be due to large clasts rising upwards due to dispersive pressures or to progressive loss of larger clasts from the lower, more strongly sheared part of the flow. Ungraded beds usually indicate high shear strength or high viscosity, thereby preventing turbulence and effective grain interaction. Normal grading usually suggests that turbulence was developed during deposition (see 3.7). **Lateral grading** of grain size is sometimes noted, especially in orthoconglomerates (Fig. 7.15). Sometimes, especially in volcanic rocks, **density** (clast composition) **grading** has to be considered separately from size grading, especially where pumice or other vesicular clasts are involved (see Fig. 7.17c).

7.4.4 Fabric

By **fabric** we mean the orientation of particular dimensions of the constituent clasts. In some deposits a strongly preferred orientation of clast long axes may occur; in others the short axes may be aligned; and in others no pattern may be discerned (Fig. 7.5).

In lithified conglomerates measurement of the three-dimensional orientation of clasts may not be practicable, and a statistical resultant may have to be estimated from measurements of apparent axes in two-dimensional horizontal and vertical surfaces (Fig. 7.5b). In the laboratory, three-dimensional data can be plotted on a stereonet (Fig. 7.6) or components measured in two dimensions can be plotted as rose diagrams (Fig. 7.5b).

The most commonly occurring fabric, which arises when the rudite is rich in discs and blades, is **imbrication** (Figs 7.7 & 8). Here the flattened surfaces all dip in an upstream direction. This is common in horizontally bedded, clast-supported conglomerates, and two variants may occur. In one, the *a*-axis is transverse to the dip direction of the clasts and the *b*-axis is parallel to the dip. In the other the *b*-axis is transverse and the *a*-axis is parallel to dip. Imbrication of either sort should be carefully distinguished from the preferred orientation of clasts in cross-bedded gravels. Flattened clasts on avalanche faces are oriented parallel to the face and therefore dip down stream, opposite to that of genuine imbrication. Other clasts roll down slope on slip faces with their long axes parallel to the slope.

Rudites with flattened clasts lacking preferred orientation suggest transport by processes in which the clasts were not completely free to move relative to one another, possibly by virtue of higher flow viscosity or a rapid rate of accumulation, which did not allow time for an organised fabric to develop.

Well organised fabrics suggest that clasts have been free to move individually and independently of one another above the bed and that they have been selectively incorporated onto the bed when they were arrested in a stable position. Clasts that landed in other than the stable, upstream-dipping orientation would be re-entrained in most cases.

Theoretically, the attitude of each particle on the bed is its response to the combined forces that acted on it during and shortly after deposition. Gravity tends to keep the particles in place, the lift force may or may not act upwards, and the drag force will try to roll the particle. The most stable position is when the forces of removal (lift and drag) are at a minimum. This occurs when the plane of flattening of the particle is tilted downwards at a small angle into the current. The drag is then minimised, the lift forces may even be negative, and the contacts of the particle lie to the side and well forward of its centre of gravity.

Of the two variants of clast orientation, the one with the long axis (*a*) transverse to flow has been ascribed, on the basis of experiments, to the rolling of clasts on the sediment surface. In view of its association with turbidites and density-flow deposits, the fabric with long axis parallel

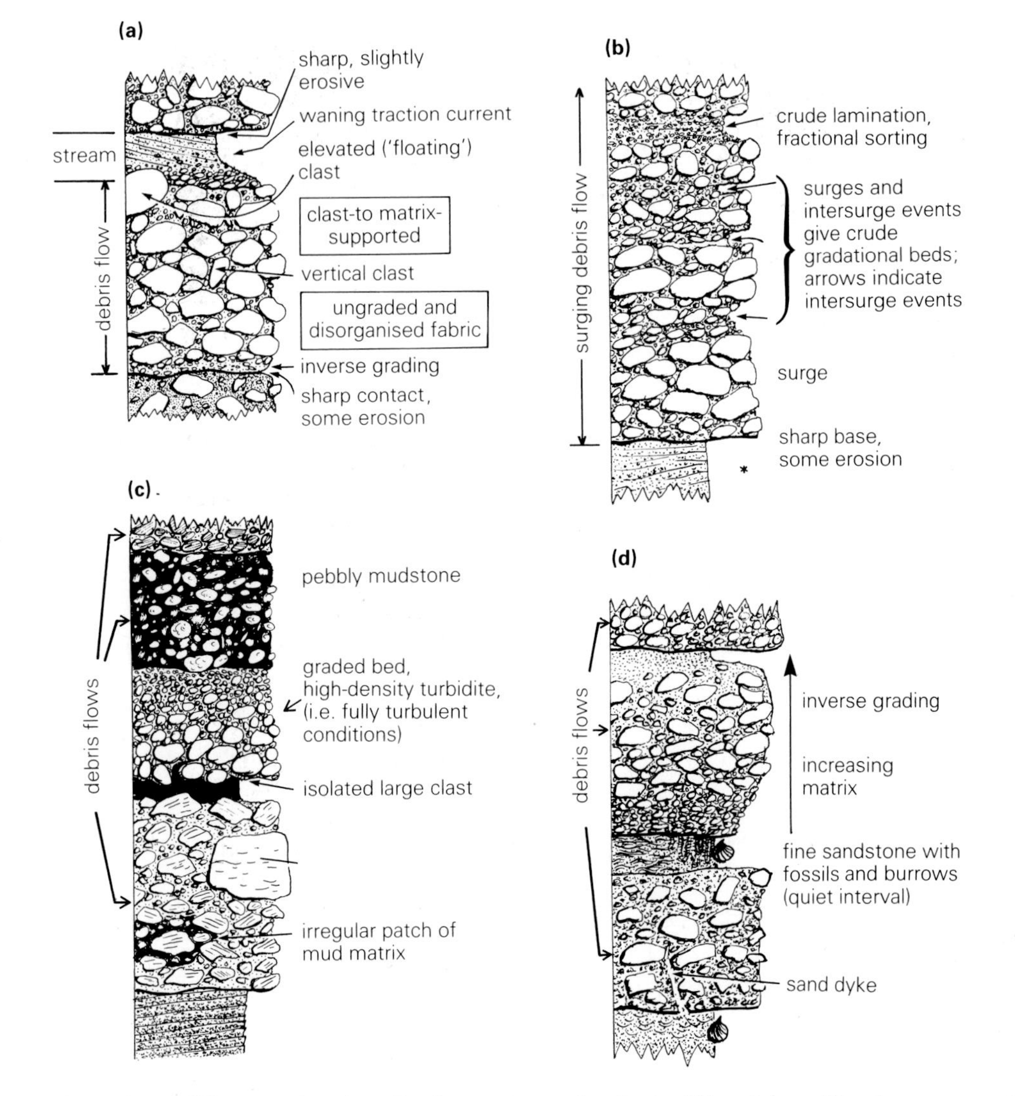

Figure 7.4 Examples of different styles of grading in conglomerates due to different depositional processes acting in a variety of typical sub-aerial (a,b) and sub-aqueous settings (c,d). Bed thickness ranges from a few decimetres to metres (after Nemec & Steel 1984, Figs 15 & 16).

to dip has been ascribed to flows that maintain large clasts above the bed, possibly through the dispersive pressures associated with intergranular collision, up to the point of deposition.

Well rounded and well sorted clasts in discrete but extensive beds may be formed by high-energy wave action. When present, imbrication commonly dips seawards on the upper and middle parts of the beach (Fig. 7.9).

Fabrics in paraconglomerates are difficult to interpret, although they have been used in attempts to differentiate between glacially emplaced tills and the products of mass flow and also between different types of till, e.g. subglacial lodgement and superglacial flow till (Fig. 7.10). Although many tills show a preferred, clast long-axis orientation parallel to ice flow, the pattern is seldom well defined and few certain generalisations seem possible. There will always be ambiguity in cases such as flow tills where later mass flow has reorganised the material of the original till.

Try to describe and interpret the data depicted in Figure 7.11.

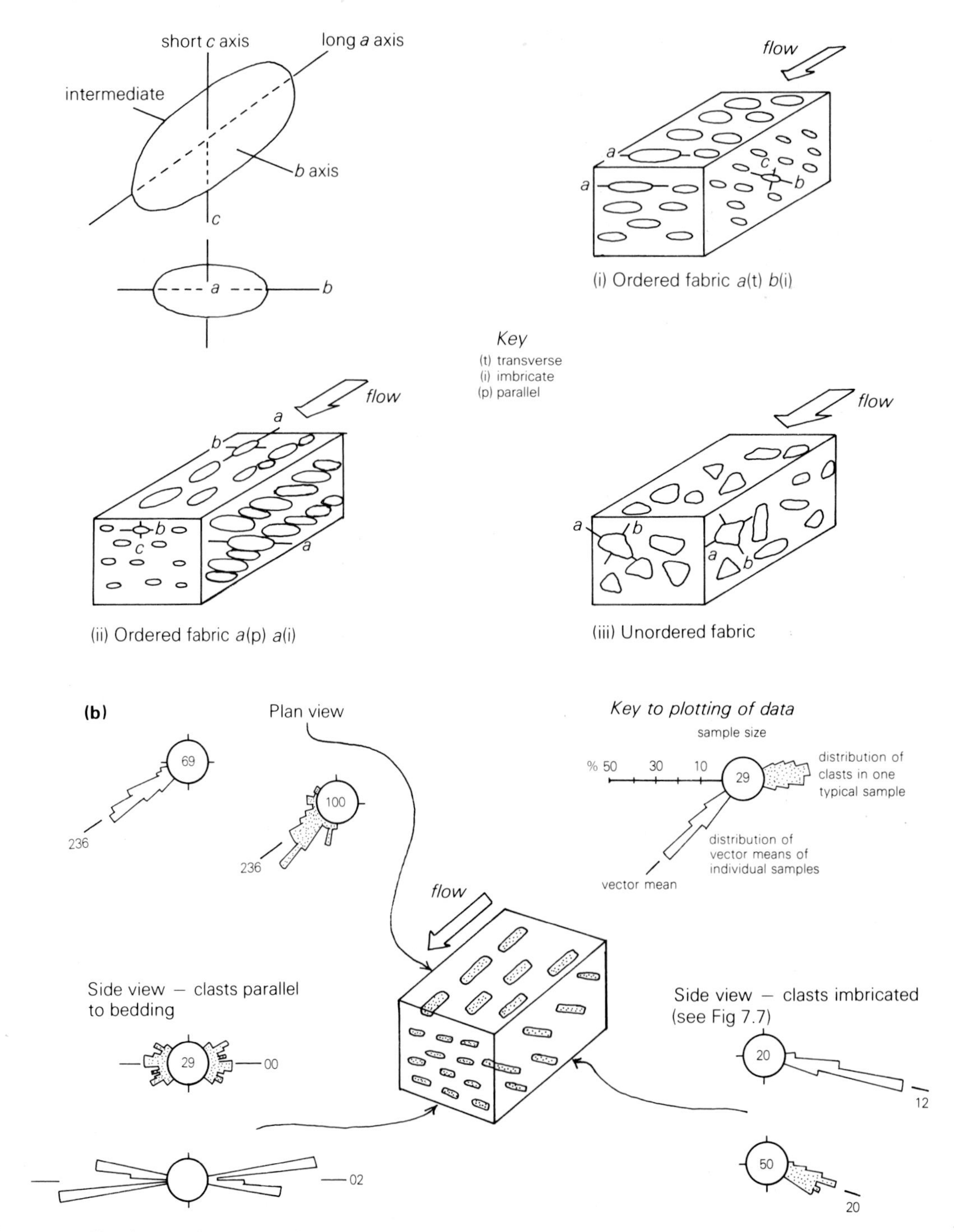

Figure 7.5 The fabric of rudites: (a) the nomenclature of ordered and unordered fabrics; (b) the plotting of fabric measurements for well cemented ordered rocks. Data collected from orthogonal faces not from three-dimensional measurements of clasts. Plan views of beds measured at 69 localities, 100 clasts at each place; imbrication measured at 20 localities on facies parallel to flow and perpendicular to bedding; measurements at 6 localities with facies perpendicular to flow and bedding. (After Davies & Walker 1974 and Walker in Harms *et al.* 1975.)

Figure 7.6 The three-dimensional disposition of rock particles shown on a stereonet projection of a lower hemisphere. In the upper diagram the elongations of the *a,b* and *c* axes of a clast centred on the plane of the projection cut the hemisphere at a' and c' and at the periphery b' It is conventional to project the axial data onto the equatorial plane of the hemisphere (lower diagram) and in particular to plot and contour the density of the projections of the '*a*' axes of 25 or more clasts from each locality in order to reveal the presence or absence of ordered fabrics and preferred orientations (see Fig. 7.9) (modified after Derbyshire, Gregory & Hails, 1979.)

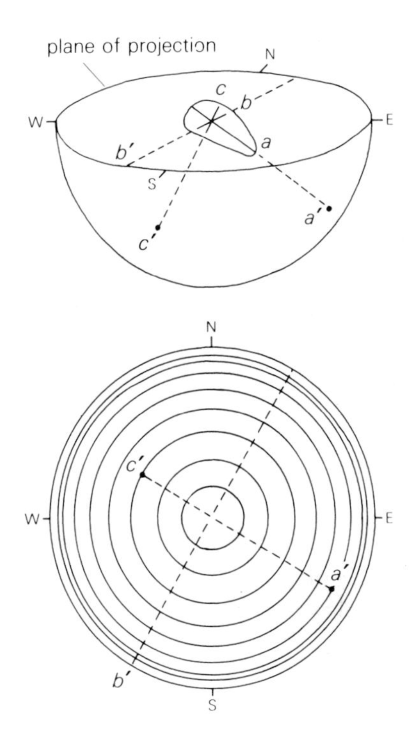

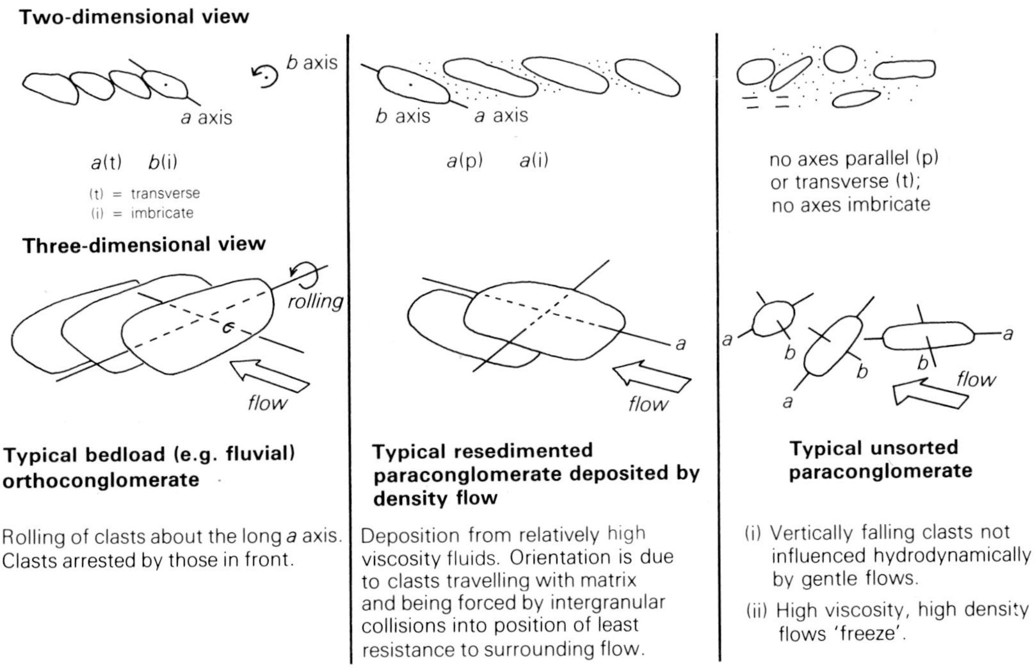

Two-dimensional view

b axis

a axis

a(t) *b*(i)

(t) = transverse
(i) = imbricate

b axis *a* axis

a(p) *a*(i)

no axes parallel (p) or transverse (t); no axes imbricate

Three-dimensional view

rolling

c

flow

a

flow

a

b

b

a

b

a

flow

Typical bedload (e.g. fluvial) orthoconglomerate

Rolling of clasts about the long *a* axis. Clasts arrested by those in front.

Typical resedimented paraconglomerate deposited by density flow

Deposition from relatively high viscosity fluids. Orientation is due to clasts travelling with matrix and being forced by intergranular collisions into position of least resistance to surrounding flow.

Typical unsorted paraconglomerate

(i) Vertically falling clasts not influenced hydrodynamically by gentle flows.

(ii) High viscosity, high density flows 'freeze'.

Figure 7.7 The nature and processes of origin of imbricated disc- and blade-like clasts.

Figure 7.8 Clast imbrication development in coarse alluvial gravels. Section parallel to flow and the clasts dipping to the left indicate flow to the right. Pleistocene, Spanish Pyrenees, Huesca Province.

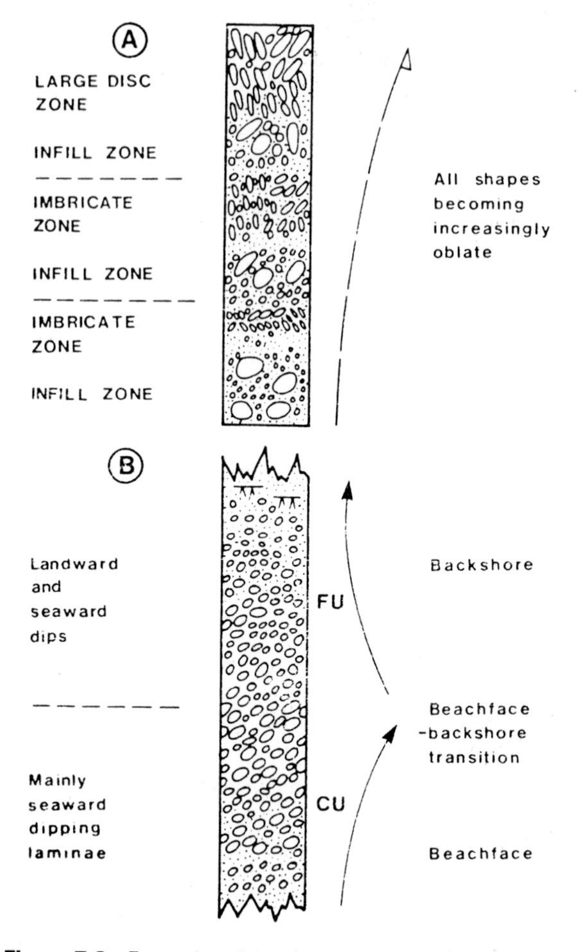

Figure 7.9 Examples of some systematic vertical changes in gravel fabric due to the progradation of pebbly beaches. Note that the patterns of change involve not only fabric but also grain size, sorting and grain shape. The changes of grain size are referred to as coarsening upwards (CU)and fining upwards (FU) where the change occurs over several beds due to gradually changing processes (after Nemec & Steel 1984, based on Bluck 1967 and Maejima 1982).

7.4.5 Stratification and cross stratification

Detection of originally horizontal surfaces is often difficult in thick 'beds' of rudite, but they can sometimes be picked out by slight changes of colour, grain size, sorting and fabric in strata that are otherwise massive or crudely bedded (Figs 2.3 & 7.2). In such situations it may be important to measure and report bed and set thickness, but this is often difficult, and a rather subjective estimate of thickness may be all that is possible. Bed contacts may be described as 'gradational', where one bed merges with the other beneath, whereas other contacts between fairly distinct beds may be described as 'amalgamated' (Fig. 7.12) where the beds have a similar matrix. Sharp bed contacts, often marked by changes in grain size, may be associated with an erosion surface (Fig. 7.13a); an extreme example being a basal conglomerate above an unconformity (Fig. 7.13b).

Multiple, distinct beds with sharp contacts are the products of a series of discrete depositional events. A basal conglomerate above an unconformity commonly marks the onset of deposition after a time interval dominated by tectonic uplift and erosion. Gravel above an erosion surface in a conformable sequence denotes a sharp change in energy or sub-environment; for example, when a river channel switches or migrates to an area of a river plain which had previously been abandoned or was inundated only at high flood. Gradational contacts suggest fluctuating and pulsating depositional processes. Amalgamated contacts between distinct beds may signify that two episodes of sedimentation were closely spaced in time, and that the energy available for the second episode partly reworked the underlying beds before they achieved significant coherence. Massive and crude bedding may involve rapidly fluctuating episodes of sedimentation in which the sediment concentration is high, 'freezing' of the load takes place, and individual depositional events are hard to distinguish.

In looking at rudites, pay particular attention to conglomerate beds that are only one clast thick. They may represent the cessation of rolling of gravel bedload. In

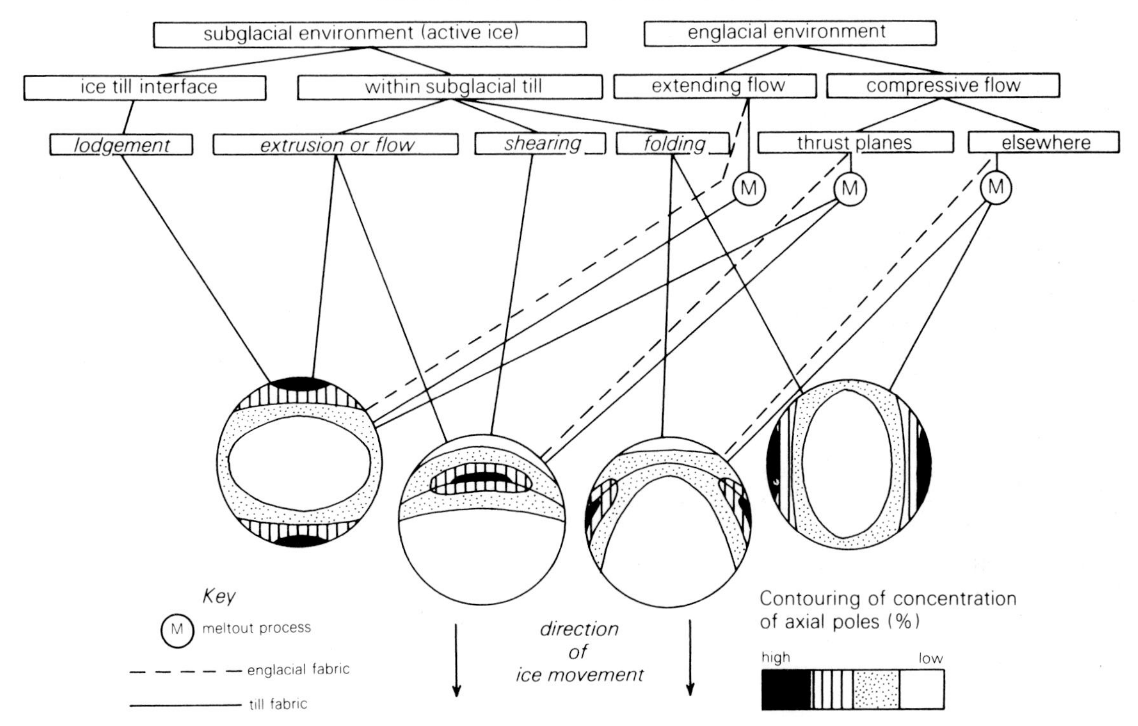

Figure 7.10 Possible relationships between fabric-forming processes and hypothetical fabric types in lodgement and melt-out tills as depicted on contoured plots on a stereonet (modified after Derbyshire, Gregory & Hails 1979).

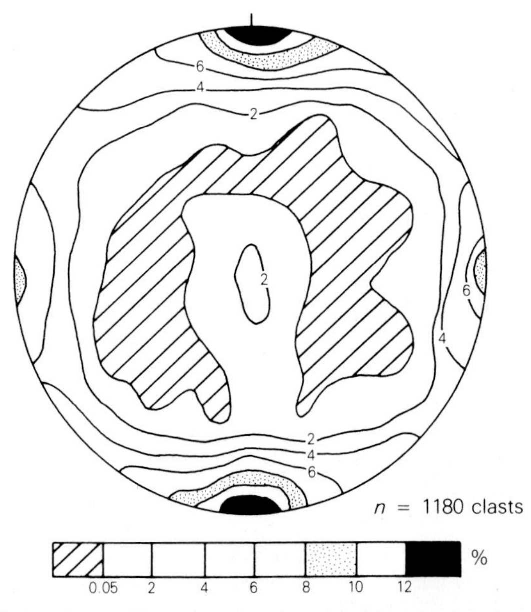

Figure 7.11 The plotting and contouring of *a*-axis fabric data (in the manner described in Fig. 7.6) for a large sample of clasts in a glacial till. Describe the pattern of organisation of the till fabric; suggest the direction of flow of the clasts; suggest the type of conglomerate present.

many cases they are accumulated 'lags' developed when a strong, erosive current winnows a gravelly sand or shell bed and takes the sand grains and hydrodynamically unstable pebble discs and shells into saltation or suspension. Pebbles and larger shells remain or 'lag' behind and are concentrated as a thin layer which may eventually armour the surface. A special case of lag conglomerate is that associated with channel migration (see Section 4.4.3, Fig. 4.20). In contrast, a layer composed of ventifacts, the majority of which are the right way up (i.e. faceted surfaces uppermost) and therefore probably *in situ*, records a setting where strong winds have winnowed away surface sand, and blasted and faceted the remaining pebbles, perhaps turning over a few in the process.

Once the likely attitude of the depositional horizontal has been established, various kinds of oblique or cross bedding may be identified. Alternatively, the early identification of cross bedding may help to establish the attitude of the original horizontal. Indistinct low-angle bedding (Fig. 7.12) has been noted in beach and fluvial gravels (Fig. 7.15) as well as in deep water conglomerates (see 7.5).

Cross stratification, with foresets at angles of 15–25° is more common in conglomerates than in breccias (Fig.

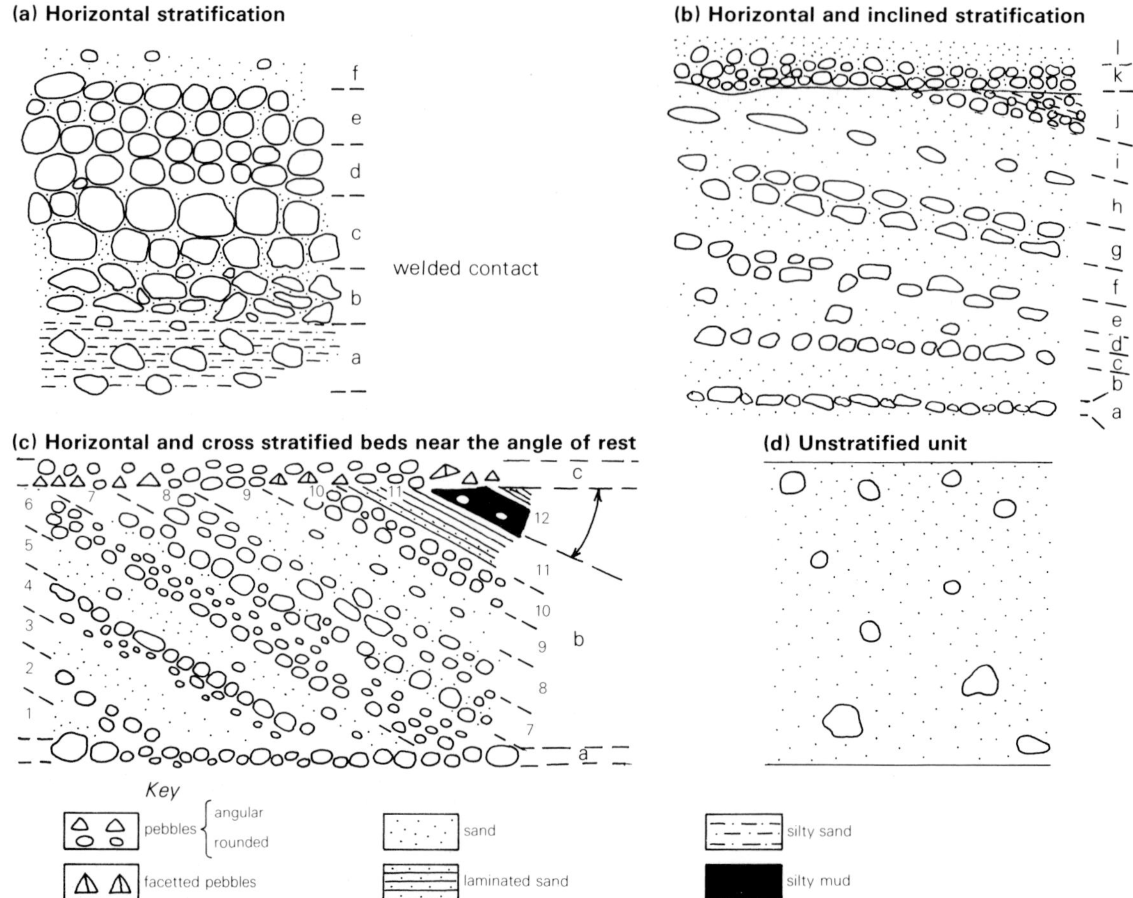

Figure 7.12 Stratification in rudites: (a) horizontal stratification with welded contacts; (b) horizontal stratification and inclined stratification; (c) horizontal and cross-stratified units near the angle of rest; (d) unstratified unit.

For each of the beds and units in each of the diagrams (a), (b), (c) and (d), describe the rock units using the terms cited in Figure 7.1 and the descriptions given in Sections 7.2.1, 7.4.1, 7.4.2, 7.4.3, 7.4.4, and 7.4.5.

Figure 7.13 (a) Channel cut and fill (Cannock Chase Pebble Beds, Triassic, Staffordshire); (b) rudites above a basal unconformity (Old Red Sandstone (Devonian) on Dalradian metamorphics, Argyll, Scotland). (After the Institute of Geological Sciences, with the permission of the Director.)

7.12). Normally, cross beds in gravel are 1–2 m thick, but isolated tabular and trough sets up to 10 m thick have been reported. Sets of tabular or trough type are difficult to identify in small exposures. Compared with cross beds in sands, the sets in gravel, particularly of tabular type, are more commonly single and isolated. Single trough sets are known to fill erosion hollows, but some troughs form cosets. Very large isolated tabular 'Gilbert'-type cross sets may be developed in gravel (see 6.2.9). Internally, foresets may be indistinct or spectacularly rhythmic and normally graded with coarse clasts at the base of each foreset layer (Fig. 7.14). Short axes of clasts tend to be oriented perpendicular to the foresets, and flattened surfaces therefore dip down current. Sandy foresets are also seen in dominantly gravelly sets, but mud-draped gravel foresets are rare. Reactivation surfaces (see 6.2.6) can be observed in larger sets.

The origin of cross beds in gravels is not always clear. By analogy with sand, the lee faces of bedforms are the most likely sites for foreset avalanching and grainfall. Echo-sounding in rivers in flood has revealed crescentic underwater dunes composed of fine-grained gravel, but their three-dimensional shape is poorly known. The suggestion that their migration gives rise to cosets of cross beds is conjectural. However, the origins of other types of set are known. Large single sets of cross beds have been observed in lake deltas and in ponds in abandoned river channels. Longitudinal and diagonal bars in braided rivers are composed mainly of flat, subhorizontal bedding, but the advance of slip faces at the downstream ends of bars may result in tabular sets of cross beds of limited extent (Fig. 7.15). Single small troughs, many metres deep and filled with crude cross strata and gravelly cosets, have been related to 'catastrophic' discharges, associated with the breaking of natural dams during glacial melts for example.

Flat subhorizontal bedding in gravel is known from many environments. For example, it forms at high discharges and relatively shallow current depths during the development of longitudinal bars in braided rivers (Fig.7.15).

DEPOSITS ASSOCIATED WITH FLAT AND CROSS BEDDED RUDITES AND THEIR SIGNIFICANCE

In successions where rudites are dominant, it is also important to record interbedded units of sandstone, siltstone or mudstone, since these reflect important changes in the sedimentation regime. Thin units of cross bedded or cross laminated fine-grained sandstones are not uncommon as interbeds, and are often the deposits of lower discharges, lower current velocities, and shallow water depths. The boundary shear stress needed to roll pebbles commonly coexists with levels of turbulence needed to maintain sand grains in suspension (Fig. 7.3). This relationship, which is not valid for high sediment

Figure 7.14 Cross-bedded fluvial rudites: (a) general view; (b) cyclical layering of foresets. Scales given by man 1.75 m and hammer 30 cm. Cannock Chase Pebble Beds, L. Triassic, Staffordshire.

concentration or high transport rates, predicts that interbedded sand and gravel may occur without radical fluctuations in flow strength. Interbedded siltstones and mudstones may, however, represent periods of protracted settling from suspension during quieter sedimentary regimes. The presence of ripple marks, mud cracks, body fossils and particularly trace fossils provides evidence of these periods, and gives additional evidence of the range of settings in which the energy fluctuations were taking place.

7.4.6 *Bed thickness and grain size relationships*

In thick, rudite-rich sequences it has become useful to measure bed thickness and maximum particle size within each bed (defined as the arithmetic mean of the ten

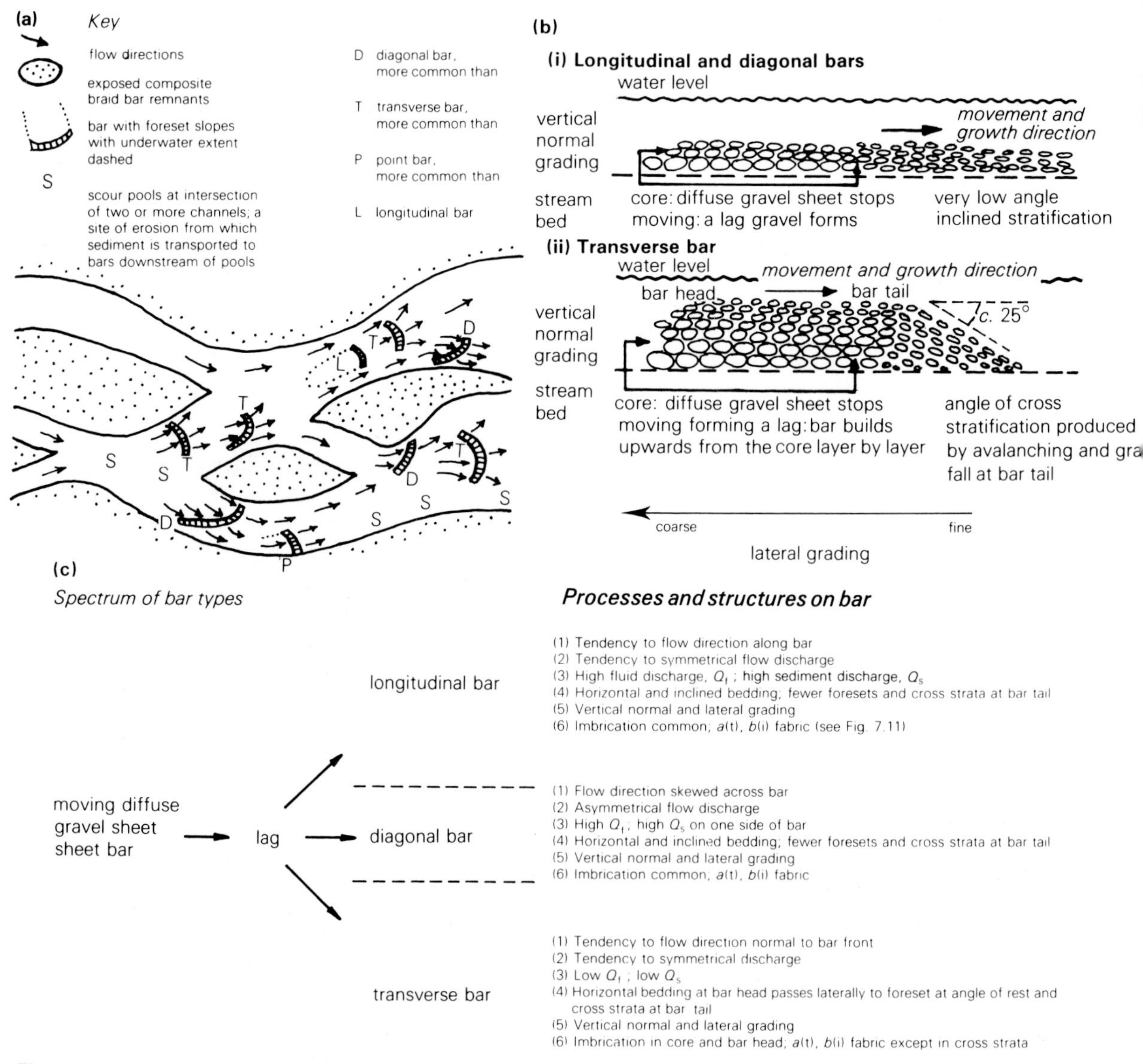

Figure 7.15 The distribution of flat-bedded inclined and cross-bedded strata in relation to processes creating fluvial bars in braided gravelly streams: (a) gravel bar types in the Kicking Horse River of British Columbia; (b) cross-sections showing the association of stratification types and processes of gravel accumulation in longitudinal, diagonal and transverse bars; (c) scheme of origin of sheet, longitudinal, diagonal and transverse bars. (Modified after Smith 1974, Hein and Walker, 1977.)

largest particles). Bed boundaries are not always easy to identify. In many such sequences boundaries may be erosive. This can sometimes lead to difficulties in estimating bed thickness on account of both amalgamation of beds (and thereby over-estimates) and removal of part of the bed (under-estimates). Particular care must therefore be taken to minimise this problem.

Plotting of these two variables will in some cases reveal a clear linear relationship; in other cases a scatter occurs with no apparent order. Cases of doubt must be resolved by the use of standard statistical tests. A correlation coefficient can be calculated. Where a linear relationship is apparent it is inferred that the beds were the products of discrete depositional events. The thickness of the bed (*BT*) is thought to reflect the thickness of the flow, while the maximum particle size (*MPS*) relates to the competence of the flow.

For carefully collected data, the value of the correlation coefficient for a group of beds can give an approximate measure of the consistency of the physical behaviour of depositional events. It also seems possible to distinguish deposition from predominantly cohesive (Bingham) from predominantly cohesionless (Newtonian) flows (Fig. 7.16, cf. Fig. 3.18), two types of debris flow which may occur in many environmental settings.

7.5 Processes of formation of mass properties and structures

We suggest that you attempt to map variations of distribution of the features described in Section 7.4 on a horizontal surface in suitable present-day environments, e.g. braided rivers at low water, beaches at low tide, debris flows and abandoned parts of alluvial fans. Try to sample from the proximal to the distal ends of extensive environments, whether they be linear or radial.

Cut banks of rivers and gravel pits often reveal short vertical sequences of pebbly strata. For work on these, it may be useful to take a series of overlapping photographs at fixed distances from a roughly planar outcrop. It is then possible to trace, from the photomosaic, gross change of distribution of the features over

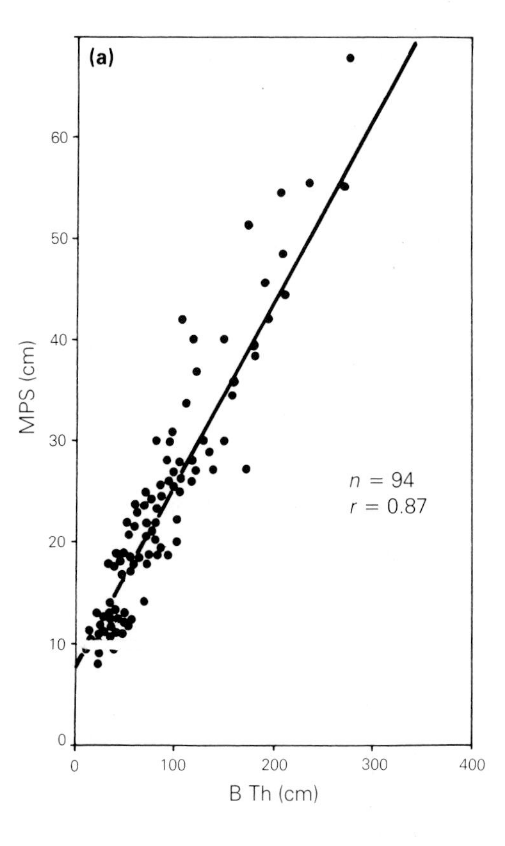

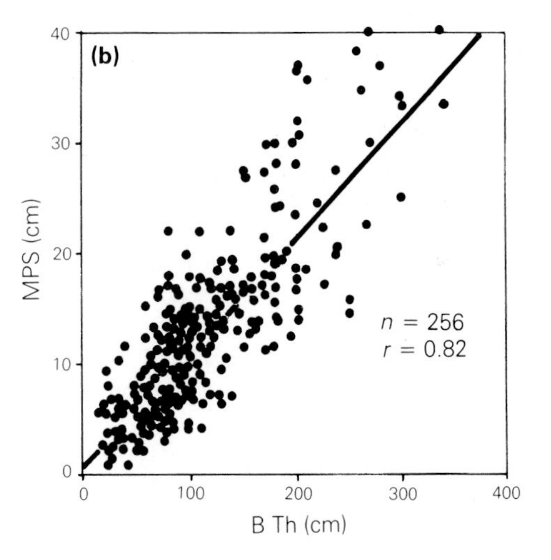

Figure 7.16 Examples of *MPS/BTh* plots from: (a) a sub-aerial fan deposit, Stornoway Group, Permo-Triassic, north-west Scotland; and (b) submarine fan delta, Ksiaz Formation, Devonian, south-west Poland. (a) shows a higher correlation coefficient, and the correlation line intersects the *MPS* axis, thus suggesting deposition by rather consistent, cohesive debris flows. (b) shows a lower correlation coefficient and the correlation line passing through the origin, suggesting slightly more variable, cohesionless behaviour (r = correlation coefficient). [After Nemec and Steel (eds) 1984 Figs 20 and 21.]

many tens of metres. This technique is equally appropriate in ancient rudites, where cementation allows the outcrop to stand proud, without degradation, for many years.

Experience suggests that it is useful to collect and plot certain data, the distributions of which can be of considerable value in interpretation. A simple example is the plotting of the mean size of the ten largest clasts against distance, either across a present-day environment or down the inferred or presumed palaeocurrent (e.g. on an alluvial fan or across a scree). Another example involves the measurement of bed thickness and the comparison of bed thickness with mean maximum clast size (see 7.4.6).

Without knowledge of the wider context of a particular succession interpretation is best limited initially to depositional process. The identification of environment may be speculated upon with success in some cases (see Ch. 10) but will generally only be suggested by a broader analysis of facies.

The descriptive features cited in Section 7.4 are observed to be combined in many different ways. Some common associations and trends are now recognised but there is a great need for extensive field observation and the collection of data from the deposits of both the present and the past.

Large-scale coarsening or fining upwards of the component units of ancient alluvial fan sequences can be interpreted in terms of energy variations related to tectonics, relief and climate. Increase in energy of depositional events, leading to progradation of fans and coarsening-upwards sequences can be caused by increasing tectonism, which enhances basin-margin relief and/or an increasingly wet climate. Decrease in energy, leading to recession of fans and fining-upwards sequences, is related to waning of tectonic activity after the initial pulses and/or a diminution of rainfall.

Limestone breccias occur quite commonly, in some cases associated with reefs. Early carbonate lithification means that any postdepositional disturbance is likely to lead to *in-situ* brecciation rather than slumping and folding (see 8.3.1 & 8.4).

Volcaniclastic sequences show varied combinations of structures and mass properties which are not always easy to interpret in view of their specialised processes of origin (see 7.3.9) and the fact that many materials are resedimented soon after initial deposition (Fig. 7.17a & b).

In the absence of other well established associations leading to useful models for present or ancient sediments, students may care to order their description, analysis and interpretation of rudites with the help of Figure 7.18 and to improve these crude relationships as experience in field, laboratory and library grows.

7.6 Uses of structures

The obvious major use of structures in rudites is in identifying processes and, eventually, environments of deposition. At the present this can only be done in an outline way despite the suggestions made in Figures 7.5, 8, 15 and 16.

An obvious target for economic geologists seeking oil, gas, water or metalliferous deposits is a thick sequence of orthoconglomerates in which porosity and permeability are high. Fluvial, shoreline and certain turbidite sediments will be more favourable in this regard than the deposits of environments where paraconglomerates are more common. Autoclastic and pyroclastic breccias are often the host rocks for primary sulphide mineralisation and zeolite formation.

Studies of palaeocurrents based on the measurement of imbrication and cross bedding may help greatly in predicting the directions in which a sequence should become thinner and finer and hence the direction in which porosity and permeability diminish.

It is easy to measure palaeocurrent directions at low water in gravelly streams, or on fans or beaches at low tide, and we recommend such exercises. Palaeocurrent measurements can be made on a variety of scales from imbrication, through cross bedding to large features such as channels. The orientation of the sides of large-scale, gravel-filled channels is likely to be parallel to the overall palaeoslope; that of large cross beds is likely to show a unidirectional trend but to vary rather more about a mean. In fluvial rocks the distribution of cross-stratal dip directions may be symmetrically bimodal about the downstream direction due to the growth of foresets at the ends of asymmetric diagonal bars (Fig. 7.15). Regional palaeocurrent patterns derived from imbrication in channels on submarine or alluvial fans are often radial. Those for beaches are mostly up or down slope, although a longshore orientation is noted in some instances.

Where pebble imbrication coexists with minor structures developed in sand on, for example, the exposed bed of a river, comparison of directions from sediments of different grain sizes can give valuable clues to changes in flow pattern as water stage falls. Sand, being more mobile, is more easily reworked by reduced flow, whereas structures of gravels tend to preserve directions of high-stage flows.

To the structural geologist, normal grading and cross stratification may help to determine 'way-up' in rocks that are steeply dipping. Furthermore, well rounded, mature conglomerates, with sphericity and roundness values that are consistent and predictable, may be

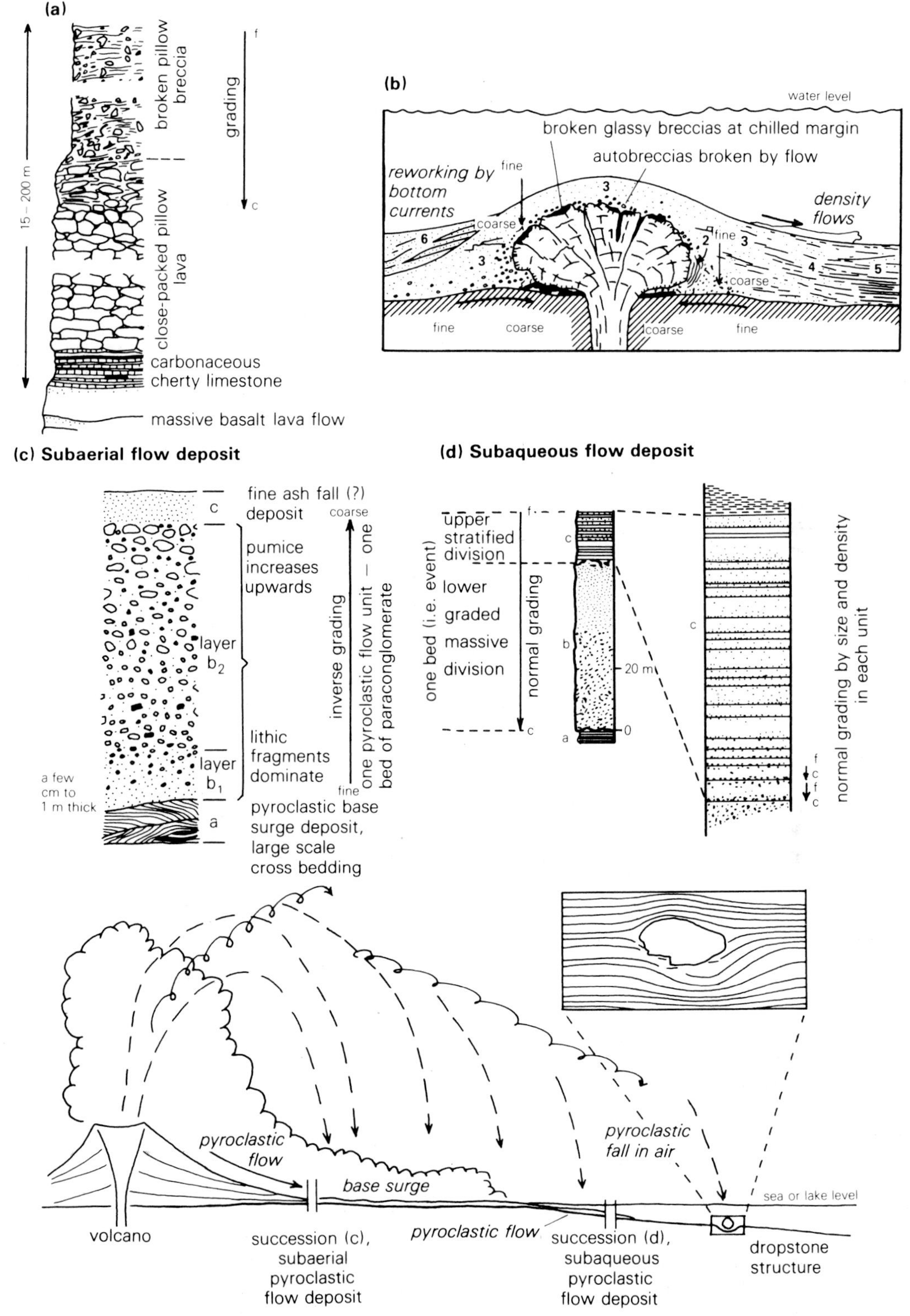

Figure 7.17 Volcaniclastic processes and rudite sequences: (a) sub-aqueous flow sequence in basic magma; typical pillow breccia – hyaloclastite sequence (modified after Carlisle 1963, Lajoie in Walker 1979); (b) sub-aqueous autoclastic and hyaloclastic flow sequences in acid rhyolitic magmas with resedimentation. A conjectural model for vertical and lateral associations of rock types and rhyolitic hyaloclastites: 1, submarine extruded dome covered by autobreccia and hyaloclastites; 2, blocky talus with remanants of ancient crust; 3, unstratified sand-size hyaloclastites *in situ*; 4, 'stratified' hyaloclastites *in situ*; 5 and 6, resedimented 'hyaloclastites' (after Lajoie in Walker 1979); (c) subaerial and subaqueous pyroclastic flow deposits. Vertical sequence of primary structures commonly present in subaerial and subaqeous pyroclastic flow deposits. (After Sparks *et al.* 1973, Fiske & Matsuda 1964 and Lajoie in Walker 1979.)

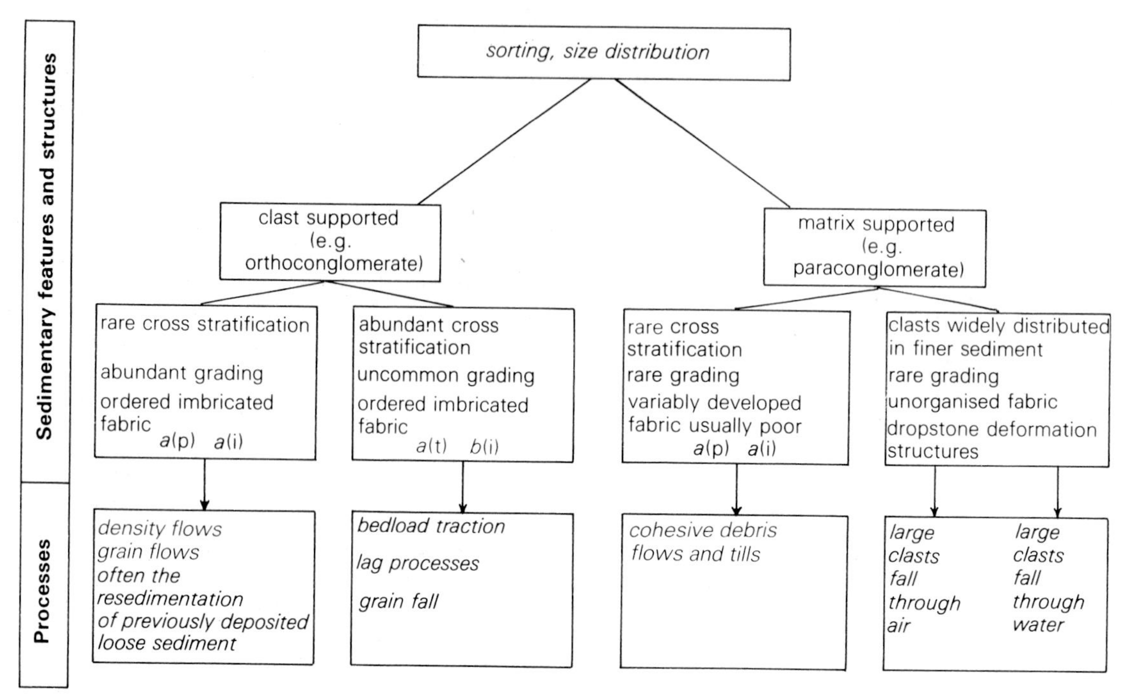

Figure 7.18 Common but not absolute associations of structures in rudites in relation to processes of origin (modified after Walker in Harms *et al.* 1975). Students are invited to consider how these structures and processes might arise in different environmental settings.

affected by tectonism. Pebbles may become stretched and flattened, and studies of the undeformed and deformed pebbles will enable a measure of strain to be estimated.

Reference and study list

References marked with an asterisk * are suitable for 16–19 year old advanced level students in schools and colleges in the UK as well as for undergraduates.

Field experience

First-hand experience: in this chapter we have suggested that your field programme should include observing and recording some of the following processes in their natural settings.

RIVER AND STREAM COURSES
Formation of frameworks, filling of frameworks, imbrication, development of lag deposits.

BEACHES
Distribution and orientation of different shaped clasts; the formation of frameworks and their fill.

DEBRIS FLOWS, SOLIFLUCTION (SOIL CREEP) AND DEBRIS AVALANCHES
Matrix-supported to clast-supported gravels; various grading types.

COASTAL OR DESERT SAND DUNES
Deflation, the formation of ventifacts and a pebble-armoured surface.

The ancient record provides well exposed sequences showing a range of features and structures that extend the experience derivable from the present record. Gravel pits in Quaternary deposits are frequently available and often repay investigation.

Laboratory experience

No films that deal exclusively with gravelly settings are known to the authors. Filming of volcanic activity is necessarily distant, but often there are after-the-event shots of the resultant products, e.g. dropstones in ashfall deposits or imbrication in ash-flow deposits. Angle of initial slip and angle of rest experiments on angular and rounded gravel are possible. The effects of introducing pebbles into otherwise sandy regimes are easily investigated in a flume. Description of texture and fabric in the main types of orthoconglomerate and paraconglomerate is easily made from hand specimens in the laboratory and is a quite basic skill.

*Anon. 1970. Film: *When the mountain moves*. 16 mm; colour, sound; 12 min. Christchurch: New Zealand Forest Service, Forest Research Institute (Protection Forestry Division PO Box 21–011, Christchurch, New Zealand). [Good coverage of mass-flow deposits.]

*British Broadcasting Corporation 1980. Film: *Anatomy of a volcano*. (The Mount St Helens eruption of 1980.) BBC Horizon Series. Colour, sound; 40 min. [Good shots of pyroclastic and autoclastic events.]

Derbyshire, E. 1981. Film: *Glaciers and sediments*. Colour, sound; 38 min. Keele: Department of Geography, University of Keele, Staffs. ST5 5BG, UK.

Dobkins, J. E. and R. L. Folk 1970. Shape development on Tahiti–Niu. *J. Sed. Petrol*. **40**, 1167–203. [This paper shows that measuring and estimating sphericity and roundness can be valuable in understanding the processes which gave rise to ancient gravel textures and fabrics.]

Johansson, C. E. 1965. Structural studies of sedimentary deposits. *Geol. Foren. Stockholm. Forhand*. **87**, 3–61. [The results of many experiments are cited.]

Jopling, A. V. 1965. Angle of repose box suitable for either class use or research. *J. Geol Education*, **13**, 143–4.

Open University 1980. Film: *Landslips*. Colour, sound. Milton Keynes: Open University Enterprises.

Light reading

*Francis, P. A. 1976. *Volcanoes*. Harmondsworth: Penguin.

*Thorarinsson, S. 1956. *The thousand years struggle against ice and fire*. Reykjavik: Boavigefa Menningarsdjods.

Essential reading

A selection from:

*Allen, J. R. L. 1970. *Physical processes of sedimentation*. London: Allen & Unwin. [Chs 4 (alluvial), 5 (shallow marine and intertidal) and 7 (glacial) give good introductions to gravel deposits.]

Allen, J. R. L. 1985. *Principles of physical sedimentology*. London: Allen & Unwin. [Ch. 9 on mass flow is useful.]

Bluck, B. J. 1967. Sedimentation of beach gravels: examples from South Wales. *J. Sed. Petrol*. **27**, 128–56.

Bluck, B. J. 1979. Structure of coarse grained braided stream alluvium. *Trans R. Soc. Edinburgh* **70**, 181–221.

Cas, R. A. F. and J. V. Wright 1987. *Volcanic successions, modern and ancient*. London: Allen & Unwin. [A good synthesis of the relationships between processes and products.]

Embleton, C. and J. Thornes (eds) 1979. *Process in geomorphology*. London: Edward Arnold.

Embry, A. F. and J. E. Klovan 1971. A late Devonian reef tract on northeastern Banks Islands, N.W.T. *Bull. Can. Petrol. Geol*. **19**, 730–81. [Terminology for carbonates with clasts > 2 mm and with greater than 10% matrix (matrix-supported floatstone) and < 10% matrix (clast-supported rudstone).]

Fisher, R. V. 1966. Rocks composed of volcanic fragments and their classification. *Earth Sci. Rev*. **1**, 287–9.

Harms, J. C., J. Southard, D. R. Spearing and R. G. Walker 1975. *Depositional environments as interpreted from primary sedimentary structures and stratification sequences*. Short course notes no. 2. Dallas: SEPM.

Johnson, A. M. 1970. *Physical processes in geology*. San Francisco: Freeman, Cooper. [Debris flow and matrix-supported conglomerate processes well described.]

Komar, P. D. 1976. *Beach processes and sedimentation*. Englewood Cliffs, NJ: Prentice-Hall. [Detailed study of the physical processes of beaches and associated sedimentary deposits.]

Koster, E. H. and R. J. Steel (eds) 1984. *Sedimentology of Gravels and Conglomerates*. Memoir no. 10. Calgary: Canadian Soc. Petrol. Geol. [Papers by Nemec & Steel and Hein are particularly valuable.]

*Lambert, M. B. 1978. *Volcanoes*. Englewood Cliffs, NJ: Prentice-Hall. [Chs 7 (products of explosion) and 8 (fragmental flows).]

Lowe, D. R. 1979. Sediment gravity flows: their classification and some problems of application to natural flows and their deposits. In *Geology of continental slopes*, L. A. Doyle and O. H. Pilkey Jr (eds), 75–82. Tulsa: SEPM Sp. Publ. 27.

Middleton, G. V. and M. A. Hampton 1976. Subaqueous sediment transport and deposition by sediment gravity flows. In *Marine sediment transport and environment management*, D. J. Stanley and D. J. P. Swift (eds), 197–218. New York: Wiley. [Description and discussion of processes of mass transport by dense media.]

Powers, M. C. 1953. A new roundness scale for sedimentary particles. *J. Sed. Petrol*. **23**, 117–19.

Reading, H. G (ed.) 1986. *Sedimentary environments and facies*, 2nd edn. Oxford: Blackwell Scientific. [Chs 3 (alluvial), 6 (deltaic), 7 (shorelines), 9 (shallow seas), 10 (shallow-marine carbonates), 12 (deep seas) and 13 (glacial) discuss the occurrence of rudites.]

Schmincke, H. E. 1974. Pyroclastic rocks. In *Sediments and sedimentary rocks* 1. H. Fuchtbauer (ed.), 160–89. Stuttgart: E. Schweizerbarte'sche.

Smith, N. D. 1974. Sedimentology and bar formation in the Upper Kicking Horse River, a braided outwash stream. *J. Geol*. **82**, 624–34.

Suthren, R. J. 1985. Facies analysis of volcaniclastic sediments: a review. In *Sedimentology: recent developments and applied aspects*, P. J. Brenchley and B. J. P. Williams (eds), 123–46. Oxford: Blackwell Scientific.

Walker, R. G. (ed.) 1984. *Facies models*, 2nd edn. Waterloo, Ont.: Geol. Assoc. Canada. [Chs 3 (glacial), 4 (volcaniclastic), 5 (coarse alluvial), 11 (turbidites), 15 (reefs) and 16 (carbonate slopes) are very relevant.]

Williams, H. and A. R. McBirney 1979. *Volcanology*. San Francisco: Freeman, Cooper. [Chs 6 (air fall and intrusive pyroclastic deposits) and 7 (pyroclastic flows and lahars) are very relevant.]

Wright, A. E. and F. Moseley (eds) 1975. *Ice ages, ancient and modern*. Geol. J. Sp. Issue 6. Liverpool: Seel House Press. [See especially papers by Boulton and Francis.]

Zingg, Th. 1935.Beitrage zur schotteranalyse. *Schweiz. Mineralog. Petrog. Mitt*. **15**, 39–140.

Further reading

Bluck, B. J. 1980. Structure, generation and preservation of upward fining braided stream cycles in the Old Red Sandstone of Scotland. *Trans R. Soc. Edinburgh* **71**, 29–46.

Boulton, G. S. 1972. The role of thermal regime in glacial sedimentation. *Sp. Publ. Inst. Br. Geogrs* **4**, 1–19.

Bull, W. B. 1977. The alluvial fan environment. *Prog. Phys. Geogr.* **1**, 222–70.

Carlisle, D. 1963. Pillow breccias and their aquagene tuffs, Quadra Island, British Columbia. *J. Geol.* **71**, 48–71. [A study of basaltic flow breccias.]

Collinson, J. D. and J. Lewin (eds) 1983. *Modern and ancient alluvial systems*. Sp. Publ. 6, Int. Assoc. Sediment. [Papers by Masari and Ramos & Sopena are good descriptions of pebbly sediments.]

Crowe, B. M. and R. V. Fisher 1973. Sedimentary structures in base-surge deposits with special references to cross-bedding. Ubehebe craters, Death Valley, California. *Bull. Geol Soc. Am.* **84**, 663–82.

Derbyshire, E., K. J. Gregory and J. R. Hails 1979. *Geomorphological processes*. London: Butterworth. [Coastal processes (pp. 107–61); cryonival and glacial processes (pp. 186–283).]

Dowdswell, J. A. and M. Sharp 1986. Characterisation of pebble fabrics in modern terrestrial glacigenic sediments. *Sedimentology* **33**, 699–710.

Eyles, N. and B. M. Clark 1985. Gravity-induced soft-sediment deformation in glaciomarine sequences in the Upper Proterozoic Port Askaig Formation, Scotland. *Sedimentology* **32**, 789–814.

Eynon, G. and R. G. Walker 1974. Facies relationships in Pleistocene outwash gravels, Southern Ontario: a model for bar growth in braided rivers. *Sedimentology* **21**, 43–70.

Hails, J. R. and A. P. Carr (eds) 1975. *Nearshore sediment dynamics and sedimentation*. New York: Wiley.

Hein, F. J. and R. G. Walker 1982. The Cambro-Ordovician Cap Enrage Formation, Quebec, Canada: conglomeratic deposits of a braided submarine channel with terraces. *Sedimentology* **29**, 309–29.

Jopling, A. V. and B. C. McDonald (eds) 1975. *Glaciofluvial and glaciolacustrine sedimentation* Tulsa: SEPM Sp. Publ. **23**. [Papers by Boothroyd & Ashley and Rust are recommended.]

Lipman, P. W. and D. R. Mullineaux (eds) 1982. *The 1980 eruption of Mount St Helens*, Washington. US Geol. Surv. Prof. Pap. 1250.

Miall, A. D. 1977. A review of the braided river depositional environment. *Earth Sci. Rev.* **13**, 1–62.

Miall, A. D. (ed.) 1978. *Fluvial sedimentology*. Can. Soc. Petrol. Geol. Mem. 5.

Postma, G. 1983. Water escape structures in the context of a depositional model of a mass flow dominated conglomeratic fan delta (Abrioja Formation, Pliocene; Almeria Basin, S. E. Spain). *Sedimentology* **30**, 91–103.

Rust, B. R. 1972. Structure and process in a braided river. *Sedimentology* **18**, 221–45.

Stanley, D. J. and G. Kelling (eds) 1978. *Sedimentation in submarine canyons, fans and trenches*. Stroudsburg, Pa: Dowden, Hutchinson & Ross.

Steel, R. J., S. Maehle, H. Nilsen, S. L. Røe and Å. Spinnangr 1977. Coarsening-upward cycles in the alluvium of Hornelen basin (Devonian), Norway. Sedimentary response to tectonic events. *Geol Soc. Am. Bull.* **88**, 1124–34. [Alluvial sequences related to synsedimentary tectonism at the basin margin.]

Steel, R. J. and D. B. Thompson 1983. Structures and textures in Triassic braided stream conglomerates ('Bunter' Pebble Beds) in the Sherwood Sandstone Group, North Staffordshire, England. *Sedimentology* **30**, 341–67.

Sugden, W. 1964. Origin of faceted pebbles in some recent desert sediments of southern Iraq. *Sedimentology* **3**, 65–74.

Walker, G. P. L. 1971. Grain size characteristics of pyroclastic deposits. *J. Geol.* **79**, 696–714. [The distinction between ashfall and ashflow deposits by grain-size characteristics.]

Washburn, A. L. 1973. *Periglacial processes and environments*. London: Edward Arnold.

Williams, P. F. and B. R. Rust 1969. The sedimentology of a braided river. *J. Sed. Petrol.* **39**, 649–79. [Description and analyses of facies in a modern gravelly river, the Donjeck in the Yukon, together with facies sequences and the generation of a general 'model'.]

Extra references relating to figure captions

Davies, I. C. and R. G. Walker 1974. Transport and deposition of resedimented conglomerates: the Cap Enrage Formation, Cambro-Ordovician, Gaspé, Quebec. *J. Sed. Petrol.* **44**, 1200–16.

Dunbar, C. O. and J. Rogers 1957. *Principles of stratigraphy*. New York: Wiley.

Fiske, R. S. and T. Matsuda 1964. Submarine equivalents of ash flows in the Tokiwa formation, Japan. *Am. J. Sci.* **262**, 76–106.

Hein, F. J. and R. G. Walker 1977. Bar evolution and development of stratification in the gravelly, braided Kicking Horse River, British Columbia. *Can. J. Earth Sci.* **14**, 562–70.

Maejima, W. 1982. Texture and stratification of gravelly beach sediments, Enju Beach, Kii Peninsula, Japan. *Osaka City Univ. J. Geosci.* **25**, 35–51.

Nemec, W. and R. J. Steel 1984. Alluvial and coastal conglomerates: their significant features and some comments on gravelly massflow deposits. In *Sedimentology of gravels and conglomerates*, E. H. Koster and R. J. Steel (eds), 1–31. Can. Soc. Petrol. Geol. Mem. 10.

Piper, D. P. and P. J. Rodgers 1980. *Procedure for the assessment of the conglomerate resources of the Sherwood Sandstone group*. Miner. Assess. Rep. Inst. Geol Sci. no. 56.

Sparks, R. S. J., S. Self and G. P. L. Walker 1973. Products of ignimbrite eruptions. *Geology*, 115–8.

Walker, R. G. 1977. Deposition of upper Mesozoic resedimented conglomerates and associated turbidites in south western Oregon. *Bull. Geol Soc. Am.* **88**, 273–85.

8 Depositional structures of chemical and biological origin

8.1 Introduction

A large part of the material weathered and eroded from land areas is transported to the seas as ions in solution. The composition of sea water has remained fairly constant throughout a large part of geological time and it follows that ions must have been taken out of solution by the precipitation of new minerals. This precipitation can be inorganic or it can be aided by or due entirely to organic agencies.

The most abundant minerals precipitated from sea water are aragonite and calcite and most of this precipitation appears to be organic. Inorganic precipitation of carbonates is possible but most inorganic precipitates are evaporite minerals, the most abundant of which are gypsum, anhydrite and halite. In non-marine settings such as saline lakes, the brine chemistry may be different from that of sea water, and different assemblages of evaporite minerals may form.

In this chapter we deal first with the structures produced by inorganic precipitation from bodies of saturated brine and then with structures due to organisms acting either to precipitate sediment or to bind existing particles.

8.2 Chemical precipitation

Inorganic precipitation of minerals from solution is largely confined to evaporite minerals, commonly gypsum or halite. For any mineral to be precipitated inorganically, an aqueous solution must be supersaturated with respect to that mineral. Whether the water body is connected to the sea or is enclosed as a lake, conditions of net evaporation must occur and this usually implies a hot, arid setting. When supersaturation is achieved, precipitation takes place so long as other ions in the solution do not interfere with crystal growth. Nucleation can occur spontaneously anywhere within the water column or on objects already on the floor of the basin. Crystals that nucleate at the water surface may float for a while, held by surface tension, and may exceptionally form rafts or crusts. Eventually they fall to the floor of the basin and the bulk of precipitation and crystal growth takes place there. For well formed crystals to develop, they need both free space and an interval of time.

Processes of nucleation and crystal growth can be modelled in the laboratory by allowing 1000 ml of a saturated solution of sodium chloride to evaporate gradually in a suitable tank. Look carefully with a hand lens at the crystals that form at the water surface and see how they develop as they fall to the bottom. Is it possible to distinguish crystals that nucleated at the water surface from those which nucleated on the floor of the tank? How do crystals continue to grow once they are on the floor? Try to monitor the temperature and rate of evaporation during the experiment. For a more elaborate experiment, try to do the same thing with about 4 l of sea water. With some chemical analysis it may be possible to study the order of crystallisation of different minerals and the changing chemistry of the remaining brine as evaporation proceeds.

8.2.1 Laminated evaporites

A common feature of many ancient evaporite-bearing sequences is a fine, millimetre-scale interlamination of different mineral phases or of an evaporite mineral and organic-rich material. Where two minerals are present, these are usually calcite ($CaCO_3$) and anhydrite ($CaSO_4$). Individual layers show great lateral continuity and may show grading (Fig. 8.1). Ungraded laminae probably record periods of settling of crystals precipitated at the water surface, possibly on a seasonal basis. In contrast, graded layers suggest reworking of previously precipitated crystals (see 6.7). This might imply episodic high energy events, such as storms, which stir up crystals and then allow them to settle out as energy wanes.

8.2.2 Fabrics due to vertical crystal growth

Growth of crystals on the basin floor commonly produces distinctive fabrics (Fig. 8.3a–c). Growth is most rapid parallel to certain crystallographic axes, commonly the *c*-axis. Crystals that precipitate within the water column fall to the bottom with different orientations.

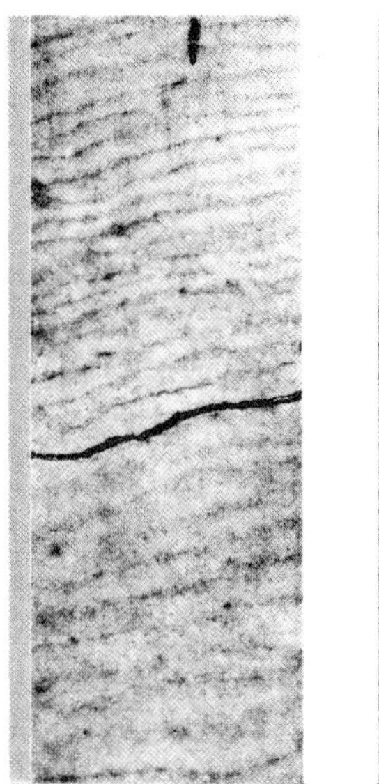

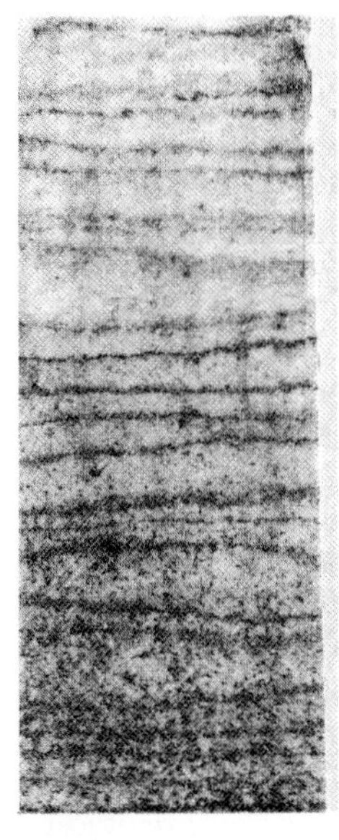

Figure 8.1 Two cores in thinly bedded anhydrite with bedding defined by slight grain-size differences and impurities. Cores are 9 cm wide. (Photograph courtesy of R. Arthurton.)

Figure 8.2 Columnar crystal growth in halite, interrupted by solution surfaces which record intervals of reduced salinity. Upper part of photograph shows randomly oriented halite cubes in fine sediment matrix. (Photograph courtesy of R. Arthurton.)

Those whose axis of preferred growth is oriented vertically will continue to grow most rapidly upwards, whereas those with the axis more inclined will eventually disappear upwards. The surviving, vertical crystals give rise to the tightly packed, columnar texture seen in many ancient deposits of halite and gypsum (Figs. 8.2, 8.3c).

During growth, the sediment surface is made up of crystal faces upon which precipitation is taking place. Records of the instantaneous position of this surface are sometimes picked out in ancient evaporites by thin layers or drapes of argillaceous sediment deposited during events such as river floods into saline lakes or storms in lagoons. Particularly severe events may lead to partial dissolution of the mineral and the truncation of the vertical crystal fabric. Above such surfaces, the pattern of vertical growth is re-established (Fig. 8.3d).

8.2.3 Pseudomorphs

Even if the solution of evaporite minerals is total, all evidence of the minerals may not be lost. Some sequences, usually of interbedded sandstone and mudstone or, less commonly, limestone and mudstone, show evidence of former evaporites, usually halite or gypsum, in the form of pseudomorphs. The original evaporite crystals have been replaced by sandstone or

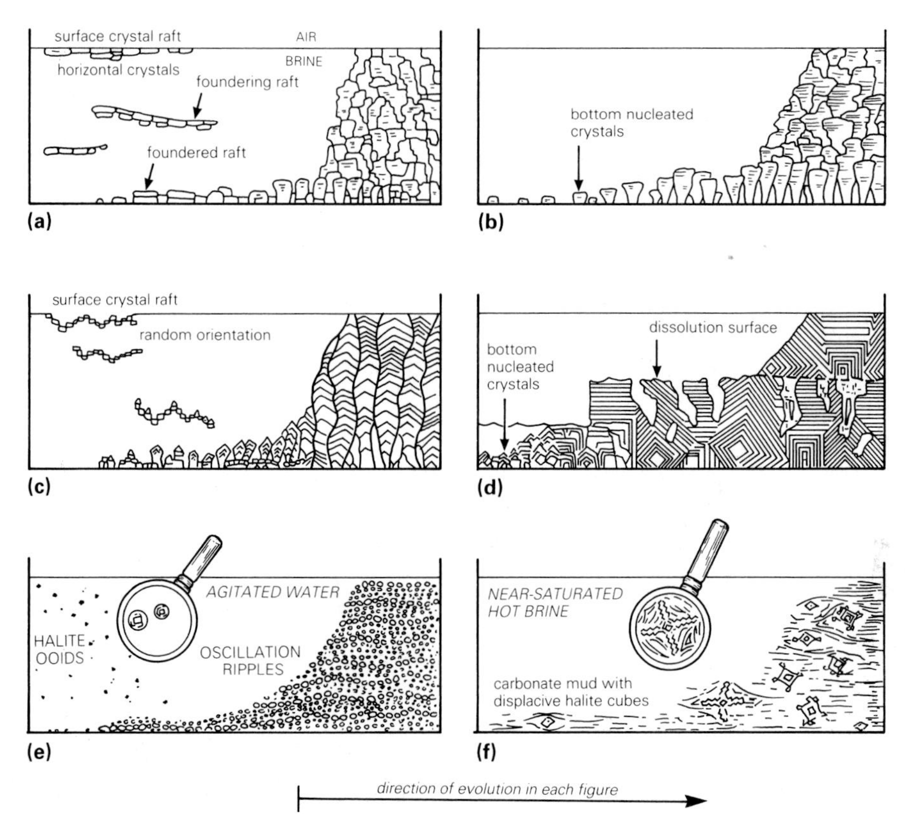

Figure 8.3 Idealised textures associated with the growth and emplacement of halite under differing conditions (after Schreiber in Reading 1986, based on Arthurton 1973, Shearman 1971, 1978, Weiler, Sass & Zak 1974).

limestone but the mineralogy of the evaporite can still be deduced from the shape of the pseudomorph. Halite pseudomorphs are cubic protrusions on the lower surfaces of sandstone or limestone beds. They occur in a variety of sizes, commonly up to about 1 cm across, and they may be isolated or may occur in lines or clusters (Fig. 8.4). Some have one face of the cube parallel to bedding and have a square appearance but, more commonly, one corner of the cube protrudes so that triangular shapes are apparent. Larger pseudomorphs commonly show a pattern of steps on their faces, giving them an indented 'hopper' form. The best shaped pseudomorphs are commonly those isolated from their neighbours.

Pseudomorphs record the former presence of evaporite crystals growing at or just below the muddy sediment surface from an overlying supersaturated brine (Fig. 8.3f). The occurrence of many, small pseudomorphs suggests rapid nucleation and possibly rapid evaporation, whereas a few large crystals may have resulted from slower, more sustained evaporation. Although pseudomorphs commonly occur in association with desiccation mudcracks, they do not themselves indicate emergence, but rather they record the existence of a shallow body of hypersaline brine which may or may not have dried out completely. A shallow water body is implied by the relatively small volumes of evaporite minerals involved.

Preservation of pseudomorphs takes place as the result of a rapid influx of sediment-laden water, probably by a flood. This dissolves the crystals on the basin floor and fills the resulting spaces with coarser sediment. This mechanism is sometimes borne out by the occurrence of small erosional tool marks with the pseudomorphs.

By occurring on the soles of sandstone beds, pseudomorphs constitute a reliable indicator of 'way-up'.

8.2.4 Diagenetic and reworked evaporites

Not all evaporite minerals occur as primary, basin-floor precipitates. Many occur as diagenetic concretions or nodules formed *within* a host sediment. The

Figure 8.4 Halite pseudomorphs on the bases of sandstone beds from a thinly interbedded sandstone/siltstone sequence: larger pseudomorphs with stepped 'hopper' faces. Proterozoic, Independence Fjord Group, N. Greenland.

textures, fabrics and structures of such evaporite deposits are dealt with in Section 9.3.1.

Some evaporites show structures due to physical processes of erosion and deposition. Small-scale scours, ripples, cross lamination and cross bedding are all quite common (Fig. 8.3e) and they record the reworking of primarily precipitated evaporites by currents, waves or even by the wind, as shown by the large aeolian dunes of gypsum found in New Mexico (see 6.3.2).

8.2.5 Spring deposits: tufa, travertine and sinter

Around many present-day springs and caves and in some deposits of Holocene and Pleistocene age, there are chemical precipitates in whose deposition evaporation played only a minor role. Two principal groups of deposit occur: calcium carbonate and silica.

Deposits of calcium carbonate are precipitated from both hot and cold springs and the precipitation may be due to cooling, evaporation, loss of dissolved carbon dioxide, or to chemical reaction, all of which may be aided by algae. One form of calcium carbonate is **tufa** which commonly occurs as a coating on plants and plant debris. Its texture is usually highly porous and spongy, and plant impressions may often be found within it. A second, more laminated and compact form of calcium carbonate is **travertine**, which occurs commonly in caves and also as the surface deposits of both hot and cold springs. Precipitates of calcite in caves result from calcium carbonate, dissolved by percolating groundwater being reprecipitated on emergence, probably as a result of loss of CO_2. Elongate vertical columns (stalactites and stalagmites) develop on the roof and floor of the cave, and vertical surfaces are coated with dripstone layers. The spectacular terraces of hot springs in Yellowstone Park, USA, or in New Zealand, are of travertine. Sections through these deposits show a fine lamination, some of which is columnar, and this could be confused with algally produced stromatolitic lamination (see 8.3.2). The lamination can be seen to be made up of layers of fibrous calcite crystals whose fibres are normal to the layering.

Deposits of silica are confined to hot springs and geysers and are known as **sinter** or **geyserite**. These occur as encrustations around geysers and springs and they develop a wide variety of surface morphologies. Continued deposition leads to a variety of types of lamination, many of which compare quite closely with those of algal stromatolites (see 8.3.2 and Walter in *Further reading*).

8.3 Precipitation and binding of sediment by organisms

Organisms are active in both the precipitation of mineral matter and the binding of sedimentary particles. Here we deal, in general terms, with organically produced structures under two main headings – reefs and bioherms, due mainly to precipitation by various organisms, and stromatolites, due to the binding of sedimentary particles by algae.

8.3.1 Reefs and bioherms

Many animals and plants living in the sea and in freshwater settings produce aragonite and calcite as skeletal or other strengthening structures (e.g. corals, echinoids and calcareous algae). After the death of the organisms, this material constitutes the main component of many carbonate sediments. Some skeletons remain more or less intact to become sedimentary particles in their own right, and others disintegrate or are broken up and abraded. The fine needles of aragonite, which make up much of the lime mud of present-day carbonate environments and which gave rise to many micritic limestones, were precipitated by calcareous algae.

Consideration of organic grains in detail is the province of the palaeontologist and the petrographer, and in this chapter we concentrate only on the larger features of limestone and dolomite sequences that were produced by colonies of organisms. Such features had topographic expression on the contemporaneous sea floor and have been termed **build-ups, reefs, mounds** or **bioherms.** They vary greatly in size, in morphology and in their organic make-up, which has changed throughout geological time as different groups of organisms became important. At the present day, organic build-ups of carbonate on the sea floor occur in a wide range of water depths, though the most prolific and spectacular examples are in fairly shallow water of tropical and subtropical coasts. For a discussion of the changing nature of reefs throughout geological time, see the books by Wilson and Toomey (see *Further reading*).

At the present day, reefs occur as barriers, running for long distances parallel to a shoreline and separating the open ocean from a more protected lagoonal area. Others are fringing reefs which encircle islands and may have a lagoon or reef flat between the reef edge and the land. Atolls are reefs that encircle a lagoon which lacks a central landmass at the present day. These larger forms commonly have horizontal dimensions measured in kilometres and they may separate areas of deep and shallow water on opposite sides. At a smaller scale, small organic build-ups or **patch reefs** occur within lagoons and in other shallow marine settings. A comparable variety occurs in the rock record where the recognition of the largest forms can require extensive geological mapping or exceptionally large exposures. Smaller reef forms can commonly be recognised in quarries and cliffs. Throughout our discussion, we use the terms 'reef' and 'bioherm' interchangeably with no implications of water depth or other conditions of deposition.

Smaller bioherms and their associated sediments are commonly divided into **reef core, reef flank** and **inter-reef** components (Fig. 8.5a). Larger reefs, which are more likely to have acted as barriers between areas of contrasting water depth, are often more appropriately divided into **fore-reef, reef** and **back-reef** components (Fig. 8.5b).

The reef core is usually a tightly bound mass of limestone, generally without any clear bedding. It may include a high proportion of framework-building skeletons or it may be difficult to see any framework organisms when the reef is due mainly to the binding effect of the organisms. The framework organisms themselves are commonly encrusted with other organisms such as algae and bryozoa, and their surfaces bored by animals, recording early stability and lithification. Early formed cavities may be partially or completely filled with sediment, often in multiple generations, while others can have fills of fibrous or blocky calcite cement. Where a cavity is partially filled with sediment and the remaining overlying space by sparry calcite, the interface approximates to the depositional horizontal. This can be used as a way-up indicator and for the measurement of dip in unbedded carbonates. These *geopetal infills* (see Fig. 2.1) can also occur in the body chambers of fossils (Fig. 8.6). *Stromatactis* is a particular type of cavity infill development in lime muds, commonly in reef settings. It is characterised by a flat floor, partially draped by sediment and an irregular roof beneath which sparry calcite

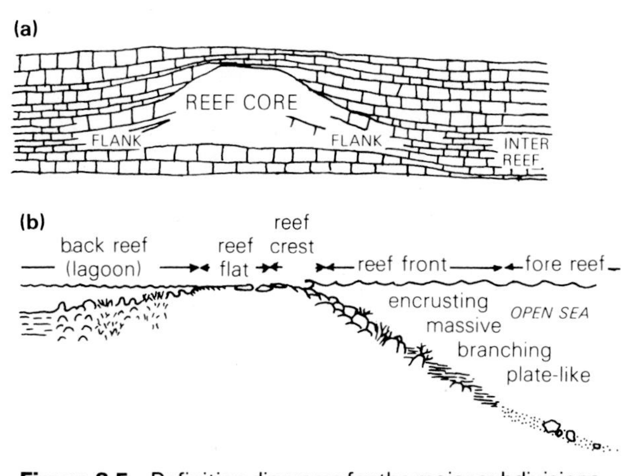

Figure 8.5 Definition diagrams for the major subdivisions of reefs: (a) isolated reef mound; (b) barrier reef at the boundary between deeper and shallower water showing main growth forms of reef-building organisms. (After James 1979.)

Figure 8.6 Brachiopod shells, with their body chambers partially infilled with sediment and partly infilled with sparry calcite. Sediment occupies the lower part of the cavity and comprises a geopetal infill. Bedding is vertical and the beds young to the left. Lower Carboniferous, Dyfed, Wales.

Figure 8.7 Stromatactis, an enigmatic structure in fine-grained limestone. Sparry calcite fills a flat-bottomed cavity which must have been kept open close to the sediment surface during early stages of diagenesis. Pentamerus Bjerge Formation, L. Silurian, Washington Land, Greenland.

infills the remaining space (Fig. 8.7). Its origin seems to relate to the compactional dewatering of the muds whereby water is trapped beneath slightly consolidated and organically bound layers. Not all the fossils of the reef core help to form a framework or to bind the sediment. Some are ordinary, detached forms which simply lived within the general reef environment. Within the reef core, a vertical change in the organic content may record the progressive development of the reef. Try always to identify which organisms reflect stages of pioneer growth, colonisation, diversification, domination, death and degradation. The flank or fore-reef deposits show clearer bedding, commonly with quite high depositional dips away from the reef core. The back-reef or inter-reef sediments show more defined, horizontal bedding. Differences in the sediments of the various zones are also reflected in the associated organisms although the fore-reef or flank deposits may include both sediment and organisms which initially formed or lived on the reef or in the back reef.

A full description of an ancient reef or bioherm should attempt to answer the following groups of questions:

Can the attitude of the original depositional horizontal be established? (See Sect. 2.1.6.)

How big was the reef? What topographic relief did the reef create when it was actively growing? Because sediment accumulates around a reef during its growth, assessment of contemporaneous relief in ancient examples is often difficult. With larger barrier or fringing reefs, there may have been significant differences in water depth on opposite sides. It may sometimes be possible to trace marker horizons from the inter-reef sediments over the reef core and these can give an indication of the topography. Bathymetry may also be suggested by larger-scale patterns of sediment distribution and by the present-day sub-aerial topography which is commonly an exhumed submarine topography (Fig. 8.8a).

What shape did the reef have in plan view? Present-day sub-aerial topography is particularly useful in some large ancient reefs where the exhumed morphology may be visible in three dimensions. However, some inferences about reef shape can be made from observations at a more local scale. Clear differences in lithology between sediments on one side of a reef and the other suggest that the reef acted as a barrier to water and sediment circulation, and the terms fore reef and back reef may then be appropriate. Small reefs or bioherms, recognisable in single outcrops are usually flanked and surrounded by similar sediment on all sides (Fig. 8.8b).

Figure 8.8 (a) Exhumed reef topography due to the removal of later softer sediments which blanketed the reef pinnacles. Jurassic, Morocco (photograph courtesy of H. S. Torrens). (b) Small patch reef overlain by thinly bedded limestones which thin as they drape the reef. Middle Ordovician, Mjøsa, Norway. (Photograph courtesy of T. L. Harland).

What organisms were involved in reef growth? Is it possible to identify a main frame-building organism? To what extent is the reef a result of frame building and to what extent is it due to sediment binding? Can any lateral or vertical variations be seen in the distribution of the types and abundances of different organisms? These questions demand palaeontological expertise and are best approached after study of a book such as that by Wilson (see *Further reading*).

To what extent did early lithification occur in the reef? This is often best judged from observation of fore-reef and reef-flank deposits. Note whether these beds are made up of bioclastic debris or contain larger blocks of reef material. Bioclastic debris often occurs in steeply dipping beds, inclined away from the reef core and commonly showing normal or inverse grading. Large blocks, sometimes of huge dimensions, occur on some steep palaeoslopes associated with dipping fore-reef beds and are interpreted as submarine scree slopes made up of blocks that fell from the reef front. Geopetal infills within large blocks may help to establish the degree of tilting of the blocks or their way-up if the infills formed in the reef prior to redeposition (see also 2.1.6).

8.3.2 Stromatolites and oncolites: structures due to algal binding

Stromatolites and oncolites are structures that show fine lamination caused by the trapping and binding of material by algae.

Stromatolites. Our knowledge of stromatolites derives largely from rocks of Precambrian age though they are also quite common throughout the Phanerozoic. Stromatolitic lamination is commonly found in mud-size carbonate sediment although coarser-grained carbonate and detrital material can also be involved. Lamination is characteristically thin, usually 1 mm or less, and has a rather delicate appearance. In some cases the lamination is irregular with small cavities filled by sparry calcite between the layers. This **birdseye structure** is the result of shrinkage of the algal layer and also the generation of gas from rotting algae. The forms which the lamination takes are extremely varied and several taxonomic schemes have been proposed for them.

For our purposes, it is sufficient to recognise some of the large- and medium-scale features of stromatolites that contribute to a good field description of them. These are illustrated schematically in Figure 8.9. At the largest scale ('mode of occurrence') the broad shape of the stromatolitic unit is described. Such units can be of any thickness from a few centimetres to several metres, and stromatolitic biostromes may extend horizontally for many kilometres.

Within bioherms or biostromes, the stromatolitic lamination may be organised into either columnar or non-columnar forms, and these in turn show a great variety of shape and scale (Fig. 8.9). The types illustrated in Figure 8.9 do not depict all possible variations and a good, scaled field drawing or photograph will usually be more valuable than words. It is important to develop a feel for the three-dimensional form of stromatolites; in addition to their vertical section (Fig. 8.10), describe their appearance in horizontal section or on bedding surfaces wherever possible (Fig. 8.11). With columnar forms, try to establish their cross-sectional shape and size. Columns tend to be circular or elliptical in plan view, the latter type sometimes having a preferred direction to their long axes. Record the direction of any such orientation as it may give a guide to current or wave directions. Bedding surfaces may show a three-dimensional relief (Fig. 8.12) and this allows visualisation of the morphology of the sediment surface during deposition.

Within a biostrome or bioherm there may be both lateral and vertical variation and a full description should include not only the size and shape of the stromatolitic unit, but also the types, scale, orientation and distribution of different types of lamination.

A note of caution is necessary here in that stromatolitic lamination is rather readily confused with some non-biological lamination, such as that in travertine, tufa and the silica deposits which form around geysers and hot springs (see 8.2.6).

Although much less widespread and varied than they appear to have been in the geological past, present-day algal mats show a range of morphological forms from flat sheets through various crinkled and pustular types to well developed columns. Present-day types take up zonal patterns of distribution which relate to the physical and chemical conditions in which they develop; for example, columnar structures with elongate cross sections have preferred elongation normal to the shore.

Stromatolitic lamination results from the trapping and binding of sediment by the mucilaginous filaments of algae which form mats growing on the sediment surface. Sediment settles from suspension onto the mats and is generally not precipitated by the algae themselves. The lamination is produced by variation in sediment supply giving more and less organic-rich layers. The algal mat re-establishes itself after an episode of rapid deposition by growing through the sediment layer, thus binding the sediments.

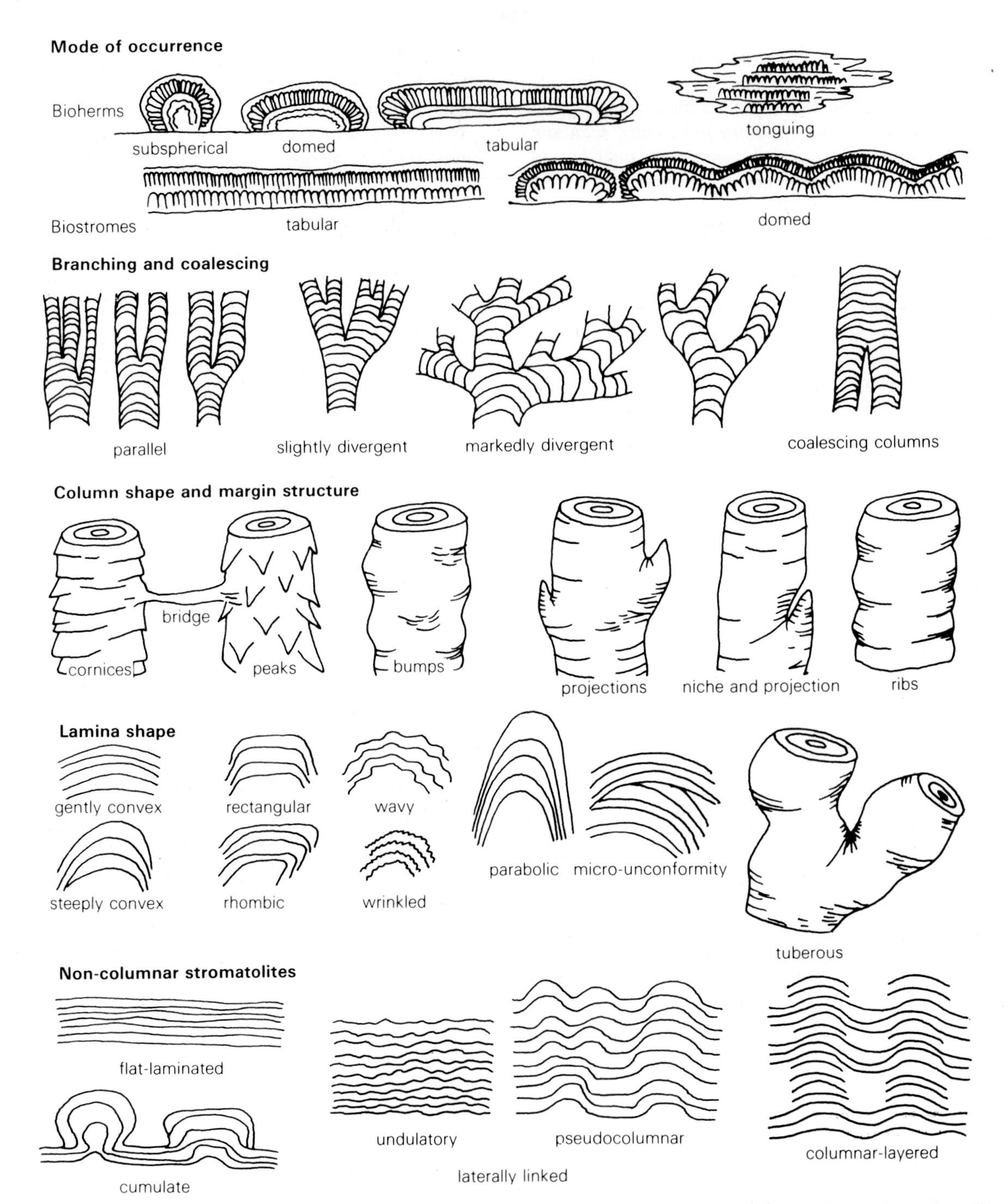

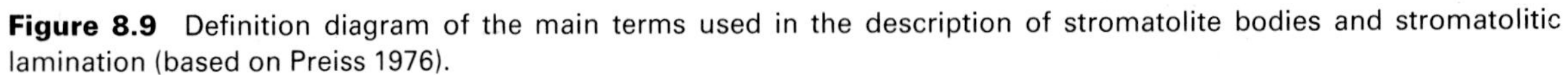

Figure 8.9 Definition diagram of the main terms used in the description of stromatolite bodies and stromatolitic lamination (based on Preiss 1976).

Figure 8.10 Examples of stromatolitic lamination seen in vertical section. (a) Morœnesø Formation, Proterozoic, N.E. Greenland; (b) Kastrup Elv Formation, L. Cambrian, Washington Land, Greenland.

Figure 8.11 Horizontal sections through stromatolite columns. Fyn Sø Formation, Upper Proterozoic, N.E. Greenland.

Figure 8.12 Stromatolite domes with their three-dimensional relief preserved on the upper bedding surface. Morœnesø Formation, Proterozoic, N.E. Greenland.

Most present-day stromatolites are associated with elevated salinities in inter-tidal and supratidal settings, although subtidal examples are also known. The more widespread occurrence of stromatolites in rocks of Precambrian age probably results from the absence of animals that grazed upon the algae. The elevated salinities associated with many present-day examples create conditions hostile to predators and allow the mats to flourish. The commonly stated view that stromatolites are indicators of intertidal conditions is misleading and, for Precambrian examples particularly, there seem to be no environmental requirements other than the availability of water and sunlight.

Attempts to use stromatolites as a basis for the biostratigraphic zonation of otherwise unfossiliferous Proterozoic sediments have only met with limited success and acceptance.

Oncolites. Oncolites are spherical or less well rounded structures, commonly up to 5 mm in diameter but sometimes bigger, often with a rather flattened shape.

Figure 8.13 Cross sections through oncolites showing internal lamination. Morœnesø Formation, Proterozoic, N.E. Greenland.

Internally they have a roughly concentric pattern of fine lamination similar to that present in stromatolites (Fig. 8.13). Careful examination of the lamination in cross section may sometimes show discontinuities.

These structures occur in both ancient limestones and in present-day lagoons and lakes. They result from the binding of sediment by algae onto isolated nuclei. Their growth, with its discontinuous pattern of lamination, suggests relatively calm conditions with occasional higher-energy episodes that turned the oncolites over and allowed growth to proceed on the opposite side. Oncolites may be confused with diagenetically formed pisoliths which occur in some carbonate-rich soil profiles.

8.4 Early cementation

In addition to the binding activity of organisms, carbonate sediment on the sea floor may be subjected to early cementation through the precipitation of aragonite or Mg-calcite from sea water. Early cements are not particularly common in seafloor sediments but are favoured by slow depositional rates, a rather stable sea floor and relatively high levels of wave and current activity. Carbonate cementation also occurs on some beaches giving **beach rock**, a phenomenon also common in beaches of detrital sand. Carbonate sediments which become sub-aerially exposed through uplift or a fall of sea level are also subjected to cementation through solution and re-precipitation due to the passage of fresh water, when the cementing mineral is always calcite. Sustained sub-aerial exposure can lead to more extensive dissolution and to the development of soil profiles.

8.4.1 *Seafloor lithification*

Cementation close to the sea bed may be patchy or lead to the development of continuous layers. Early cemented patches take the form of **concretionary nodules** around which differential compaction can take place, especially if the host sediment is fine-grained. This is recognised in ancient sediments through tracing laminae from the surrounding compacted sediment into and across the nodule. When seen in the rock record, such nodules commonly follow particular bedding horizons.

Laterally continuous cemented layers commonly form a few centimetres below the sea floor and tend to have rather flat top surfaces and rather irregular bottoms. Continued precipitation may lead to buckling and breakage of the layers, development of anticlines and the thrusting of layers one over the other to give **tepee structures** (Fig. 8.14). In plan view the upwards buckled ridges have polygonal forms. These structures occur on the floors of present-day marine lagoons as well as in the rock record.

When the cemented layers, which probably developed a few centimetres below the sea bed, are swept clean of loose sediment, the lithified surface may become encrusted and bored by marine organisms and be mineralised by phosphatic and manganese-rich minerals. These bored and encrusted surfaces are called **hardgrounds** and their occurrence in the rock record provides evidence for intervals of non-deposition. The recognition of borings is dealt with in Chapter 9 (see Fig. 9.36).

Early cementation of carbonate sediments on submarine slopes may lead to the spectacular development of breccias through the secondary movement of the lithified or partially lithified material as mass gravity flows. Breccias made up of rather tabular clasts are especially characteristic (Fig. 8.15).

More general comments on preferential cementation and the development of nodules and concretions are given in Section 9.3.

8.4.2 Sub-aerial exposure

Sub-aerial exposure of carbonate sediments leads to their rapid lithification in most circumstances. Percolating rainwater dissolves carbonate and reprecipitates it as calcite cement. Sustained exposure leads to further modification which may take one of two forms.

Sustained dissolution associated with abundant fresh water will lead to the development of **karstic features**. Such features are not just the product of recently exposed carbonate sediments but also develop very extensively in older limestones which have been subjected to abundant rainfall. Karstic features developed at the landform surface give very characteristic morphologies and landscapes (Fig. 8.16). They include patterns of sharp ridges and deep clefts due to the solutional widening of joints and steep-sided funnels and pipes. Within limestone, solution produces cavities which may grow to the size of major caves. Precipitation of carbonate as various types of travertine deposit (see 8.2.5) fills or partly fills such voids. In some instances the cavities may collapse, leading to the addition of **collapse breccias** to the fill. Such units tend to be irregular in shape and bear no relationship to bedding. The breccia clasts are all of clearly local derivation, irregular and angular in shape and with a very mixed range of sizes. Spaces between them may be

Figure 8.14 Tepee structure developed in limestone. Scale given by portion of hammer head at bottom centre. Triassic, Glamorgan, Wales, (photograph courtesy of M. E. Tucker).

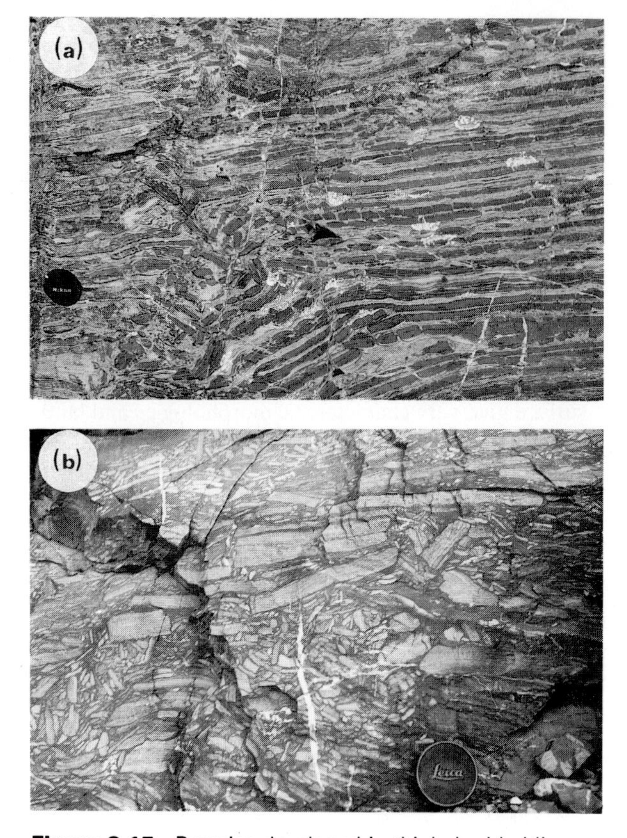

Figure 8.15 Breccias developed in thinly bedded limestones as a result of secondary movement following partial lithification. (a) Undisturbed beds passing laterally into fragmented rotated and folded beds. Upper Cambrian, Nyboe Land, North Greenland (photograph courtesy of J. R. Ineson and Geological Survey of Greenland). (b) Fragments of thin limestone beds showing disaggregation and rotation within a sandy matrix which was not lithified at the time of movement. Biri Formation, Late Precambrian, Mjøsa, Norway.

Figure 8.16 Large palaeokarstic solution pipe in limestone infilled by sandstone. The feature developed on the surface of the limestone, soon after its deposition and early lithification, during a period of lowered sea level and emergence. The base of the pipe is at the level of the hammer and the cave below is of recent origin. Lower Carboniferous, Anglesey, Wales.

partially or totally infilled with travertine deposits. In studying any ancient limestone sequence, it is very important to try to distinguish those karstic features due to relatively recent (Quaternary) activity from those which developed during the early postdepositional history of the limestone. This is often far from easy and the careful application of principles of cross-cutting relationships will be needed (see 2.1.3).

Continued solution and precipitation close to the surface, commonly under conditions of lower rainfall, may lead to more extensive and organised occurrence of calcite in **calcrete** or **caliche** soil profiles. Such profiles also occur in host sediment other than carbonates and they are dealt with in Chapter 9.

8.5 Other bedding phenomena in limestones

A large proportion of carbonate sediments are made up of sand, silt and mud-grade material which is subjected to erosion, transport and deposition similar in most respects to that experienced by detrital particles. The resulting sedimentary structures are very similar therefore to those described in Chapters 4–6. There are, however, a few structures and bedding styles which seem to be unique to carbonates and which reflect either the different grain types that are present in lime sands or the fact that carbonate minerals are, in geological terms, readily dissolved and precipitated.

8.5.1 Large-scale sigmoidal cross bedding

Certain shallow-marine limestones show sigmoidal cross bedding at the scale of several metres, often with complex smaller-scale structures superimposed. The sediment is usually oolitic and/or bioclastic limestone and the cross-bedded units may be very extensive laterally. They are thought to reflect the growth of large oolite and lime sand shoals, probably developing as a result of strong tidal currents in an area starved of detrital supply.

8.5.2 Parallel-bedded limestones

Many limestones seen at outcrop show clear parallel and horizontal bedding. Close examination, however, commonly reveals no obvious control in the form of grain size and compositional differences between beds. The limestone seems to record very uniform conditions of deposition with no primary differentiation, yet bedding is apparent. In some cases bed-parting surfaces (bedding planes) may show some signs of dissolution in the form of stylolites (see 9.3.2), suggesting that the bedding is, at least in part, postdepositional in origin. A variety of detailed processes may have operated and not all cases need have a similar explanation. Careful documentation of the bed-parting surfaces in homogeneous, bedded limestones may lead to original insights into a poorly understood phenomenon.

Reference and study list

References marked with an asterisk * are suitable for 16–19 year old advanced level students in schools and colleges in the UK as well as for undergraduates.

Field experience

First-hand investigation of structures in evaporites is possible in lagoons and in trenches cut in sabkhas. Ancient deposits are not normally exposed at outcrop, where collapse breccias overlying wet rockhead are all that is normally seen, but visits to mines may be possible. Hot springs, springs, caves and lakes (with deposits of tufa and sinter) may be visited in many parts of the world. Field excursions are increasingly possible to present-day carbonate environments wherein reefs, bioherms and stromatolites are found, but diving equipment and training are necessary in order to examine many subtidal processes. On the other hand, examination of ancient reefs, bioherms and stromatolites poses few problems, for they are frequently found in the geological record in many parts of the world and are generally well exposed. Equipment needed to monitor processes in present-day areas includes the normal field equipment plus Eh and pH meters, sampling bottles for salinity measurements and plankton counts, current velocity meters, boats, diving equipment, etc.

James, N. P., C. W. Stearn and R. S. Harrison 1977. *Field guide to modern and Pleistocene reef carbonates, Barbados, W.I.* Miami Beach, Florida: Atlantic Reef Committee, University of Miami, Fisher Island.

Kaplan, E. H. 1980. An annotated list of marine stations suitable for field courses in carbonate geology and tropical marine sciences. *J. Geol Education* **28**, 186–9.

Multer, H. G. and L. C. Gerhard 1974. *Guide book to the geology and ecology of some marine and terrestrial environments, U.S. Virgin Islands.* (Sp. Publ. no. 5). St Croix US VI: West Indies Laboratory.

Roberts, H. H. 1977. *Field guidebook to the reefs and geology of Grand Cayman Island, B.W.I.* Miami Beach, Florida: Atlantic Reef Committee, University of Miami, Fisher Island.

Woodley, J. D. and E. Robinson 1977. *Field guide book to the modern and ancient reefs of Jamaica.* Miami Beach, Florida: Atlantic Reef Committee, University of Miami, Fisher Island.

Laboratory experience

Structures of evaporites may be available in borehole cores. Simple experiments to produce textures and structures from saturated solutions of sodium chloride are well within the capability of most students.

*Arthurton, R. S. 1973. Experimentally produced halite compared with Triassic layered halite-rock from Cheshire, England. *Sedimentology* **20**, 145–60.

Murray, J. 1969. Film: *Abu Dhabi, 1969 – a modern limestone environment.* Colour, sound; *c.* 20 min. Bristol: University of Bristol, Department of Geology.

*Open University 1980. Film: *Surface processes: some aspects of limestone deposition.* Colour, sound. Milton Keynes: Open University Enterprises.

Open University 1980. Film: *Crystals.* Colour, sound. Milton Keynes; Open University Enterprises.

Light reading

Darwin, C. H. 1842 and 1851. *Structure and distribution of coral reefs.* London: John Murray.

Ford, T. D. and C. H. D. Cullingford 1976. *The science of speleology.* London: Academic Press. [Deposits and structures in caves.]

Essential reading

A selection from:

Bathurst, R. G. C. 1975. *Carbonate sediments and their diagenesis*, 2nd edn. Amsterdam: Elsevier. [Tidal flat structures, especially pp. 178–209, 217–30, 517–43.]

Dean, W. E. and B. C. Schreiber 1978. *Marine evaporites.* Short Course Notes no. 4. Tulsa: SEPM. [Section 2 (pp. 6–42), Evaporites of coastal sabkhas, by D. J. Shearman, contains many useful ideas on the origin of evaporitic structures and, particularly, textures and structures of halite and rock salt.]

Glennie, K. W. 1970. *Desert sedimentary environments.* Amsterdam: Elsevier. [Coastal and inland evaporites of deserts.]

Kirkland, D. W. and R. Evans (eds) 1973. *Marine evaporites.*

Stroudsburg, Pa: Dowden, Hutchinson & Ross. [A compilation of papers with useful contextual introductions, mostly to do with calcium sulphates and halites.]

Laporte, L. F. (ed.) 1974. *Reefs in time and space*. Tulsa: SEPM Sp. Publ. 18. [Seven case studies of modern and fossil reefs.]

Newall, N. D. *et al.* 1953. *The Permian reef complex of the Guadelupe Mountains region Texas and New Mexico*. San Francisco: W. H. Freeman. [A classic report.]

Park, R. K. 1977. The preservation potential of some recent stromatolites. *Sedimentology* **24**, 485–506.

Purser, B. H. (ed.) 1973. *The Persian Gulf*. Berlin: Springer-Verlag. [Modern carbonate–evaporite tidal flats and sabkhas.]

Scholle, P. A., D. G. Bebout and C. H. Moore (eds) 1983. *Carbonate depositional environments*. AAPG Memoir 33. [A beautifully illustrated and wide ranging description of carbonate environments and the structures which occur in them.]

Scoffin, T. P. 1987. *An introduction to carbonate sediments and rocks*. Glasgow: Blackie. [Many useful descriptions and illustrations of structures in carbonates at all scales and in their environmental contexts.]

Stoddart, D. R. 1969. Ecology and morphology of recent coral reefs. *Biol. Rev.* **44**, 433–98.

Tucker, M. E. 1985. Shallow-marine carbonate facies and facies models. In *Sedimentology: recent developments and applied aspects*, P. J. Brenchley and B. P. J. Williams (eds), 147–69. Geol. Soc. London, Sp. Publ. 18.

Walker, R. G. (ed.) 1984. *Facies models*, 2nd edn. Geol. Assoc. Canada. [Chs 14 (carbonates), 15 (reefs), 16 (carbonate slopes) and 17 (evaporites) all have relevant material.]

Walter, M. R. (ed.) 1976. *Stromatolites*. Amsterdam: Elsevier. [Several useful papers including ones on tufa.]

Wilson, J. L. 1975. *Carbonate facies in geologic history*. Berlin: Springer-Verlag. [The book deals mainly with carbonate facies and large structures in the geological record.]

Further reading

Assereto, R. L. A. M. and C. G. Kendall 1977. Nature, origin and classification of peritidal tepee structures and related breccias. *Sedimentology* **24**, 153–210.

Braitsch, O. 1971. *Salt deposits, their origin and composition*. Berlin: Springer-Verlag.

Degens, E. T. 1965. *Geochemistry of sediments*. Englewood Cliffs, N.J: Prentice-Hall. [Geochemistry of evaporites and carbonates.]

Flugel, E. (ed.) 1977. *Fossil algae: recent results and developments*. Berlin: Springer-Verlag. [Useful papers on both growth forms and microstructures of stromatolites.]

Frost, S. H., M. P. Weiss and J. B. Saunders 1977. *Reefs and related carbonates: ecology and sedimentology*. AAPG Studies in Geology 4. [Several useful papers, mainly concerned with modern reefs of the Caribbean area.]

Ginsburg, R. N. (ed.) 1975. *Tidal deposits*. Berlin: Springer-Verlag. [Articles include that by E. W. Mountjoy on intertidal and supratidal deposits within isolated Upper Devonian build-ups in Alberta (pp. 387–97).]

Hardie, L. A. (ed.). 1977. *Sedimentation on the modern carbonate tidal flats of north-west Andros Island, Bahamas*. The Johns Hopkins University Studies in Geology no. 22. Baltimore: Johns Hopkins University Press.

Kinsman, D. J. J. 1969. Modes of formation, sedimentary associations and diagnostic features of shallow water and supratidal evaporites. *Bull. Am. Assoc. Petrol. Geol.* **53**, 830–40.

Logan, B. W. *et al.* 1974. *Evolution and diagenesis of Quaternary carbonate sequences, Shark Bay, Western Australia*. AAPG Mem. 22.

Lowenstein, T. K. and L. A. Hardie 1985. Criteria for the recognition of salt-pan evaporites. *Sedimentology* **32**, 627–44.

Park, R. K. 1976. A note on the significance of lamination in stromatolites. *Sedimentology* **23**, 379–93.

Pratt, B. R. and N. P. James 1982. Cryptalgal–metazoan bioherms of Early Ordovician age in the St George Group, western Newfoundland. *Sedimentology* **29**, 543–69.

Reading, H. G. (ed.) 1986. *Sedimentary environments and facies*, 2nd edn. Oxford: Blackwell Scientific. [Chs 8 (arid shorelines and evaporites: Schreiber) and 10 (Shallow marine carbonates: Sellwood) are helpful.]

Riding, R. 1979. Origin and diagenesis of lacustrine algal bioherms at the margin of the Ries crater, Upper Miocene, Southern Germany. *Sedimentology* **26**, 645–80.

Scoffin, T. P. 1971. The conditions of growth of the Wenlock reefs of Shropshire, England. *Sedimentology* **17**, 173–219.

Shearman, D. J. 1970. Recent halite rock, Baja California, Mexico. *Trans Inst. Mining Metall.* **79B**, 155–62.

Simpson, J. 1985. Stylolite-controlled layering in an homogeneous limestone: pseudo-bedding produced by burial diagenesis. *Sedimentology* **32**, 495–505.

Southgate, P. N. 1982. Cambrian skeletal halite crystals and experimental analogues. *Sedimentology* **29**, 391–407.

Toomey, D. F. (ed.) 1981. *European fossil reef models*. Tulsa: SEPM Sp. Publ. 30. [Nineteen papers on European reefs of all ages.]

Extra references relating to figure captions

Preiss, W. V. 1976. Basic field and laboratory methods for the study of stromatolites. In *Stromatolites*, M. R. Walter (ed.), 5–13. Amsterdam: Elsevier.

Shearman, D. J. 1971. *Marine evaporites: the calcium sulfate facies*. Am. Assoc. Petrolm. Geol. Seminar, 65 pp. University of Calgary, Canada.

Shearman, D. J. 1978. Evaporites of coastal sabkhas. In *Marine evaporites*. W. E. Dean and B. C. Schreiber (eds), 6–42. SEPM Short Course no. 4. Tulsa, Oklahoma.

Weiler, Y., E. Sass and I. Zak 1974. Halite oolites and ripples in the Dead Sea, Israel. *Sedimentology* **21**, 623–32.

9 Structures due to deformation and disturbance

9.1 Introduction

Any sediment may be disturbed after deposition, but disturbance is most common in sands and finer-grained material. Earlier structures may be disrupted and distinctive new structures may form as a result of physical, chemical and biological processes. It is often difficult to tell when physically and chemically induced disturbances took place. In some cases, they occurred soon after deposition at, or close to, the contemporaneous surface, and in other cases they were associated with later burial and lithification.

Many deformational structures are valuable as 'way-up' indicators and all tell us something about conditions after deposition within the sediment or at its surface.

9.2 Physically induced soft-sediment deformation

This results from physical forces, commonly gravitational, acting upon physically weak sediment, usually silts and fine-grained sands, at the sediment surface or soon after burial. There is no established or unambiguous way of classifying these structures. Here we follow a broadly morphological scheme, based upon the mode of occurrence. However, several structures can be observed in both vertical section and on bedding surfaces, and might, therefore, be placed in more than one category. They are described under only one heading, usually their most common mode of occurrence.

Most types of soft-sediment deformation depend on unconsolidated sediment being in a weak condition. The resistance of sediment to deformation is most commonly expressed by its shear strength τ, which is a function of grain cohesion C, intergranular friction and the effective pressure between the grains:

$$\tau = C + (\sigma - p) \tan \phi \qquad (9.1)$$

where σ is pressure normal to shear, p is excess pore-fluid pressure and ϕ is the angle of internal friction.

For a sediment to fail after deposition and be deformed, its shear strength must be reduced or the applied shear stress increased. This can be achieved by a loss of cohesion, by a readjustment of packing to reduce $\tan \phi$, or by increasing the pore-fluid pressure p. Cohesion is the least readily changed property as it is largely controlled by grain size. A shock applied to waterlogged, loosely packed sediment can change the packing and, in the process, increase the pore-fluid pressure to the extent that the sediment undergoes temporary **liquefaction**. In this condition, sediment and water behave as a liquid, deforming very readily until the pore-water pressure falls due to escape of excess water and the sediment takes on a closer packing and grains make frictional contact with one another. The shocks that cause liquefaction may be externally generated, as by earthquakes, or they may be local, for example a rise in water level or an episode of sudden deposition.

This effect is illustrated by jumping up and down on a sandy beach close to the water's edge. The surrounding sediment liquefies and then water will escape to the surface. Once this has happened, the same patch of sand is not easily liquefied again.

In addition to shock and repacking, excess pore-fluid pressure can be produced by rapid deposition of fine-grained sediment. Their low permeability prevents pore-fluid escape and sediment compaction at a rate that balances the increasing overburden. **Overpressured** or **undercompacted** conditions are then said to obtain, in which state the sediment is highly susceptible to deformation.

Liquefaction of a sediment may be total, so that all grain contact is broken and the mass of sediment and water flows freely. In such cases, original lamination is destroyed, giving massive or 'slurried' bedding. In other cases, where liquefaction is only partial, deformation is more limited and original lamination may be preserved, although distorted. A mass of liquefied sediment will remain mobile or weak until the excess pore-fluid pressure is dissipated either by general intergranular flow of pore water, usually upwards, or by water escape along restricted pathways. If vigorous enough, the upward escape of fluid may lead to the **fluidisation** of sediment within escape pathways. Rapid fluid movement between the grains causes a loss of strength and increased pore space. The relative movement of grains and fluid during fluidisation allows some grain sorting to take place, usually by upwards removal of fines. In a liquefied sediment, fluid and grains move essentially together giving little scope for sorting.

9.2.1 *Features visible on bedding surfaces*

Load casts and flame structures. These structures occur most commonly on the lower surfaces of beds of sandstone interbedded with mudstones, i.e. they are a type of sole mark (see 4.2). They also occur within sandstone units and are commonly recognised in vertical section. Load casts on soles of sandstone beds are rounded, rather irregular lobes of variable size and relief. Small examples are measured in millimetres and large ones may be tens of centimetres or even metres in diameter. They seldom occur in isolation and usually cover a whole bedding surface (Fig. 9.1).

Figure 9.1 Loadcasts on the base of a sandstone bed from an interbedded sandstone/mudstone sequence. Bude Formation, Upper Carboniferous, North Cornwall.

Upward-pointing fingers or wedges of the underlying unit occur between the sandy lobes. These are **flame structures** and they are an inevitable accompaniment of load casts (Fig. 9.2). Although many load casts are simple protrusions on the sandstone base, some are more globular, being attached to an overlying sandstone by a thin neck or even being totally detached from it. In some cases there is no sign of an overlying sandstone and the **isolated load balls** or **pseudonodules** 'float' in the mudstone, which commonly shows a disturbed 'slurried' texture (Fig. 9.3).

Internally, load casts show contorted lamination. Close to the edges, lamination parallels the margin, but towards the centre contortion becomes more intense. Where lamination can be seen in the layer beneath the loaded surface, it tends to follow the margins of the flame structures, becoming contorted towards the centre of the flame. In relatively rare examples, load casts are centred on original ripples so that deformed laminae were originally cross laminae.

Figure 9.2 Vertical section through load-casted, interbedded sandstones and mudstones. The loads on the base of the sandstone have 'flames' of mudstone squeezed upwards between them. Bude Formation, Upper Carboniferous, North Cornwall.

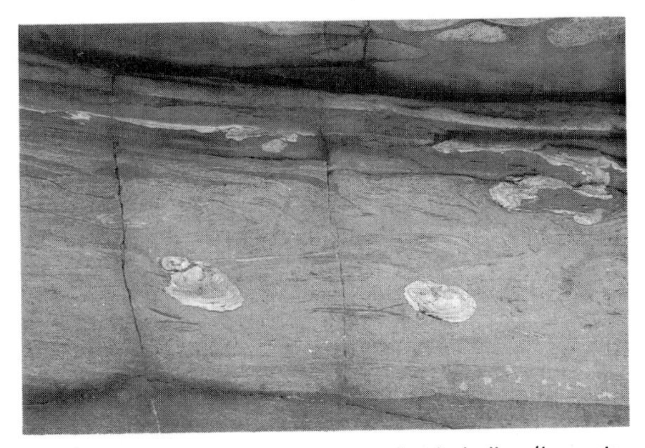

Figure 9.3 Isolated sandstone load balls ('pseudonodules') in a siltstone with a 'slurried' texture. The original sandstone bed has totally foundered and collapsed into the finer sediment. Load balls are about 5 cm diameter. Bude Formation, Upper Carboniferous, North Cornwall.

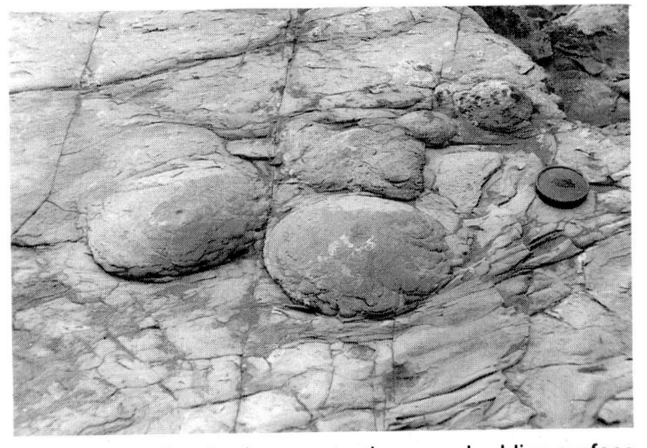

Figure 9.4 Sand volcanoes on the upper bedding surface of a sandstone bed. Note the small lobes of sand which flowed down the side from the central vent. Ross Formation, Upper Carboniferous, Co. Clare, Eire.

The basic mechanism of formation is gravity acting on beds which were unstable due to their high porosity and lack of compaction and coherence, and to differences in density between the beds. Muds commonly have a high depositional porosity (60–70 %) whereas the porosity of even rapidly deposited sands is considerably less (30–40 %). Thus if a sand layer is rapidly deposited on a mud layer, it will be denser than the mud and if both sediments are weak it will tend to sink into the mud by loading.

In loading, the mud may have failed due to excess pore-fluid pressures generated by deposition of the overlying sand layer. For isolated load balls or pseudonodules, a relatively thick mud bed must have lost its strength. The sudden deposition of the sand could cause this, but externally generated shocks are equally possible. This effect can be simulated in the laboratory by violently jarring a waterlogged sand and mud sequence.

Whatever the reason, the combination of density inversion and temporary weakness leads to the sinking of one bed into the other, either randomly or at localised pre-existing thickness differences (e.g. scours or ripples). The size of load casts correlates roughly with the thickness of the sandstone bed. For isolated load balls, the whole sand bed must sink, but preservation of internal lamination within load balls implies that the sand was not totally liquefied.

As 'way-up' indicators, load casts are generally unambiguous. Their downwards convexity and their association with flame structures are diagnostic. Contortions within large load balls can be similar to those of slumps (see p. 147), but the latter involve lateral movement which is shown by a preferred orientation of folds. With loading, the dominantly vertical movement gives a random fold orientation. In addition, loading is normally confined to one pair of beds, whereas slumping may involve many or several beds.

Sand and mud volcanoes. These relatively rare structures occur most commonly in sandstones, often interbedded with mudstones, the volcanoes themselves usually being of medium- or fine-grained sand though in some cases of silt and mud. They occur on upper bedding surfaces and are also seen in vertical section. Volcanoes are often underlain by units showing extensive post-depositional disturbance such as loading, convolute bedding, sand and mud dyke intrusion and evidence of slumping and sliding.

On upper bedding surfaces, sand volcanoes are conical or dome-like, ranging in diameter from 10 cm to several metres and up to 50 cm high. They may have a crater-like depression in their centres and their flanks may carry radially arranged sand lobes up to a few centimetres wide and with rounded ends (Fig. 9.4). Vertical sections show internal layering parallel to the flanks. In the central zone, a plug or pipe of structureless sand may underlie the crater and may link with a sand-filled dyke or tube below (Fig. 9.5).

They result from liquefied sand being extruded through a local vent at the sediment surface. The volcanoes are, in effect, localised equivalents of the transposed sand sheets associated with sandstone dykes (see p. 141). Sand volcanoes commonly reflect release of pressure from a liquefied unit, possibly following a shock.

Lobes on the flanks record the flowage of liquefied sand. The convex downslope ends show that the sand

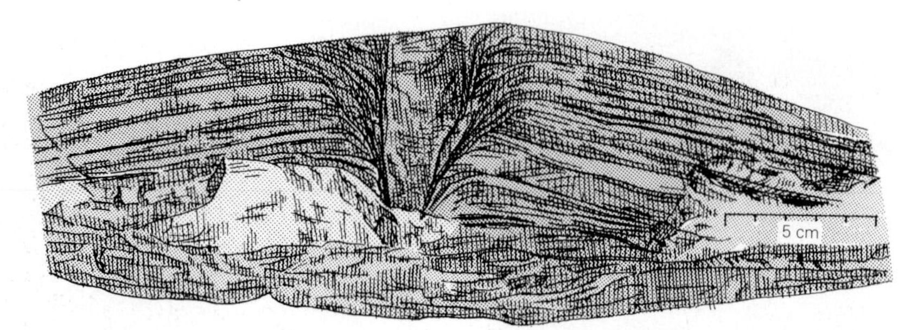

Figure 9.5 Internal structure of a sand volcano showing the central plug and the inclined bedding which could, without care, be mistaken for cross bedding. Drawn from photograph. Ross Formation, Upper Carboniferous, Co. Clare, Eire.

must have been **dewatering** as it flowed. Preservation of these lobes suggests that the extrusion of sand volcanoes took place in quiet conditions, otherwise the sand would have been reworked. In examples extruded in settings with higher energy, reworking and partial erosion may be identified.

In vertical section, the lenticular shapes of volcanoes and the inclined lamination could be mistaken for depositional bedforms and the inclined layering for cross lamination.

Patterns of cracks. Patterns of cracks with a variety of scales, shapes and origins occur on present-day sediment surfaces and on both upper and lower bedding surfaces in rocks. Four types of crack, each of a rather different origin, are discussed.

Desiccation mudcracks. These are common on the floors of dried-up ponds, lakes and playas, on river floodplains and in intertidal and supratidal areas where they are often open fissures or are only partially filled by other sediment. In rocks they occur on the bedding surfaces of interbedded, sandstone–mudstone sequences and less commonly in thinly bedded carbonates.

In rocks, the cracks occur in muddy sediment and are infilled by coarser-grained material, usually sandstone. The cracks commonly form polygons from centimetres to metres in diameter, with different sizes sometimes present on the same surface (Fig. 9.6). While there is a tendency towards a hexagonal pattern, many polygons are quadrilaterals or triangles. The cracks are parallel-sided in plan, and in vertical section they usually taper downwards. Their shape in vertical section can sometimes be complicated by folding (Fig. 9.7). Crack widths range up to several centimetres and depths up to several decimetres. On both present-day surfaces and on bedding surfaces, the areas between the cracks are commonly gently concave-upwards. The surface of a present-day mud layer may be curled up into a highly concave shape.

Drying out of an exposed muddy sediment causes contraction, giving an isotropic, horizontal, tensional stress field which diminishes downwards from the surface. The stress is released by the development of vertical cracks that taper downwards to the level of no effective stress. On slopes steeper than about 5° the crack pattern tends to be rectangular, with one set of cracks parallel to the contours. Clearly, gravity acts to give an anisotropic stress field. Cracks may be also generated from earlier disturbances such as footprints.

With homogeneous material, the depths of cracks and the diameters of polygons are directly related, the thicker the cracked layer, the larger the polygons. Thin, surface layers of mud, which dry out rapidly, commonly give small cracks superimposed upon the larger ones which develop during more sustained desiccation.

Filling of cracks takes place later, for example by the influx of a sediment-laden flood. Equally, wind-blown sand may become trapped in cracks. Mudflakes, derived from surface mudlayers, may become mixed with the sand. On burial, compaction of muds is greater than that of the sand infills, which respond by folding (Fig. 9.7).

The concave-upwards surfaces between desiccation mudcracks and their downward-tapering form are useful indicators of 'way-up'.

Sub-aqueous shrinkage cracks (synaeresis cracks). These occur in mudstones interbedded with sandstone and also in some clay-rich carbonate sediments, usually where the beds are thin. They are also common on the floors of salt pans in present-day salt marshes. The cracks are most common either as positive relief features on bases of sandstone beds or in vertical sections through muddy layers where their downward-tapering sandy fills may be contorted by small-scale compactional folding. They also occur as negative relief features on upper bedding surfaces.

In plan, synaeresis cracks tend to have irregular or radiating patterns, sometimes cross-cutting one another. Individual cracks are lenticular, pinching out rather than joining with other cracks (Fig. 9.8).

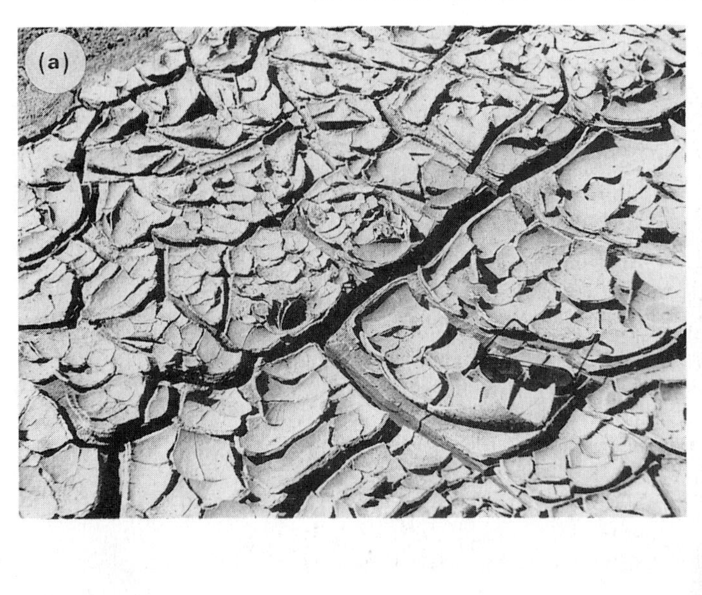

Figure 9.6 Desiccation cracks. (a) On a present-day mud surface crack patterns of different sizes have different depths of penetration and the thin surface mud layer is rolling up. Such mud roll-up clasts may be subsequently reworked by water or by the wind. Sunglasses, bottom right for scale. (b) Ancient crack pattern on a bedding surface of interbedded sandstones and mudstones. Polygons are 20–30 cm diameter. West Bay Formation, Lower Carboniferous, Nova Scotia, Canada.

Sub-aqueous shrinkage cracks result from loss of pore water from the sediment because of a reorganisation of originally highly porous clays, either due to flocculation, or because of salinity-induced changes of volume of certain clay minerals. These processes are referred to as **synaeresis**. The conditions under which they take place are not well known, but they occur in a variety of environmental settings and water depths. Marginal marine settings may be particularly favourable because clays there are likely to be subjected to changes in salinity. However, cracks of this type occur in ancient sediments of both marine and non-marine origin. The irregular distribution of cracks may sometimes be due to earlier inhomogeneities, such as burrows, in the host sediment. These cracks can be indicators of 'way-up' and can suggest postdepositional conditions. As 'way-up' indicators, they are not as good as desiccation mudcracks because intercrack areas are flat.

Figure 9.7 Sand-filled mudcracks, folded because of the compaction of the mudstone. Independence Fjord Group, Proterozoic, N.E. Greenland.

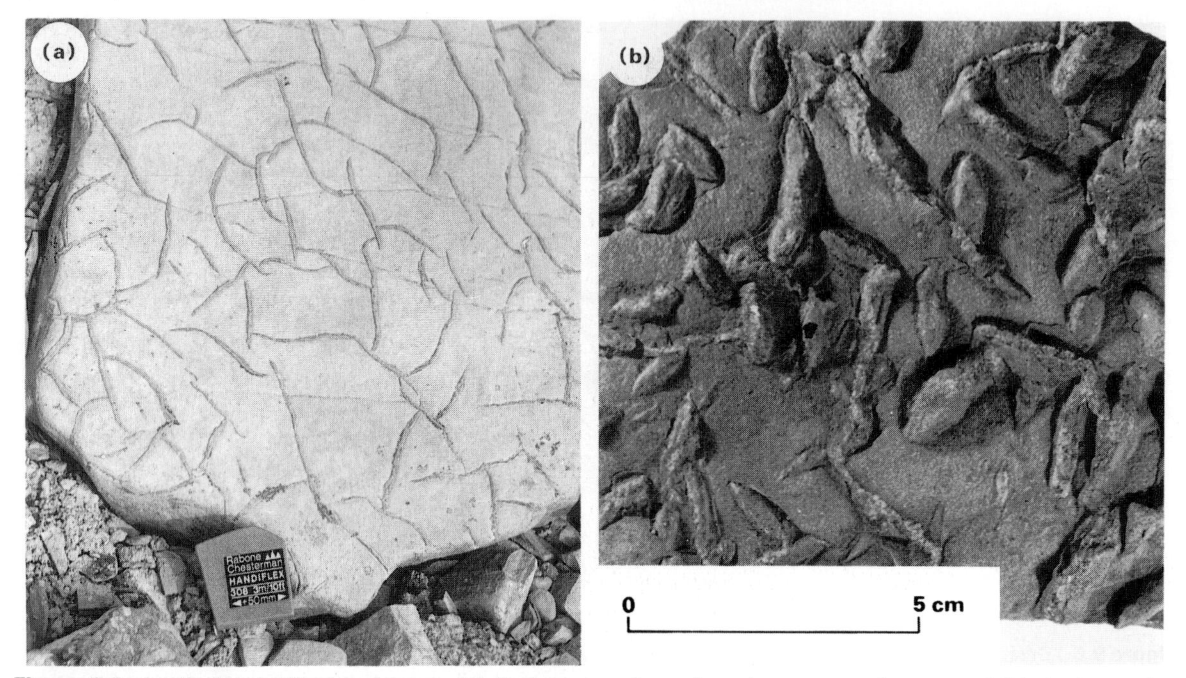

Figure 9.8 Lenticular shrinkage cracks on: (a) the upper surface of a calcareous mudstone; and (b) the base of a sandstone bed. Such cracks are generally thought to be of sub-aqueous origin. (a) Cass Fjord Formation, Cambrian, N.W. Greenland; (b) Independence Fjord Group, Proterozoic, N.E. Greenland.

Synaeresis cracks may be confused with desiccation cracks and with sandstone dykes, and they have commonly been mistaken for trace fossils, particularly in rocks of late Precambrian age. The most common misidentification is as a burrow, their lenticular shape in plan being compared with the base of a U-shaped burrow. The compactional folding has also led to confusion with organic traces.

Sandstone and mudstone dykes and transposed sand sheets. With these structures, crack-filling material is most commonly sand but the host sediment may be anything from mud to coarse-grained gravel. Dykes, though relatively uncommon, occur in a wide variety of settings from deep-water sandstone/mudstone sequences to sub-aerial mass-flow conglomerates. They are seen on both bedding surfaces and in vertical section. Transposed sand sheets, which are sometimes associated with the dykes, can easily be mistaken for normally interbedded sandstones and their certain identification is possible only with very good three-dimensional exposure.

Dykes occur in a range of sizes, with widths up to several tens of centimetres and vertical extents ranging up to several metres. As with other cracks, the fills may be folded, particularly if the host sediment is finer-grained. Dykes can sometimes be traced downwards to link with underlying beds of similar lithology. Above they may be truncated by erosion or they may link with an overlying sandstone.

On horizontal surfaces, dykes can occur as positive or negative features depending upon weathering of the dyke and host lithologies. Dykes tend to be straight and rather parallel-sided (Fig. 9.9) and, although usually random in orientation, they may show linear trends or polygonal patterns. Internally they have a variably developed lamination parallel to their sides which tends to be less clearly developed towards the centre.

In good exposure some sandstone dykes may be traced upwards into horizontal sandstone sheets. These may be laterally extensive and, at first sight, can be mistaken for sandstone beds laid down above the dyke-intruded unit. These sheets tend to be rather featureless although they sometimes show gently undulating surfaces. Locally, more intense folding may occur (Fig. 9.10). Internally the sandstones may show an irregular 'slurried' texture of weakly defined and rather disturbed lamination.

Sediment dykes result from the injection of sediment from an underlying, or more rarely overlying, source bed during a short-lived, postdepositional event when both a buried source bed and the host layer were in a weakened condition. Where the intruded host sediment is fine-grained (e.g. sand or silt), and was laid down in

Figure 9.9 Sandstone dyke cutting through a poorly sorted conglomerate of probable mudflow origin. A weak lamination parallel to the walls of the dyke is produced by shearing of the liquefied sand during its intrusion from below. Moraenesø Formation, Proterozoic, N. Greenland.

Figure 9.10 Folding in a sandstone dyke (bottom right) associated with its transition into a transposed sand sheet (top left). Moraenesø Formation, Proterozoic, N. Greenland.

relatively quiet conditions an external shock may have been needed to liquefy temporarily the source bed. Where the host sediment is coarser-grained, its sudden emplacement by mass flow, for example, could have created the conditions necessary to liquefy the source bed. The host sediment commonly has a fairly high content of fine-grained material which would have reduced its permeability. Sediment and water would then be expelled from the liquefied layer through fissures with sufficient force to carry the liquified sediment, in some cases, to the free surface. Lamination in the dyke fills reflects shearing of the liquefied sand as it moved along the fissures. When a fissure reaches the sediment surface, liquefied sand may be extruded and flow laterally before it loses excess water and mobility. Such transposed sand sheets occur on some modern debris flows and are sometimes preserved in the rock record. Their undulating bases reflect a tendency of the extruded sand to sink back into the mobile host sediment by loading.

In poor exposure, sandstone dykes could be confused with either type of shrinkage crack although this is less likely where the host sediment is coarse-grained. Internal lamination is good evidence of injection, but confusion could arise from similarity with ice wedges of periglacial areas.

Ice-wedge polygons. Large areas of present-day permafrost show patterns of polygonal cracks of variable plan shape. Similar patterns of crack are also widely recognised on aerial photographs of areas subjected to permafrost conditions during earlier periods of Quaternary glaciation. Cracks are also seen in vertical sections through glacial and proglacial sediments in these areas. More rarely they are recognised in association with ancient glacial deposits (tillites) in the rock record.

In plan view, polygons range in diameter from about 3m up to several tens of metres and individual cracks are from a few centimetres up to several tens of centimetres wide. Cracks forming at the present-day are commonly bordered by ramparts of host sediment which have been pushed upwards (Fig. 9.11).

In vertical section, ancient examples commonly penetrate vertically for several metres and show an upward-flaring, wedge shape which tapers downwards to a fairly sharp lower termination. The cracks usually show quite complex patterns of filling, often with several phases of contrasting lithologies occurring in zones roughly parallel to the sides of the cracks (Fig. 9.12). The fill also contrasts to some extent with the host sediment, but coarse clasts, derived from the host, are commonly recognised in the crack fills. Where pebbles are included in the fills, they tend to have their long axes parallel to the sides of the cracks.

The cracks result from the thermal contraction of the host sediment in the extremely cold conditions. The tensile strength of the host sediment is exceeded and a crack pattern develops. Water or loose sediment filters downwards into the crack from the active surface layer which thaws out in the summer. Sediment may help to wedge open the crack, and water collecting there may help crack development by expanding on subsequent freezing. Many wedges in present-day settings have ice masses below them.

The recognition of ice wedges is of considerable palaeoclimatic significance, and care should be taken to distinguish these features from other types of cracks.

Figure 9.11 Present-day icewedge with ramparts on either side. Washington Land, N. Greenland.

Raindrop impressions. Some upper bedding surfaces in ancient mudstones, siltstones and sandstones and also some present-day muddy or sandy surfaces show patterns of shallow pits. These are commonly associated with desiccation mudcracks. Pits may be widely separated or may completely cover the surface. They are circular or, rarely, elliptical in shape, up to about 1 cm in diameter and up to a few millimetres deep (Fig. 9.13). Where they completely cover a surface, they are polygonal. They have a slightly raised rim and the floor of each pit gives a smooth, concave-upwards crater-like form.

Large raindrops and hailstones impact with considerable force and produce small craters on damp sediment. Preservation in the rock record is only likely when the sediment is muddy, as this will have the strength to retain the impression when it dries out. Sandy surfaces, on drying, are reworked by the wind or water.

Rain pits could be confused with trace fossils or with gas-bubble escape features but the underlying sediments show no traces of disturbed lamination.

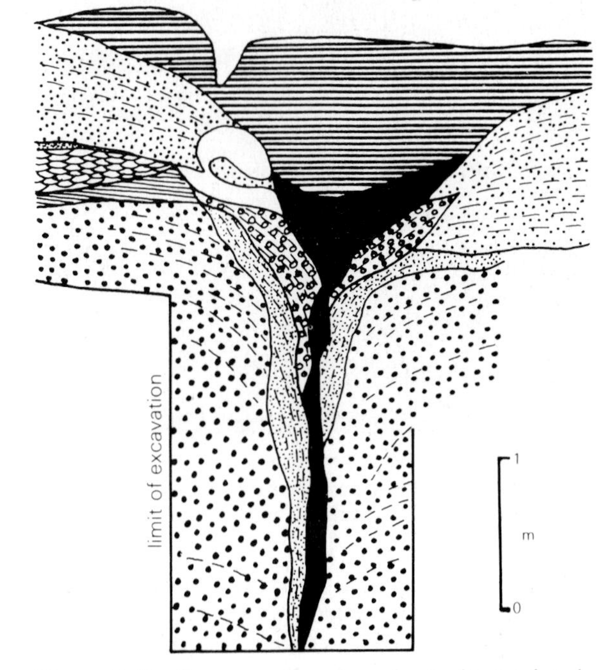

Figure 9.12 Cross section through an ice-wedge in Pleistocene sediments, exposed in a gravel pit. Note the downward tapering of the wedge and the various zones of infill. (After Gruhn & Bryan 1969.)

9.2.2. Disturbance within individual beds

Although some of them have expression in plan view, these structures are most common in vertical section. The structures include those due to deformation of primary depositional lamination as well as new structures developed by postdepositional activity. Thickness of disturbed units ranges from centimetres to metres, and all or only part of a bed may be affected.

Oversteepened and overturned cross bedding. This is common in medium-grained or finer sandstones laid down in a variety of aqueous environments. Deformation in aeolian sands and laboratory experiments relating to it is dealt with in Section 6.3.4.

Overturned or oversteepened cross bedding occurs within single sets. The deformation ranges from foresets that dip more steeply than the angle of rest in the upper parts of sets (oversteepened) to extensive overturning of foresets (Fig. 9.14) into recumbent folds. The

Figure 9.13 Raindrop impressions (rainpits) on the upper bedding surface of a siltstone. West Bay Formation, Lower Carboniferous, Nova Scotia, Canada.

Figure 9.14 Cross bedding with overturned foresets. Kap Holbœck Formation, Lower Cambrian, N.E. Greenland.

overfolding is always in the direction of the original foreset dip, and the intensity of oversteepening or overturning usually increases upwards through the set. The position of the fold axis within sets is variable and axial planes are inclined to the horizontal.

In addition to simple folds, some single sets show more complex folding which may also involve normal and reverse faulting, loss of definition of the foreset lamination, and enclosure of deformed blocks in structureless masses.

For simple oversteepened or overturned foresets, the processes seem fairly straightforward. A shear force acting on the upper surface of a bedform, and in the same direction as the current that produced the cross bedding, deformed the sediment which had been weakened. The shear force seems usually to have been the water current that produced the cross bedding and the sand was probably weakened by partial liquefaction which is quite easily achieved in rapidly deposited sands. If the whole set were partially liquefied either by shock or spontaneously, the repacking of grains into a more stable configuration would begin at the base and move upwards through the set as a **front of reconsolidation.** The upper parts of the set would, therefore, be subjected to the deforming shear stress for a longer time than the lower parts and this would give rise to the fold shapes observed.

Where the fold pattern is more complex, the above mechanism may have played a part, but forceful upward escape of water may also have caused deformation into more upright folds. Internal buckling, similar to that which produces convolute lamination (see below) may also have been involved. When deformational structures include minor faulting and patches of structureless sand, the original cross bedding may be aeolian (see 6.3.4 and 6.3.5).

Convolute bedding and lamination. These structures occur commonly in single beds of sand or silt in a wide range of environmental settings. They are most commonly recognised in vertical section, but are also seen on bedding surfaces and may be associated with water- and sediment-escape structures such as sand volcanoes.

Convolute bedding and convolute lamination are size-related terms for similar features, the former at the scale of decimetres or bigger, the latter at the scale of centimetres. However, the terms are used loosely and there is no agreed or physically significant size limit. The structure involves folding of lamination, commonly into upright cuspate forms with sharp anticlines and more gentle synclines (Fig. 9.15). Overturning of fold axes is sometimes seen, often with a preferred orientation. It is usually possible to trace laminae through the folds and it may sometimes be possible to detect original cross lamination within the folded sediment. Convolution usually increases in intensity upwards through a bed from undisturbed lamination at the base. At the top it may either die out gradually or be sharply truncated. On upper bedding surfaces, convolute lamination commonly takes the form of a complex pattern of basins and ridges.

Figure 9.15 (a) Convolute lamination within ripple cross-laminated sandstone (photograph courtesy of G. Kelling). (b) Convolute bedding within cross-bedded sandstone. The sharp upward-pointing folds suggest upwards fluid escape. Roaches Grit, Upper Carboniferous, Staffordshire.

Convolution involves plastic deformation of partially liquefied sediment soon after deposition. The common occurrence of convolute lamination in turbidite sandstones, and just below the sediment surface in present-day river floodplains and tidal flats in seismically quiet areas, suggests that liquefaction can be spontaneous as well as externally triggered. On tidal flats liquefaction may be aided by breaking waves during emergence of the bed or by the rise and fall of the water table through the sediment. Where axial planes of folds have a preferred direction of inclination, this often coincides with the palaeocurrent, suggesting that convolution formed during deposition.

The main use of convolute lamination is as evidence of rapid deposition. It has limited potential as an indicator of 'way-up' and as a rather uncertain palaeocurrent indicator when folds have a preferred direction of overturning.

Dish, pillar and sheet dewatering structures. These structures have been described only in the past fifteen years and their true significance was recognised even more recently. Although they were originally recorded from thick turbidite sandstones (and this remains their most common occurrence), they are now recognised in shallow-water sandstones and in volcanic ash layers. The host sediment ranges from coarse silts to coarse, even pebbly, sand and fine gravels. Dish structures also require a certain amount of clay for their formation. The sandstones and siltstones are usually discrete beds at least decimetres or even metres thick, in which dish and pillar structures are commonly the only structures present. The beds elsewhere appear massive. In some cases, the structures deform and cut across earlier lamination. Dish and pillar structures are usually poorly defined and they need rather exceptional exposure and weathering to stand out clearly. Only after seeing good examples is it realistic to look for these structures in less favourable exposures. Sheet dewatering structures are rather more obvious and common.

Dish and pillar structures commonly occur together in the same bed. In vertical section, dish structures appear as thin, roughly horizontal zones, often flat but more usually concave upwards (Fig. 9.16). Each is defined by a dark, clay-rich zone 0.2–2 mm thick which contrasts with thicker, paler sandstone on either side. Plan views show the three-dimensional shape to have the form of shallow dishes a few centimetres across, defined by concentrations of clay and platy mineral grains. The darker zones defining these dishes commonly have a vertical spacing of a few millimetres up to about 10 cm.

Flat zones lacking the strong dish shape, have upturned ends associated with penetration of layers by **pillar structures** (Fig. 9.16). These extend vertically for several centimetres and may pass through several of the horizontal zones and dishes. They have a core of cleaner sand with poorly defined, darker, clay-rich fringes and in plan are circular with a diameter of a few centimetres.

Sheet dewatering structures are continuous and more linear in plan. These subvertical sheets, up to a few millimetres wide, are commonly arranged in a parallel fashion and usually occur near the tops of structureless sandstone beds (Fig. 9.17). They are not necessarily associated with dish structures.

All these structures may coexist within a bed, but they may sometimes show a crude vertical zonation. Flat layers tend to underlie the more curved dishes, and pillars may extend upwards beyond the dishes. The sequence may end upwards in convolute lamination.

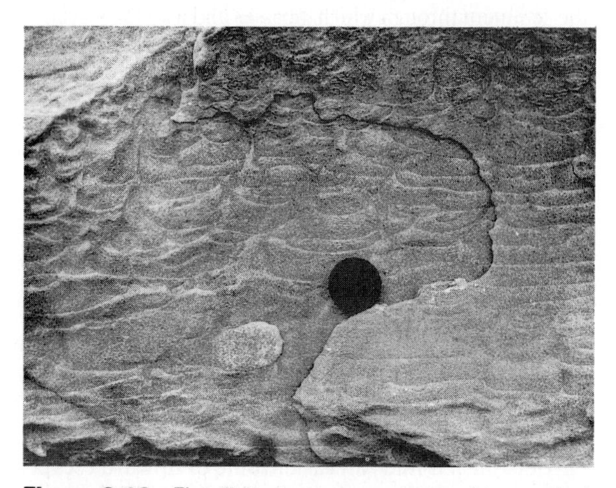

Figure 9.16 Flat dish structures penetrated by vertical 'pillar' structures which acted as conduits for vertical water escape. Eocene Flysch, San Sebastian, Spain.

Figure 9.17 Sheet dewatering structures in otherwise structureless poorly sorted sandstone, Skipton Moor Grits, Upper Carboniferous, Yorkshire. (Photograph courtesy of J. G. Baines.)

The three types of structure are all due to the post-depositional escape of water from the sand. The vertical structures (the sheets and pillars) are conduits of water expulsion, similar to sandstone dykes and the feeder pipes of sand volcanoes. Pillars and sheets, however, record the movement of water with only restricted and selective movement of sediment. The force of upward water movement causes local fluidisation of the sediment through which it passes and also the selective transport of fine particles. This results in the slightly cleaner sand of the pillars and sheets.

The mode of formation of dish structures is less obvious and is inferred from the structures themselves. Their intimate association with pillars suggests a related origin. They are thought to be produced by slower and, to some extent, horizontal water movement, restricted and controlled by semi-permeable, flat-lying barriers that probably started as weakly defined depositional laminations. The upward movement of water through the loosely packed sand soon after deposition is retarded by the laminations and further fine particles are added to them, further reducing their permeability. Some of the escaping water is forced horizontally below barriers until it finds an easier route of upward escape, at a pillar or the upturned edge of a dish. This water movement probably takes place very soon after, or even during, deposition. Rapid deposition probably gives rise to very loose grain packing and a high initial porosity.

The rarity of these structures limits their use as way-up indicators although, when present, the upward concavity of dishes could be helpful. The lack of dish and pillar structures in many massive sandstones may reflect a lack of suitable clay rather than a lack of rapid dewatering.

Dish structures could be confused with trough cross lamination but the lack of foreset laminae in each 'trough' and the different plan shape should help avoid this pitfall. Pillar and sheet structures could be confused with vertical burrows but upward turning of the flanking layers is not common in burrows.

9.2.3 *Disturbance affecting several beds*

These larger-scale structures may be recognisable at outcrop but some examples may require a larger scale of observation, possibly involving mapping. With structures such as slumps, which result in part from lateral mass movement, it would be equally valid to regard them as being of depositional or, at least, 'redepositional' origin. With larger-scale disturbances there is often a difficulty in establishing the timing of deformation and in distinguishing tectonic from sediment-induced deformation.

Slumps: sedimentary folding. Units of folded sediment attributable to slumping usually occur in interbedded sequences containing a substantial proportion of fine-grained sediment. The sequences may be composed of detrital or carbonate material and the fine-grained sediment may be clay or lime mud. Slumped units occur on or at the bases of contemporaneous slopes in a variety of environments.

Slump-folded units vary from less than 1 m up to tens or even hundreds of metres in thickness. They are usually bounded above and below by undisturbed sediment and this distinguishes them from tectonically disturbed beds (Fig. 9.18). Within the slumped unit, folds may have preferred orientations which when properly plotted on a stereonet, help to indicate the

Figure 9.18 Interbedded sandstone/mudstone sequences within which a group of beds have been folded as a result of slump movements. The folding is seen to be syndepositional by the undisturbed nature of the bedding above and/or below. (a) Upper Carboniferous, Cantabrians, northern Spain; (b) Ross Formation, Upper Carboniferous, Co. Clare, Ireland.

direction of the palaeoslope. It is important to record the style and scale of the folding, the thickness of the deformed unit as a whole as well as the thicknesses of the beds within it. Systematic recording and comparison of the orientations of fold axes and axial planes with those of tectonically produced structures (folds, cleavage, etc.) will help to distinguish between tectonic and sedimentary disturbance.

Unconsolidated sediments resting on a slope may become unstable, possibly due to elevated pore-fluid pressure in a particular layer in the sediment pile. They may then move down slope under gravity as a coherent mass. In some cases a whole slab of material may detach and move, whereas in other cases the downslope end may stay fixed as the upslope end moves towards it. In both cases there are significant differences between the behaviour at the upslope and downslope ends. The upslope end is subject to a dominantly tensional stress regime, while at the downslope end compressive stress dominates. However, the deformation recorded in slumped sediments is generally complex and must be interpreted with some caution. Slump folds often reflect a compressive regime in a downslope position with fold axes normal to the direction of movement. However, folds also result from lateral compression and also from internal shearing between parts of the slump moving at different rates. Internal shearing commonly leads to the fold axes being rotated so that they lie parallel to the movement direction. Even where folds seem to be compressive, it is important to remember that most slumps are lobate in plan and that fold axes can be expected to show a spread of directions. Upslope zones of slumps, dominated by tensional stress, commonly show a different suite of structures described below.

Slump folding helps our understanding of processes at or soon after deposition. The deformation produced by genuine slumping, involving lateral displacement of material, may be confused with that due both to vertical sinking (i.e. loading) and to tectonic deformation. Vertical sinking, as a rule, produces little or no preferred orientation of fold axes; with tectonic deformation, the folds will commonly relate to or mirror the larger-scale structures of undoubted tectonic origin.

Rotation and displacement of coherent blocks. Displaced relationships between blocks of internally coherent sediment occur on a variety of scales, in a variety of settings and in sediment of virtually any composition. Although many tectonic and sedimentary breccias could be placed under this heading, with many large-scale examples grading into tectonic structures, we deal here with small- and medium-scale examples where movement has been confined to discrete slip planes. Such structures can all be categorised as **synsedimentary faults** whose throws can vary from centimetres to many hundreds of metres. Here we deal with faults in water-lain sediments. Displacement of cohesive masses in aeolian sandstones is dealt with in Section 6.3.4.

Features of this type are so variable that no comprehensive account is possible. A systematic field description of any suspected example should be made with several questions in mind. However, do not expect that they can all be answered fully and unambiguously in every case.

(a) *Is the displacement of early, postdepositional or syndepositional origin or is it tectonically produced after lithification?* When syndepositional, the disturbance may be confined between undisturbed units above and below, and the planes of movement will be sharp with little or no brecciation. Faults produced after lithification usually have quartz or calcite veining and associated brecciation.
(b) *Did the movement take place close to the contemporaneous sediment surface and thereby create a topographic feature?* Draping of a topographic step by overlying sediment, lateral changes of thickness and the gradual upward elimination of relief should be looked for.
(c) *What is the shape, size and spacing of the surfaces of displacement?* Pay particular attention to the vertical and lateral extents of the surfaces. Are they planar or concave upwards? Many small-scale synsedimentary faults are planar with small throws, whereas larger faults, particularly in muddy or silty sequences, are concave-upwards and pass down dip into bedding-plane faults.
(d) *Is there any evidence that movement took place during deposition or was the displacement a discrete event, followed by subsequent deposition?* Here it is necessary to look carefully at patterns of thickness change across the fault (Fig. 9.19).
(e) *Have the various blocks undergone any rotation in the course of displacement?* Careful comparison of dips on either side of surfaces of displacement will help (e.g. Fig. 9.20).
(f) *Are there any minor structures such as drag folds and smaller faults which help to elucidate the nature of the movement?*

Be sure to distinguish those folds related to fault movement from those which pre-date the faulting. Some slumps, which develop folds during ductile deformation, are cut by faults due to later brittle failure.

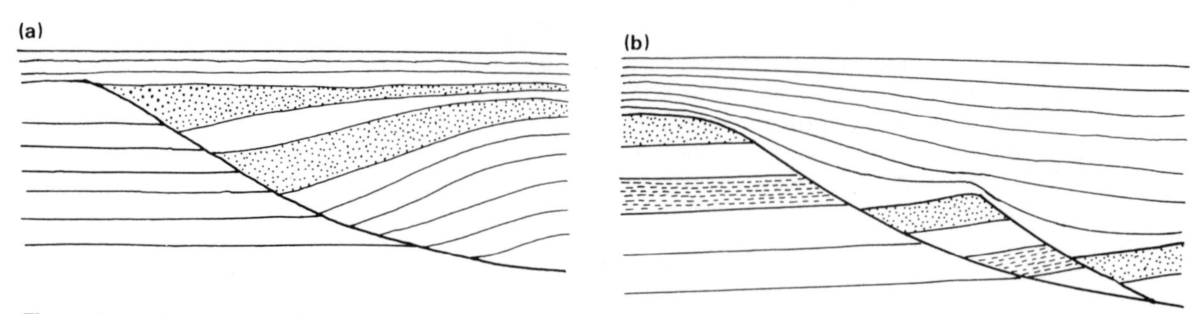

Figure 9.19 Schematic sections through synsedimentary faults: (a) the fault movement has continued during deposition; (b) movement has taken place to the extent that a topography developed on the sediment surface. This was then draped by later deposits.

Synsedimentary faults occur for a variety of reasons. Most have a normal throw indicating at least local tensional stress. It is not possible to give an exhaustive account of the ways in which these structures operate but the following illustrations may be helpful.

Slump scars (9.19b). It was suggested earlier that the upslope end of a slump sheet would be a zone of overall tension. If the sheet moved off in its entirety, it would leave an empty slump scar which would then be draped by later sediment (Fig. 9.21). However, at many slump scars a series of slip surfaces develops, and packets of sediment between these surfaces move only short distances, often rotating to dip up slope in the process (Fig. 9.20).

Collapse features. Small-scale faulting in sands and sandstone can often be due to the collapse of some buried objects in the sand. Logs, masses of vegetation or blocks of ice may rot or melt and the sand then collapses into the resulting space.

Growth faults. Most growth faults are on a scale too large to see at outcrop and their recognition normally requires mapping or geophysical investigation. Small examples have, however, been recognised in extensive exposures (Fig. 9.22). The typical pattern of displacement and thickness change is shown schematically in Figure 9.19a. Faults of this type are typically found in deltaic deposits, particularly those with a high proportion of fine-grained sediment. High depositional rates lead to high pore-water pressure in buried sediment and the consequent sliding of material towards the basin. Collapse

Figure 9.20 Large rotational slide of a thick sandstone bed. The thick sandstone on the left and its underlying fine-grained sediments are *in situ*. The block on the right has slid along a narrow, concave-upwards shear zone and has rotated back into the zone. Figure (arrowed) for scale. Lower Cretaceous, Kvalvågen, Spitsbergen.

Figure 9.21 Slump scar within a sequence of mudstones and graded siltstones. Note how the sediment above the scar is identical with that below and how it drapes the surface. Proterozoic, Kongsfjord Formation, Finnmark, Norway.

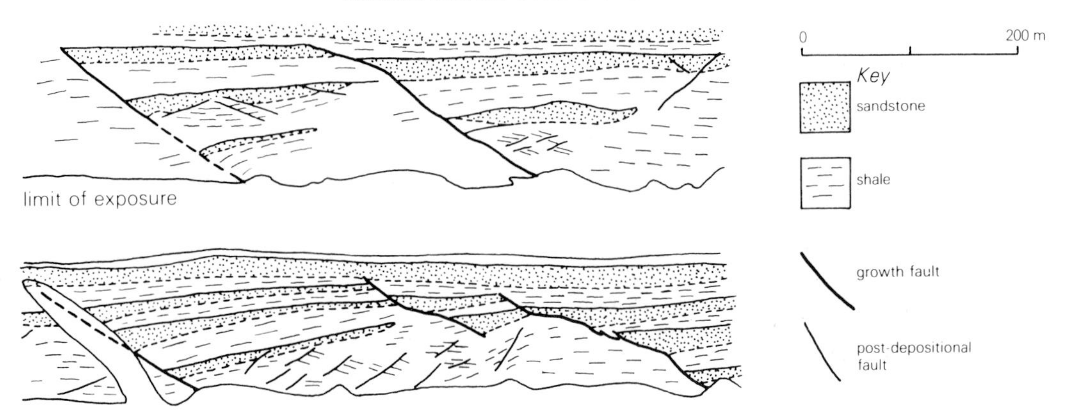

Figure 9.22 A system of growth faults exposed in cliff sections on Edgeøya, Svalbard (after Edwards 1976).

of the pile of delta-front sediments takes place along curved fault planes which pass down slope into bedding-plane faults. At the depositional surface the foundering accommodates a thickening of sediment on the down-thrown side of the fault. When movement ceases, the fault line will be draped and a more uniform thickness pattern will be re-established.

9.3 Chemically induced disturbance

Structures produced by postdepositional chemical activity are variable in their occurrence, mineralogy, and in the nature and timing of the chemical reactions. Three main types of process can be ideally envisaged: precipitation of minerals from pore waters, reactions between host sediment and pore waters, and solution of sediment by percolating water. Although this subdivision is theoretically acceptable, there are, in practice, real difficulties in distinguishing between the products of precipitation and reaction, and the structures produced by these processes are best considered together.

9.3.1 Products of precipitation and reaction (nodules and concretions)

We are not concerned with general, large-scale lithification and cementation but with the local chemical precipitation and reaction within sediments which create structures commonly referred to as **nodules** or **concretions.** These two terms are used interchangeably. Nodules and concretions occur in host sediments of virtually any composition. They commonly stand out clearly because of a contrast in cementation or composition between the concretion and the host sediment. They range in size from large bodies, metres in diameter, down to small bodies of 1 mm or less. Shapes and patterns of distribution of concretions are highly variable. If we can understand the processes of formation of nodules, these may tell us about chemical conditions within a sediment after deposition.

There are five main sets of questions to have in mind when observing nodules:

(a) Is the structure the product of direct precipitation into original pore space or is it the result of reaction between the host sediment and pore waters?
(b) What is the mineral composition of the nodule or concretion?
(c) What is causing the concretionary material to be localised? Is it following particular layers or beds? Is it associated with organic traces such as roots or burrows, or with body fossils? Is it randomly distributed? Is there any systematic variation in the vertical distribution of concretions?
(d) Has the concretion been precipitated to enclose grains of the host sediment (i.e. poikilitic growth)? Has it grown by pushing aside the host sediment (displacive growth)? Has it been precipitated in a large void space of primary or secondary origin?
(e) When did the processes occur?

Precipitation versus reaction. Resolution of this question is not easy, particularly in the field, and in practice it is not terribly important, as most concretions result from a combination of processes. Concretions or nodules in clastic sequences are, on balance, more likely to be formed by precipitation, whereas those in carbonate and other chemical sediments (e.g. evaporites, ironstones) are more likely to have formed through reaction. There are few clear guidelines for making the distinction in the field, and laboratory examination of thin sections will nearly always be needed.

Mineralogy of concretions and nodules. The mineralogy of the material forming or cementing a concretion is our most important indication of the chemistry of pore water during nodule growth. For a particular mineral to be precipitated, the pore waters must be supersaturated with the constituent ions. Other chemical conditions, notably acidity/alkalinity (pH) and oxidation/reduction potential (Eh) must also be appropriate. Figure 9.23 shows the general conditions under which different minerals form, but some caution should be used in its application. Recently, the study of stable isotopes (especially of oxygen and carbon) has allowed the conditions of precipitation of some concretions to be more closely determined. Below we comment briefly on the occurrence and significance of some common, concretion forming minerals.

Calcite ($CaCO_3$). Calcite is one of the most commonly occurring minerals. The calcium carbonate may have been available within the sediment, for example, as shell fragments, or it may have been introduced from outside. Many calcite concretions in clastic sediments are of early diagenetic origin, forming from alkaline pore waters. Calcite also commonly infills larger cavities, particularly in limestone.

Dolomite–ankerite–siderite ($CaMg(CO_3)_2$–$Ca(MgFe)(CO_3)_2$–$FeCO_3$). This family of carbonate minerals forms a continuous series with varying iron content and it commonly occurs in concretions in mudstones and siltstones. Alkaline conditions are required and precipitation of siderite and ankerite favoured by reducing conditions (Fig. 9.23). Evidence of early diagenetic (precompaction) origin is common, suggesting that the requisite chemical conditions obtained soon after deposition. In some coal measure mudstone sequences, siderite occupies a substantial proportion of the thickness, commonly occurring as laterally coalescing nodules or 'beds' which were formerly mined as low-grade ores, the so-called 'clay-ironstones'.

Pyrite and marcasite (FeS_2). Pyrite and marcasite are very similar sulphide minerals that occur in nodules in both clastic and carbonate rocks. They are particularly common in fine-grained, dark-coloured mudstones usually associated with preserved organic matter.

Both minerals reflect strongly reducing conditions within the sediment after deposition, or even at the

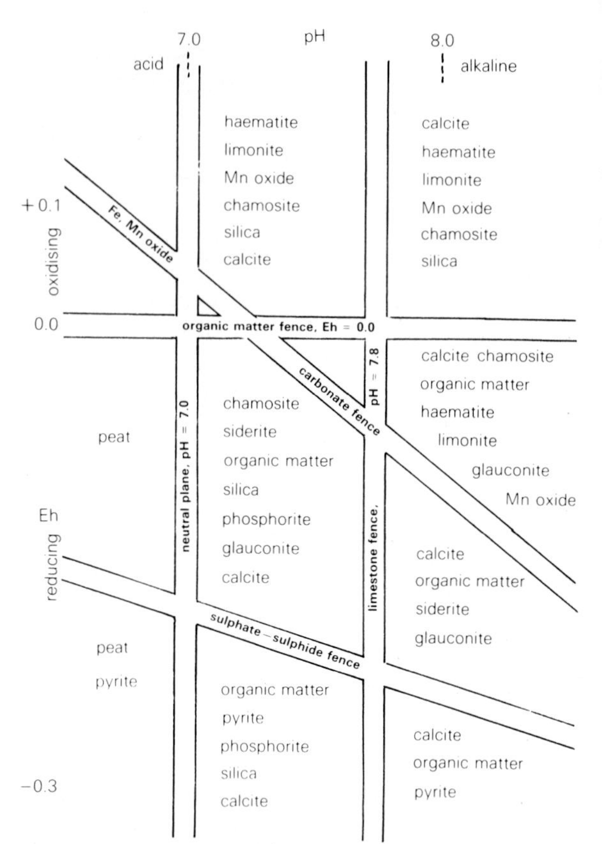

Figure 9.23 Stability fields of commonly occurring authigenic minerals found in sedimentary rocks, in terms of the prevailing Eh and pH. For ground water with an ionic composition comparable to that of sea water. (After Krumbein & Garrels 1952.)

sediment surface in a so-called 'euxinic' environment. Within the sediment, restricted mixing of pore waters with overlying oxygenated water and the action of oxygen-reducing bacteria use up free oxygen. Sulphate-reducing bacteria then take over, producing free sulphide ions which are fixed by iron to give finely disseminated pyrite or marcasite. This gives a black colour to sediment only a few centimetres below the surface in finer-grained parts of present-day tidal flats. Larger pyrite concretions commonly show an internal radial pattern of crystal growth.

Silica (SiO_2). Siliceous nodules or concretions occur both in carbonates, where the silica is commonly cryptocrystalline (chert or flint), and in sandstones, where it occurs as overgrowths on detrital quartz grains. Silica precipitation seems to require weakly alkaline conditions, but the precise controls and processes of flint and chert formation are poorly understood. In carbonates, chert is usually picked out by its darker colour and it is invariably a replacement of the carbonate, not an intergranular precipitate (cf. Fig. 9.25). In sandstones, silica-cemented concretions usually contrast strongly with less well cemented sandstone around them. Usually it is not easy to establish the timing of silica diagenesis, but some nodules in sequences of continental origin compare with the **silcrete** nodules of some present-day soils and, by inference, are thought to have formed at an early stage. In addition to occurring as concretions chert also occurs in thinly bedded units from the lithification of primary siliceous oozes.

Evaporites. Gypsum ($CaSO_4.2H_2O$) and anhydrite ($CaSO_4$) both occur as nodules of early diagenetic origin in highly alkaline conditions at the present-day (Fig. 9.24), for example beneath the surfaces of ephemeral lakes and supratidal flats in hot, arid settings. Comparable structures are common in ancient sediments. The host sediment is usually carbonate, but evaporite nodules also occur in clastic sediments, particularly in muddy siltstones. With these diagenetic evaporites, the problem of precipitation versus reaction is particularly acute, as a carbonate host sediment is relatively reactive. Gypsum can occur poikilitically, enclosing grains of host sediment, and anhydrite usually develops displacively as nodules and layers. In some examples, gypsum precipitation follows plant roots, giving soil profiles called '*gypcretes*'.

In ancient sequences, the textures, fabrics and structures developed by the growth of evaporite minerals may still be recognised even though the evaporite minerals have been replaced by pseudomorphs of more stable phases such as silica.

Figure 9.24 Nodular anhydrite, which grew displacively in a carbonate host sediment. Ordovician, Washington Land, N.W. Greenland.

Haematite (Fe_2O_3). Haematite occurs in nodular forms as well as in its more familiar state as the pigment in red beds. In each case it requires oxidising conditions, although it can persist in slightly reducing alkaline conditions (Fig. 9.23).

Nodular haematite usually occurs in red or partially reduced sequences and commonly the nodules occur in profiles attributable to ancient soil development.

Barite ($BaSO_4$). Barite is quite common as a local cement, particularly in red sandstones where well formed crystals poikilitically enclose the sand grains. In modern deserts **sand roses** form close to the sediment surface. Oxidising conditions and a supply of barium in solution are needed for barite to form in this way.

Limonite ($2Fe_2O_3.3H_2O$). Although many iron minerals weather to limonite, the mineral also appears to form concretions and nodules in its own right, if oxidising groundwater conditions prevail (Fig. 9.23). It most commonly occurs as concretions in sands and sandstones, often cementing the only lithified parts of otherwise unconsolidated sands.

Form and location of concretions. A commonly recurring question is: Why has the concretion formed at this place and in this particular shape? Sometimes the answer is obvious. In other cases only general explanations are possible. The question implicitly assumes that something in the host sediment caused chemical conditions to be suitable for a particular mineral at one place and not at another. Here we outline the most obvious controls, although many concretions appear to be randomly located.

Concretions which roughly follow bedding. In many fine-grained clastic sequences, in chalk and in limestones, concretions or nodules occur in zones parallel to bedding. Sometimes a slight lithological contrast is seen between the concretionary horizons and those which contain fewer or no concretions. Individual concretions tend to be rather flattened parallel to the bedding and, in extreme cases, the concretions coalesce laterally into more or less continuous 'beds'. Common examples of this type are siderite concretions in siltstone and mudstone sequences, the flints in chalks and cherts in more homogeneous limestones and early diagenetic evaporites (Fig. 9.25).

Subtle differences in, for example, organic content or permeability, may control the development of such concretions. Compositional differences may allow a slightly different pore-water chemistry to develop, whereas permeability controls the rate at which pore

Figure 9.25 Diagenetic nodules following particular bedding horizons. (a) Chert nodules in limestone, seen in cross section; Portfjeld Formation, Lower Cambrian, northeast Greenland. (b) Calcite cemented concretions in siltstone seen in three-dimensional exposure; Lower Jurassic, Yorkshire.

water passes through the sediment. The nucleation of individual concretions must depend on even more subtle inhomogeneities in the sediments.

Concretions which follow burrows. Elongate and irregularly shaped concretions, particularly those that show branching patterns and cut across bedding, usually follow burrow traces. Often it may be possible to extract the cemented burrow from loose sediment. Concretions centred on burrows are commonly of flint in chalk and of limonite in poorly consolidated sands.

Concretions which follow rootlets. In seat earth palaeosols in coal measures, concretions are commonly associated with rootlets which penetrate the bedding. The concretions tend to be elongate normal to bedding. Mostly they are composed of siderite which weathers to limonite, but often small pyrite crystals fringe the concretion and are also scattered within it. The concretions follow the traces of thicker roots, whereas the more common thin rootlets are preserved as carbonaceous films. The association of thin carbonaceous rootlets usually enables these root concretions to be distinguished from those that follow burrows.

In other fossil soils, concretions of limonite, silica, gypsum and/or haematite or simply a general colour mottling follow the root traces of plants.

Concretions centred on body fossils. There are three main ways in which fossils provide a locus for the development of concretions. First, the fossil itself can become a concretion when concretionary material replaces and exactly replicates the fossil as, for example, in pyritised ammonites or in flint echinoids in chalk. Secondly, the fossil forms a nucleus for the precipitation. When broken, many ellipsoidal or more irregular carbonate concretions show fossils in their centres. The rotting organism provided local chemical conditions in the pore water which favoured precipitation. Thirdly, and more rarely, a concretion can occur

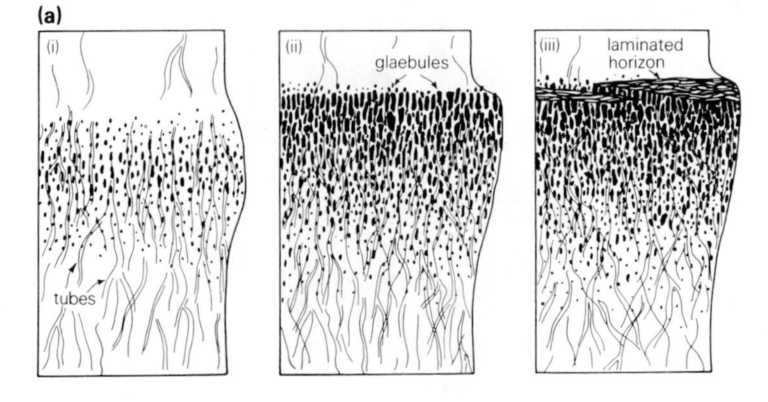

Figure 9.26 Carbonate concretions developed as a result of soil-forming processes to give calcrete or caliche profiles. (a) Stages in the development of a mature profile; (i) is the earliest stage and (iii) represents a mature profile. Examples are preserved in the rock record at all stages (after Allen 1974). (b) A fairly mature profile corresponding roughly to stage (ii) of figure (a). (c) A profile showing the development of veins linking nodules and showing the catenary pattern known as 'pseudoanticlines'. Both (b) and (c) are from the Lower Old Red Sandstone, Llanstephan, Dyfed, Wales.

around the position occupied by the soft parts of an animal while the hard parts remain undisturbed. Irregular masses of pyrite occur sometimes at one end of a belemnite guard, in the position where the animal's soft parts would have been. Rapid deposition or a rather inhospitable sediment surface are needed to allow the soft parts to become buried before scavengers consume them. Once buried, anaerobic rotting would lead to the reducing conditions necessary for pyrite precipitation.

Concretions in distinct vertical profiles. Concretions, particularly of calcite, sometimes occur within red siltstones and sandstones in distinct vertical profiles ranging in thickness from decimetres to several metres. Similar profiles also occur in limestones, where they are often associated with karstic features (see 8.4.2). The concretions, which are commonly a few centimetres in dimensions, have irregular shapes but are, in many cases, elongated vertically. Some profiles have only scattered nodules (Fig. 9.26a), while others show an upwards increase in nodule size, abundance and coalescence, often as meshes and networks with linking veins (Fig. 9.26b). In rather rare examples, the upper part of the profile is a more or less continuous bed of limestone with a crude horizontal lamination. These carbonate-rich layers tend to be laterally continuous and to maintain their character over long distances. As such they are often useful for correlation at least over short and intermediate distances. In some examples, the upper surface of the profile shows a relief of broad cuspate curves which also may relate to radial patterns in the subvertical concretions and to convex-upwards veins within the carbonate network. These structures have been called **pseudoanticlines**.

These profiles and associated features compare closely with those of present-day soils of certain semi-arid areas, the so-called **calcrete** and **caliche** soils. These result from the vertical movement of water through the sediment due both to the downward movement of rainwater and the upward movement of ground water under dry evaporating conditions. Some of the vertical nodules (or **glaebules**) may reflect original plant roots, while others may follow fractures. The more continuous limestone layers with lamination record the development of **hard-pan** conditions in a fully mature profile. For calcretes developed in siltstones and sandstones, there is a possibility that the calcium was introduced as wind-blown dust, as either carbonate or sulphate, while an internal origin is clearly likely in limestones. The development of calcretes requires that the sediment surface is exposed, with little or no sedimentation for a considerable period of time, probably measured in thousands of years. The maturity of the profiles reflects the relative durations of the non-depositional intervals within any particular sequence. The correct identification of this type of profile is clearly important in environmental and palaeoclimatic reconstruction. Not only do the profiles indicate prevailing conditions but they also enable us to detect fluctuations in depositional rate. Other similar profiles involve silica, haematite and gypsum nodules giving **silcrete, ferricrete** and **gypcrete** soils respectively. (For more detailed discussion see Allen and Goudie in *Study list*.)

Figure 9.27 Cone-in-cone structure in carbonate-cemented mudstone. Probably Upper Carboniferous Coal Measures, location unknown. Scale is 5 cm long.

Mode and timing of growth of concretions. Deductions about the way in which concretions grew within the host sediment will be made mainly from looking at internal features of the concretions. They may also give information on the timing of concretion development and on the degree of compaction that the sediment has subsequently undergone. There are four main ways in which concretions and nodules develop.

Primary pore filling. Evidence for this is the occurrence within the concretion of original sediment grains and original bedding features. Such concretions occur mainly in clastic sequences. In mudstones, it may be possible to see the original lamination running through the concretion and linking up with lamination in the host sediment on either side. If this is observed, compare the thickness of the lamination inside the concretion with that in the host sediment and also note how the lamination in the host sediment behaves around the concretion. The lamination is commonly much thicker within the concretion than outside it and the lamination in the host sediment drapes around the concretion for some distance above and below. Clearly a certain amount of compaction has taken place after the concretion formed and comparison of lamina thicknesses can give a measure of its magnitude. When compaction is demonstrably large, it is a fair inference that the concretion formed quite soon after deposition.

In some concretions in mudstones, particularly those with a carbonate cement, the central parts of concretions show a pattern of irregular lenticular cracks filled with coarse calcite crystals. These are **septarian nodules** and the cracking reflects a synaeresis-type contraction (see p. 148) due to the dewatering of a gel-like mass of clay minerals. In other concretions a curious angular pattern of fractures is found around the margins involving stacked conical fracture surfaces. This is **cone-in-cone** structure (Fig. 9.27), which reflects the stress field set up by the growth of the concretionary cement.

In sandstones, early carbonate concretions often weather out as holes due to a later phase of general cementation by silica which renders the carbonate more susceptible to weathering than the host. It is often possible to obtain a clearer impression of the original shape and packing of the sand grains in partly weathered carbonate concretions where the silica cement is not present.

Displacive nodule growth. Here the host sediment is physically pushed aside as the nodule grows, and little or none of it is incorporated in the nodule. Internally, this may sometimes be apparent from the crystal structure of the nodule as with those pyrite nodules with a radial fabric. One of the most common examples of displacive growth is that of gypsum and anhydrite nodules which occur both as layers and as more irregular forms. As the nodules grow, they push apart the host sediment until only thin remnants remain between the nodules, leading to the so-called **chicken-wire texture** (Fig. 9.28). Growth of anhydrite as layers sometimes sets up stresses that produce highly contorted folding within the layer.

Nodule growth by replacement. This is not always easy to recognise. The replacement is, in some cases, associated with displacive growth, but in other cases, details of the internal structure of the host sediment are preserved even though total replacement has taken place. In addition to showing vestiges of the original lamination, replacement nodules commonly have reaction rims where the process of replacement has not gone to completion. If the replacement is complete, concretions formed by reaction between host sediment and pore water can have many of the attributes of displacive nodules and the two processes are not mutually exclusive.

Concretions due to cavity infill. Cavities within sediment may be of primary or secondary origin and are most common in limestones. An example of a primary cavity is the body chamber of a shelled organism where the strength of the shell maintains the cavity until the fill has been precipitated (cf. Fig. 8.6). A secondary cavity could result from the solution or rotting of some object, probably after the host sediment was partially lithified so as to support the cavity. Infills of voids are characterised by well formed, pure crystals growing inwards from the walls. Concentric zones may show crystals of increasing size from wall to centre. Two types of cavity fill of particular interest, **geopetal infills** and **stromatactis** were dealt with in Chapter 8 (see Section 8.3.1 and Figs 8.6 & 7).

Figure 9.28 Anhydrite nodules which have grown displacively to push aside the host sediment into thin veneers between the nodules, giving so-called 'chicken-wire' texture. Cape Webster Formation, M. Ordovician, Washington Land, Greenland.

9.3.2 Products of dissolution

Two types of dissolution effect occur commonly in limestone sequences.

Figure 9.29 Stylolites in chalk. The crenulated surfaces are developed normal to the principal compressive stress and usually follow bedding. They commonly show concentration of insoluble residues. Chalk, Upper Cretaceous, Yorkshire.

Stylolites. On certain bed partings, particularly in limestones but also rarely in sandstones, a highly crenulated contact is seen (Fig. 9.29). In vertical section it may be possible to detect that the surface is defined by a thin layer of clay. Relief is usually a few millimetres and seldom more than a few centimetres. When such a surface is exposed in plan, the small-scale relief is seen to be highly irregular.

Stylolites result from dissolution of both upper and lower beds at a particular bedding surface. The beds grow into each other due to the vertical compressive stress. The relief of the irregularities gives a minimum measure of the thickness of material removed in solution. This can sometimes represent a high proportion of the original rock. The clay parting that occurs on some surfaces represents the insoluble residue from the solution of the limestone.

Collapse breccias. In sequences that contain or may have contained evaporites, brecciated horizons sometimes occur. Angular blocks may be of any size, but they are similar in composition to other rocks in the sequence. Often blocks will be shifted only slightly relative to one another, and it is possible to restore them mentally to their original positions, as in a jigsaw puzzle. Such breccias may often be the result of the wholesale removal in solution of a thick evaporite unit, so that the only evidence we have of its former existence is the breccia caused by the collapse of overlying sediment.

9.4 Biogenic sedimentary structures: trace fossils

9.4.1 Introduction

The study of trace fossils (**ichnology**) is concerned with understanding disturbance of sediment by living organisms, i.e. **biogenic** sedimentary structures. Apart from the consistent recognition of vertebrate (e.g. dinosaur) footprints from 1828 onwards, trace fossils were at first grouped as 'fucoids' (fossil seaweeds), and their algal origin was hotly debated. Until quite recently, trace fossils have been ignored in most geology courses. At outcrop they have commonly been dismissed as 'burrows' or 'worm traces', suggesting that they have little to contribute to the elucidation of Earth history. In fact the reverse is often true. Trace fossils, in contrast to many body fossils which are rolled and derived, are records of life and events that took place *in situ* during or soon after the deposition of the sediment. They often occur where no body fossils have been preserved, for example in non-marine red beds, or where organisms were entirely soft bodied. Trace fossils record behavioural, ecological and sedimentological events which body fossils and other sedimentary structures cannot highlight directly. Their study may alter one's view of a problem or turn an investigation in a new direction. Sometimes they provide the key to what, at first, appear to be purely sedimentological problems, for example the origin of massive beds.

Even where body fossils and sedimentary structures are abundant, trace fossils should be studied with equal rigour, for they yield information against which to test all manner of conjectures and speculations. They may inspire working hypotheses that would otherwise not be considered. They encourage the study of sediments from biological, ecological and biochemical standpoints, so complementing the study of the physical origin of structures. Students working on certain sedimentological projects should not be afraid to consult colleagues with an ichnological background, rather than remain narrow and ignore a major source of information and ideas.

The beginnings of such an approach lie in gaining experience of traces produced in sediments by animals and plants that are living at the present day. Many worthwhile projects are possible at the coast and in rivers, but failing this, certain films (e.g. Farrow) and specialised books may help (e.g. Schaefer, Frey, Basan *et al.*, Reineck & Singh). The record of past events provided by trace fossils in rocks is, however, often more informative than that derived from the present. Present-day traces have not yet been studied sufficiently widely and critically and it is commonly easier to see certain trace fossils (e.g. those in subtidal and abyssal settings) in solid rocks.

9.4.2 Observation and recording in the field and in the laboratory

The study of trace fossils requires one to try to relate fragmentary two-dimensional patterns to complex three-dimensional records of behaviour left by a diverse range of organisms. A wide range of techniques has been developed. We concentrate here on cheaper, simpler techniques which rapidly enlarge experience.

In the field. In present-day subaerial and intertidal environments direct and 'after-the-event' observation is possible: in subaqueous settings observation is more costly as diving equipment and underwater cameras or television may be needed. Whatever the environment, students should define a problem and plan an appropriate programme of sampling, description and analysis of the environment. Estuaries provide fairly safe and accessible locations for a variety of case studies. Such exercises can also develop skills such as plane-tabling, aligning transects, siting quadrat surveys, sampling sub-environments for sediments as well as for organisms and the records of their activity, photographing evidence to scale, orientating data, drawing scaled diagrams and collecting and curating samples (see Ch. 10 for further details). Other useful techniques include the making of box cores, the taking of vertical and horizontal peels using lacquer, polyester resin and expoxy resin, and the casting of burrows, both subaerially and underwater (see Appendix 2).

In dealing with trace fossils in rocks, many of these techniques are inappropriate. Others are comparable, but may need to be augmented and adapted. The drawing of scaled field diagrams and the photographing of trace fossils may be helped by outlining inconspicuous features with a felt-tipped pen or chalk; burrows may be accentuated by wetting a rock surface by glycerine and water, paraffin or light mineral oil or smearing it with small quantities of ink (whereupon uptake of stain is controlled by differences in porosity); delicate scratches and fine detail should be whitened with powdered chalk or ammonium chloride, and photographed in strong oblique light.

Graphic logs of sections should include data on occurrence and distribution of fossils (Fig. 10.6).

In the laboratory. The following procedures may be appropriate in the laboratory, particularly with respect to borehole cores:

(a) The making of peels from box cores (Fig. 9.30).
(b) Designing and carrying out experiments suggested by field observations to test ideas on the origins of marks of uncertain origin. Even small laboratories can maintain fresh- or sea-water aquaria.
(c) Staining of fine-grained carbonate and clay mineral-rich rocks by organic dyes such as alizarin red, methylene blue, or Indian ink.
(d) Making acetate peels by polishing a cut surface and etching it with acid, then applying acetone and covering this with an acetate sheet which, when adherent, can be peeled off.
(e) Subjecting 1 cm-thick sawn blocks, whether naturally cemented or impregnated, to X-radiography or infrared and ultraviolet photography. These techniques are helpful in revealing at least traces of structures where none appears to exist. Massive beds may show intense bioturbation, i.e. maximum rather than minimum organic activity; this, supplemented by diagenetic effects, enhances their apparent homogeneity. Infrared photography is cheap in that it only requires a special film and filter, although the cutting of thinner slabs (0.5 cm), which gives the best results, is difficult. Exposure time should be proportional to the organic content of the rocks, arenaceous ones being more transparent than the argillaceous. Ultraviolet photography is best applied to limestones which contain little iron.
(f) Artificial weathering of apparently homogeneous rocks for five minutes using sand-blasting equip-

Figure 9.30 An epoxy resin peel of a trace fossil. (After Farrow in Frey 1975).

ment with an abrasive of unsorted sand slightly finer than the grain size of the rock.

(g) Making thin sections of impregnated sediment or rock. These should be made larger (about 5 x 5 cm) and slightly thicker (0.04 mm) than normal, whereupon they can be mounted in a slide projector. Thin sections may be stained to good effect (see (c), above).

(h) Hand specimens may be photographed in ways appropriate to work in the field. They may be further described with respect to the features which form the basis of the classifications set down in Sections 9.4.3, 9.4.4, 9.4.5.

9.4.3 *Classification of trace fossils*

We emphasise a practical rather than an academic approach to the complex task of describing and interpreting outcrops that contain trace fossils. Nevertheless, there are three main approaches to classifying trace fossils, using:

(a) morphological aspects: the taxonomic approach;
(b) preservational–sedimentological aspects: the toponomic, stratinomic approach;
(c) behavioural–environmental aspects: the ethological approach.

Taxonomic classification. A beginner may be bewildered by not being able to classify a set of trace fossils satisfactorily, even though he has access to appropriate journals and the relevant *Treatise.* Systematic ordering of trace fossils according to the taxonomy of the organisms producing them would be highly desirable, but even experts disagree on how to do this. The problem is that one trace fossil 'genus' may cover traces made by several different organisms, and several 'genera' may be made by the different activities of one organism. International codes of biological nomenclature are to be modified in order to accommodate the naming of trace fossils (see *Further reading*). Some workers despair even of this. In this connection, always try to remember that the basic unit of nomenclature is the **ichnogenus** defined morphologically, and that the lower category, the **ichnospecies,** often reflects size, preservational or minor behavioural variations of the basic form.

More worthwhile questions to ask in the field usually relate to the nature and process of preservation of the traces, and to the possible behavioural significance of the features.

Classification according to mode of preservation. When trace fossils are analysed sedimentologically, it may become clear that their morphology can be explained in terms of animal behaviour and also be related to relatively few sedimentary and diagenetic processes. The study of the preservation of trace fossils is termed **toponomy.**

You should, therefore, gain some experience of describing the common modes of preservation, while trying to relate these to possible interpretations of plant and animal activity. Diversity of forms of traces arises from the activities of differently shaped animals with different behaviour patterns. Remember to look systematically under, within and on top of the bed and for features that cross-cut the bedding, then decide which way is way-up (cf. Fig. 9.34). Bear in mind, however, that the time relationships which hold for the generation of sedimentary structures by physical processes do not necessarily hold for trace fossils; features at the base of a bed or within it may be caused by burrowers that postdate the deposition of the bed, and the timing of events may be hard to establish.

Key questions about the morphology and mode of preservation of trace fossils. These sets of questions should encourage students to think of each set of trace fossils in *three* complementary ways and hence should help them to observe, describe, measure and record them most effectively. In doing so, students should conjecture and hypothesise about the possible organisms and processes that gave rise to them. Figure 9.31 presents a lower bedding surface and a cross-section of what is apparently a set of trace fossils. Try using the following questions to help understand these features further.

The first set (Q.1) encourages description of the morphology of preserved structures; the middle group (Q.2 & Q.3) the mode of preservation, and the last group (Q.4–Q.8) the position and process of preservation.

Q. 1 *What is the morphology of the trace fossil?* Are there identifiable shapes of organisms or parts of them? Is the trace best described as:

(a) a single shape (e.g. a print or track made by a foot);
(b) several similar shapes repeated to form a pattern (e.g. a track made during locomotion);
(c) a trail (i.e. a continuous groove made during locomotion);
(d) a radially symmetrical shape developed in a horizontal plane (e.g. by the resting of a starfish);
(e) a tunnel or shaft caused by a burrower seeking food and/or refuge;

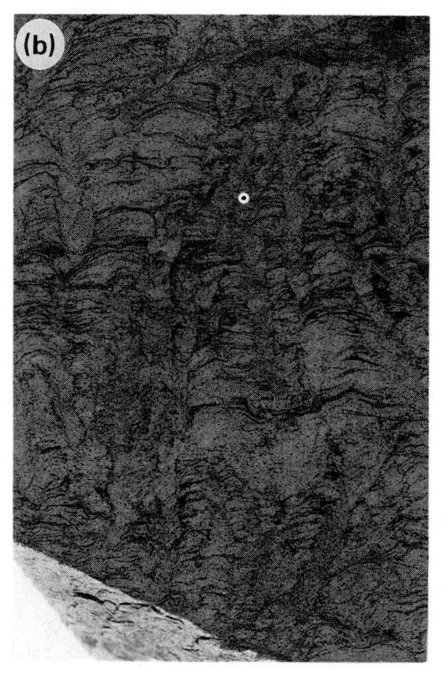

Figure 9.31 Two views of a set of suspected trace fossils; (a) on a lower bedding surface and (b) in vertical cross section. Traces are about 7 mm wide. Haslingden Flags, Upper Carboniferous, Central Lancashire (see Fig. 9.39(o) for names of traces).

(f) a series of spreiten, which are U-shaped, closely related, concentric laminae caused by an animal shifting the location of its burrow as it grows or moves upwards, downwards, forwards and backwards by excavating and backfilling (Fig. 9.32).
(g) a pouch shape, for example caused by the resting of bivalves;
(h) a network pattern.

For further descriptive details relating to each of these categories, see Table 9.1. Attempt to relate the trace fossils depicted in Figures 9.39 and 9.40 to these categories.

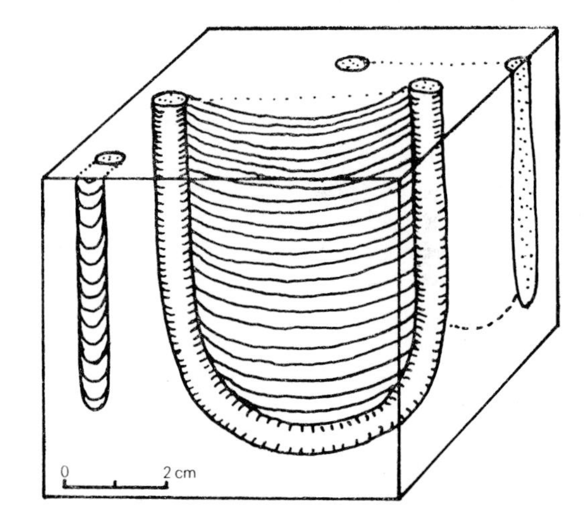

Figure 9.32 A trace fossil with spreiten. The U-tube is the dwelling burrow of an organism. The arms of the U indicate that the organism lengthened and deepened the tube by removing sediment from the floor of the burrow and plastering it against the ceiling, so forming a concentric 'spreite'. The spreiten are protrusive, the present burrow being the last formed and underlying all previously formed ones.

Q. 2 *Is the trace fossil preserved as a cast or mould?* The best way to appreciate the difference between these two modes of preservation is to make moulds and casts in the laboratory using plasticine and a fill of dental plaster; a mould and cast of a hand- or footprint for example. Is there evidence that the fill was **passive**, i.e. by normal sedimentation, or active by, for example, the back-filling action of a burrower (see Fig. 9.33)?

Q. 3 *Is the trace fossil preserved as a diagenetic concretion? Chondrites, Rhizocorallium, Thallassinoides* and *Ophiomorpha* (Figs 9.39 & 40) are often preserved as calcite and siderite nodules in shale or limonite nodules in sand. Small-diameter burrows are often preserved in pyrite, which oxidises to red-brown goethite, in flint or chert (e.g. crustacean burrows in the Chalk, or in collophane-cemented nodules (e.g. shrimp

Table 9.1 A morphological classification of some common invertebrate trace fossils (after Simpson in Frey (ed.) 1975 p.43). Examples which are illustrated in the figures are marked by an asterisk *.

Rank IV	*Rank III*	*Rank II*	*Rank I*	*Examples*
(a) Track-like trace on bedding plane		a. 'Prods' or 'scratches' all alike	(1) Clustered 'scratches'	*Paleohelcura*
			(2) Rows of 'prods'	*Tasmanadia*
		b. 'Prods' or 'scratches' of different kinds	Rows of 'prods'	*Kouphichnium* phyllopod tracks*
(b) Trail-like trace on bedding plane	1. Freely winding	a. Simple trail	(1) No ornament	*Gordia*
			(2) Transverse ornament	*Climatichnus*
		b. Bilobed trail	Transverse ornament	*Cruziana**
		c. Trilobed trail	Transverse ornament	*Scolicia**
	2. Windings in contact with one another; pattern on bedding plane	a. Simple trail	No ornament	*Helminthoidea**
		b. Bilobed trail	Transverse ornament	*Nereites**
(c) Radially symmetrical in a horizontal plane	1. Without axial vertical structure	a. Five-rayed	Rays are grooved	*Asteriacites**
		b. Multirayed	Club-shaped rays	*Asterosoma*
	2. With vertical axial structure	a. Circular outline	Conical depression	*Histioderma*
		b. Multirayed	Radial branches	*Lennea*
(d) Tunnels and shafts	1. Of uniform diameter	a. Vertical	(1) Isolated	*Tigillites*
			(2) *En masse*	*Skolithos**
		b. Horizontal	Winding	*Planolites**
		c. U-shaped		*Arenicolites**
		d. Regularly branching		*Chondrites**
	2. Variable diameter	Irregular network		*Thalassinoides**
(e) Forms having a spreite		a. U-shaped	(1) Vertical plane	*Diplocraterion**
			(2) Horizontal plane	*Rhizocorallium**
		b. Spiral	Inclined plane	*Zoophycos**
		c. Branched	Vertical	*Phycodes**
(f) Pouch shaped		a. Smooth surface		*Pelecypodichnus**
		b. Transverse ornament		*Rusophycus**
(g) Miscellaneous		Net pattern		*Paleodictyon**

burrows) in present-day conditions. These features are often distinguished by burrow margins with different chemical and physical compositions. They have been produced by the ingestion of clay-sized silicates and secreted as colloidal organic compounds, rich in Ca, Mg, Na and traces of Cu and Fe, which are made to bind the quartz sand grains of the burrow wall. Alternatively, in dark shales produced in reducing conditions, the water-pumping activities of animals may give rise to lighter-coloured 'haloes' around the nodular traces. Despite this, it is often difficult to distinguish burrows from traces of plant roots.

Q. 4 *Is the trace fossil preserved in an interfacial position on the top of the casting medium as an epichnial trace like a ridge (positive feature) or a groove (negative feature)?* (Fig. 9.34). What is its composition? Are there any marks on the top or bottom of the ridges and grooves?

Q. 5 *Is the trace fossil preserved in an interfacial position on the bottom of the casting medium as a hypichnial trace, e.g. a ridge or groove?* (Fig. 9.34). If so, is there any evidence that this was a sediment/water interface? Are only the sub-interface laminae deformed? Was the trace fossil preserved at a sediment/sediment interface, possibly between contrasting lithologies, possibly at a concealed junction? What is the composition of the underlying and overlying beds? Are the underlying *and* overlying laminae deformed?

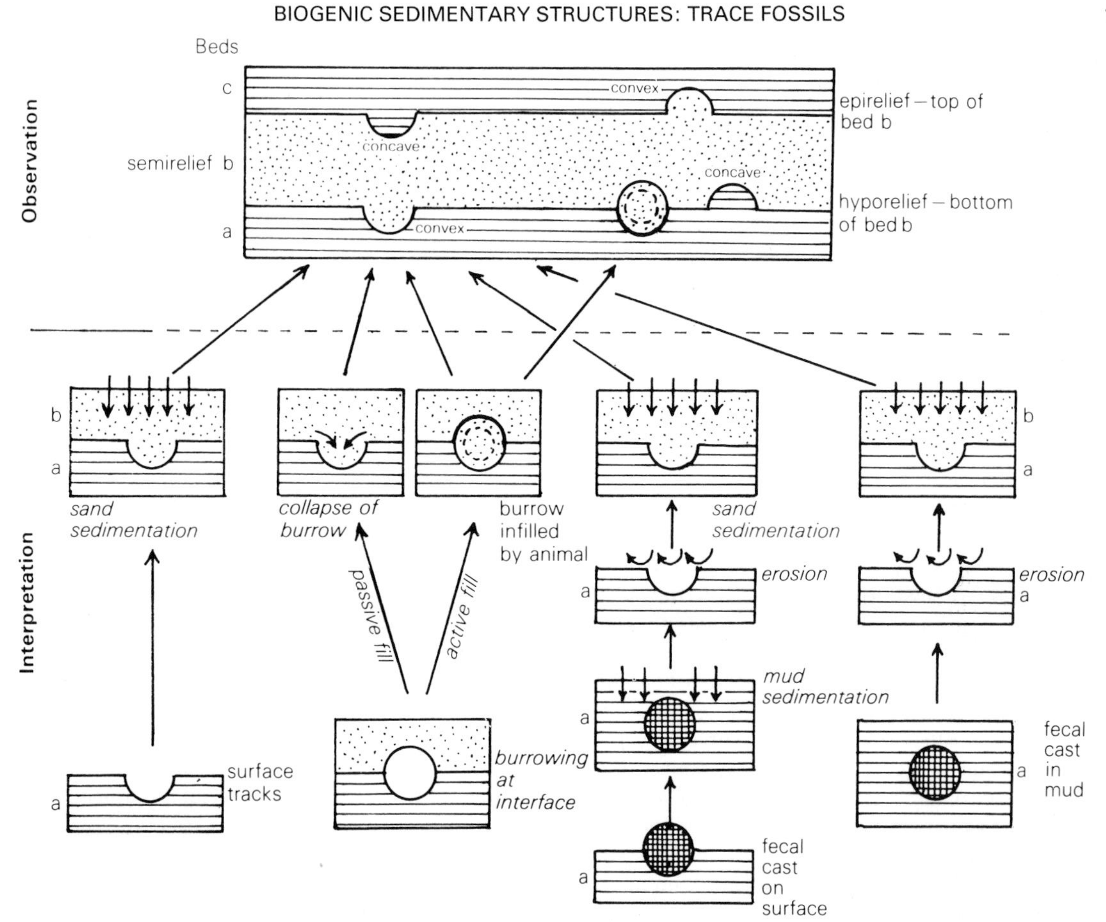

Figure 9.33 A classification of trace fossil preservation types and their interpretation (modified after Seilacher (1964), Webby (1969) & Hallam in Frey (1975)).

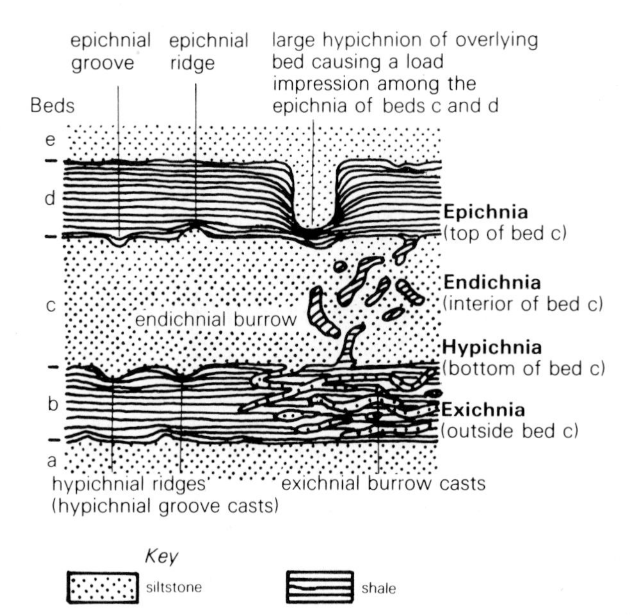

Figure 9.34 A diagrammatic classification of the modes of preservation based on the nature of the main casting medium according to Martinsson (1970). The four key terms to the right of the diagram refer to the trace fossils relating to bed c (modified after Basan in Basan 1978).

Q. 6 *Is the trace fossil preserved within a bed but outside the main body of the casting medium as an exichnial trace?* (Fig. 9.34). Here the traces of one lithology (e.g. sandstone) are isolated in a different lithology (e.g. shale). A sharp upward termination of the fill might suggest a former connection of the burrow fill to a bed of sand which has subsequently been removed by erosion, i.e. a concealed bed junction.

Q. 7 *Is the trace fossil preserved in an internal position with the main body of the casting medium as an endichnial trace?* (Fig. 9.34). *What kind of 'ichnofabric' is present?* Are the burrows very densely distributed and interpenetrating? If so, the sediment should be referred to as having a **bioturbate texture**. Are the burrows common but distinct? If so, the term **burrow mottling** may be more appropriate (see Fig. 9.35a). Are the structures preserved in full relief? Is the wall of the cast of different composition from the body of the cast, as when a burrow in sand is lined by a layer or layers of mucus and/or faecal pellets made of mud. Does the trace contain internal structures, e.g. spreiten? Try to distinguish cross-cutting relationships and work out whether the fabric is best interpreted in terms of the evolution of the burrows with time or different communities living contemporaneously at different depths. Try to quantify the numbers of individual trace fossil 'species' and the amount of bioturbation.

Q. 8 *Is the trace preserved by burial following erosion, i.e. is it a derived trace fossil?* This arises when, after burrowing, erosion takes place and currents winnow away a soft matrix leaving the mucus-bound burrow linings as sediment-filled 'gloves'. These can be covered by later, possibly different, sediments (Fig. 9.37). Alternatively currents may scour out burrows made in mud and afterwards fill them with sand. Bored pebbles and pieces of bored wood may be reworked as clasts into younger sediment.

In order to consolidate the ideas developed so far, attempt to use combinations of the terms introduced above to describe and interpret the features depicted in Figure 9.36.

Preservation potential. Processes of sedimentation strongly influence the trace-fossil assemblage which is preserved, thereby producing a physically induced bias. The majority of biogenic traces, particularly the epichnial ones on the upper surface of beds at the present day, have almost zero fossilisation potential. Trace-fossil associations dominated by surface or near-surface traces indicate more or less continuous sedimentation; those dominated by traces made at depth below the sediment surface indicate discontinuous sedimentation with interspersed periods of erosion. **Protrusive spreiten**, marking sets of burrow fills where the last-formed burrow underlies all earlier ones, reflect adjustment of the animal's position in response to erosion at the sediment surface or to growth. **Retrusive spreiten**, or nested cones marking sets

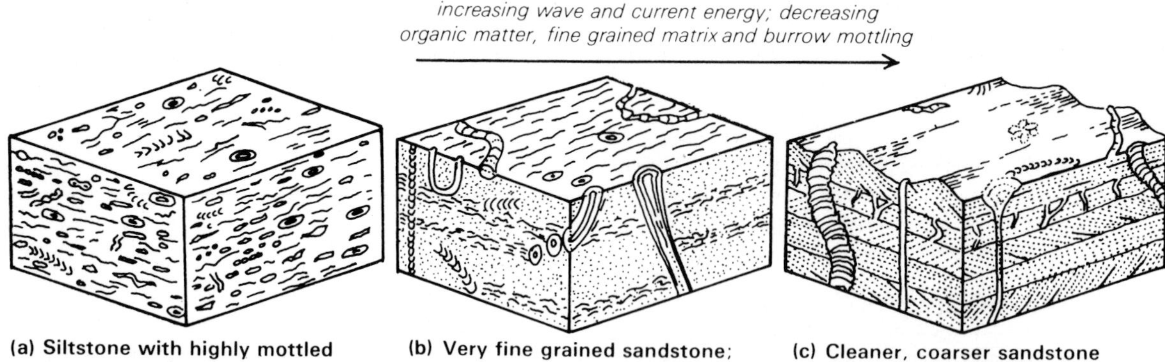

Figure 9.35 Diagrams to show the changing pattern of bioturbation in response to increasing energy in the depositional environment from (a) to (c). (After Howard in Frey 1975 and Howard in Basan 1978.)

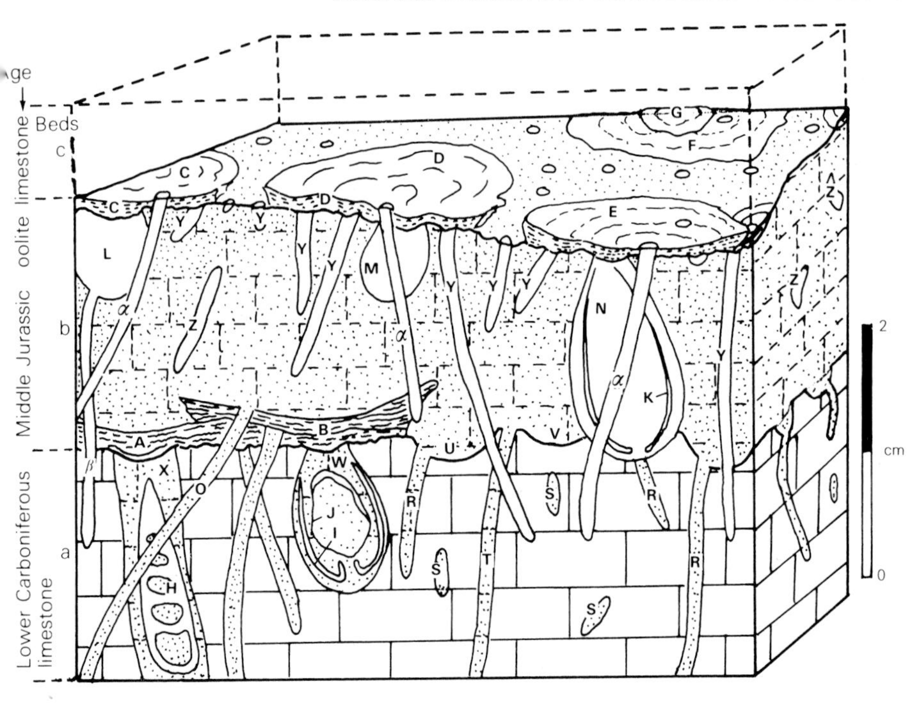

Figure 9.36 An exercise on describing the mode of preservation of some trace fossils depicted in a diagram (after Bromley in Frey 1975) which shows a vertical cross section of a hardground. 1, identify the body fossils to group level; 2, describe the morphology of the trace fossils (Table 9.1); 3, describe the modes of preservation of the trace fossils using the classifications of Seilacher-Webby and Martinsson (Figs 9.33 & 34); 4, describe the mode of behaviour of the trace fossils (Fig. 9.38); 5, write a stratigraphic history of the rocks depicted in the diagram using the letters a–c to identify the beds and A–Z, and α–γ, to identify the body or trace fossils.

of burrow fills where the last-formed burrow overlies all previous ones, reflect adjustment to increased sedimentation as the animal moves upwards. Preservation of delicate individual structures made by some organisms depends on a lack of later, more general, bioturbation. The preservation potential of shallow or near-surface traces is greatest in low-energy settings.

Attempt to describe and interpret the broad sequence of events that gave rise to the preservation of the traces shown in Figure 9.37, including the recognition of protrusive and retrusive spreiten.

Classification according to behaviour (an ethological classification). Certain types of trace-producing behaviour are common to several groups of organisms. Five or six basic patterns of behaviour are recognised and these are shown diagramatically in Figure 9.38. Although there is some overlap between the basic behavioural categories, the scheme is useful in understanding the origins and interrelationships of fossil and recent traces.

Key questions concerning the behaviour of trace-producing organisms. Further sets of questions (Q.9–Q.15) may be asked about assemblages of trace fossils in order to encourage conjectures and hypotheses about possible behaviour patterns. These can be tested against experience of how present-day organisms graze, crawl, rest, burrow, feed at depth and escape.

Q. 9 *Could the traces be the result of animals temporarily interrupting their locomotion on or above the sediment surface to rest or seek refuge?* Isolated, shallow, trough-like depressions may record the outline or morphology of the undersurface of an animal, or marks caused by its settling into sediment, or burying itself just under the sediment surface, or by its leaving the sediment. Such resting traces, called *Cubichnia,* are made by epibenthic and nektobenthic mobile animals (starfish, bivalves, arthropods like crabs, and flat fish, among others). Examples are *Asteriacites, Pelecypodichnus, Rusophycus* (Fig. 9.39). Traces are transitional to crawling, dwelling and escape structures (Fig. 9.38).

Q.10 (a) *Could the traces be produced by locomotion of animals on the sediment surface or along an interface?* Features which might suggest this are footprints, trackways, trails and horizontal burrows. Their linear, sinuous or branched sculpture reflects movement of walking limbs, bristles or other appendages, or the muscular movements of a body, or the dragging of a shell. Could different, but apparently related, traces be made by animals walking, galloping, hopping, crawling, half-walking, half-swimming? Could the traces be caused by an animal being drifted by

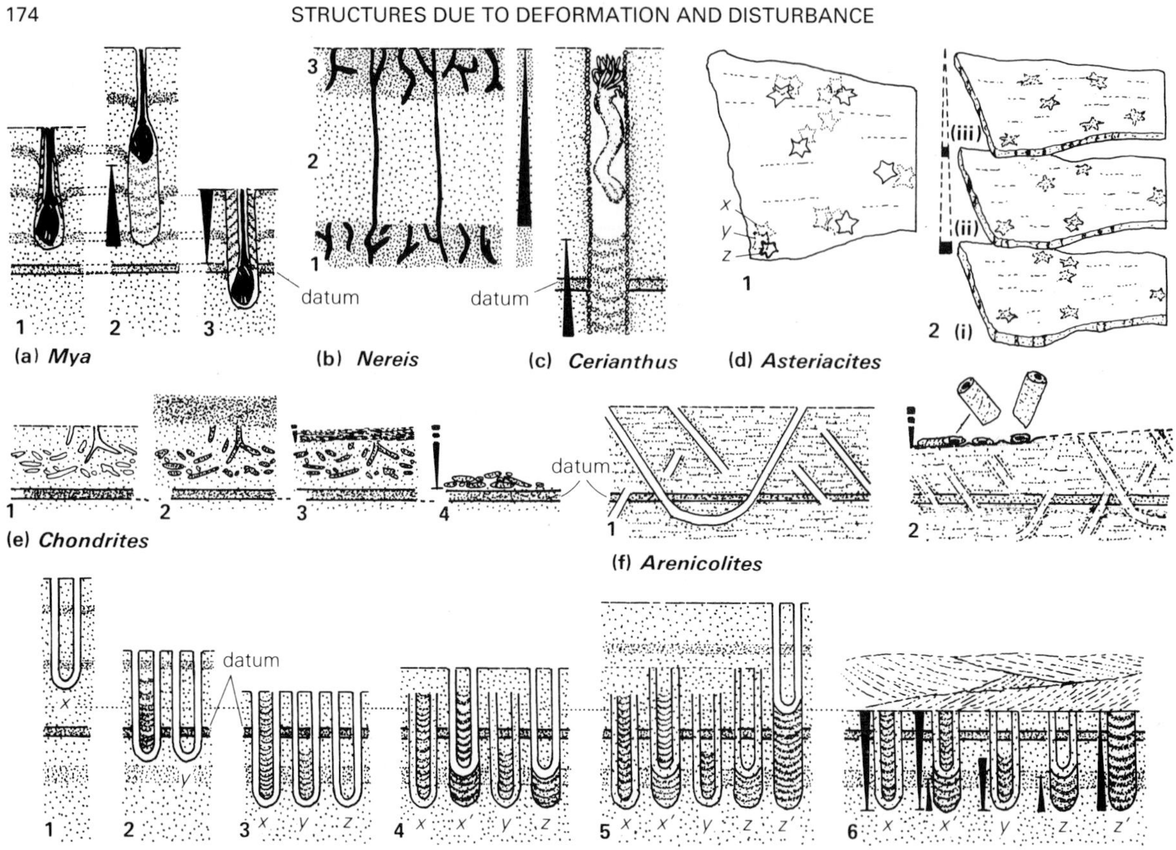

Figure 9.37 Exercise to reconstruct the amounts of sedimentation and erosion in clastic rock sequences. The diagrams represent single or closely related sequences of vertical sections. In (a–g), explain the sequence of events that has given rise to the preservation of each of the sets of trace fossils. In many cases a lithological datum-plane (time-marker) is shown. After working out the sequences, explain the significance of the tapering columns in the diagrams. The movement patterns relate to: (a) the bivalve *Mya* which has a single siphon; (b) the polychaete worm *Nereis*; (c) the sea anenome *Cerianthus*, an organism that dwells in a single tube and produces a pattern similar to traces such as *Skolithos* or *Monocraterion* which are not attributable to known organisms; (d) a resting trace of a starfish *Asteriacites*. Exposure 1: a bedding plane with groups of traces; exposure 2: three bedding planes, oldest (i); (e) the preservation patterns of *Chondrites*, believed to be the work of sediment-feeding organisms; (f) the preservation pattern *Arenicolites curvatus*, believed to be the work of a suspension-feeding organism, possibly a worm; (g) the movement pattern *Diplocraterion yoyo* (after Goldring 1964). (The whole drawing is modified after Howard in Basan 1978.)

a cross current, while it first lost and then regained a direction of movement in the face of hostile currents? Examples are provided by arthropod and tetrapod tracks, e.g. *Diplichnites, Kouphichnium* and *Chirotherium*.

(b) *Could the traces be made just below the surface of sediment, especially in silt–sand?* Such surface and subsurface traces are known as *Repichnia*, and are made by benthos and nektobenthos, predators, scavengers and deposit feeders (such as snails). Examples are *Gyrochorte, Cruziana* and *Aulichnites*. Repichnial traces are often transitional to resting traces and surface grazing traces (Fig. 9.38).

Q.11 *Could the traces be due to carefully organised surface or horizontal interface grazing?* Are the traces on the surface? Do they show discontinuous, systematically patterned meanders, loops, spirals and networks, or (occasionally) branched patterns? Are they non-overlapping, curved to tightly coiled, grooves, pits and furrows which may have delicately constructed spreiten? Could they result from 'strip mining' by an animal to ensure the economical exploitation of

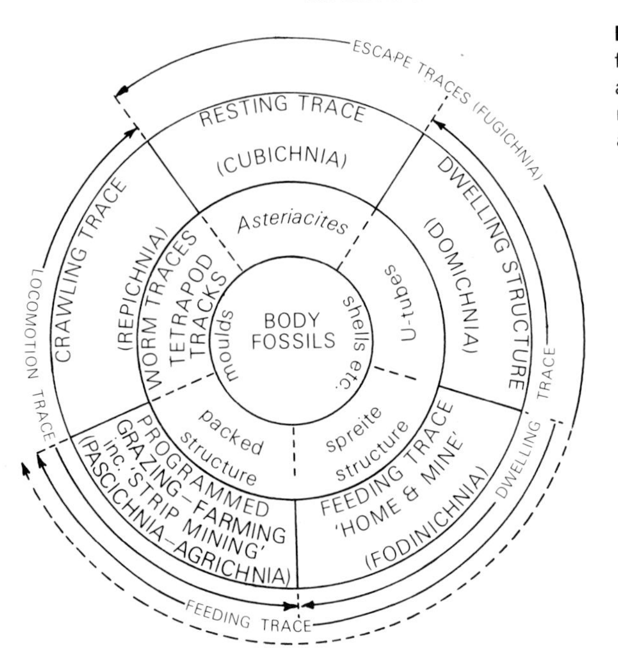

Figure 9.38 The behavioural classification of trace fossils, modified after Seilacher (1953), Osgood (1970) and Simpson in Frey (1975). The diagram shows the six main behavioural categories and their relations to one another and to body fossils.

nutritious sediment? Such surface grazing traces, wherein the complete form is commonly preserved, are made by mobile epibenthos (e.g. grazing gastropods such as limpets, worms, echinoids and arthropods) and are known as *Pascichnia.* Examples are *Helminthoida* and *Nereites* and planar types of *Zoophycos* (Figs. 9.39 & 40).

Q.12 *Could the traces result from animals systematically 'farming' the flora and fauna on the sides of the tunnel system?* Are the traces burrows which are systematically patterned and non-overlapping? Could they be preserved at the top surface of a stratum because subsequent erosion has been able to cut down only as far as the plane of the indurated burrow fills? The loop, net and hexagonally, polygonally branched patterns (e.g. *Paleodictyon*) are likely to fit here. Such traces are referred to as farming traces: *Agrichnia* (Fig. 9.39(i)).

Q.13 *Could the traces be burrows excavated while animals search for food within the sediment?* Features which might suggest this are single, branched or unbranched, cylindrical to sinuous, radial shafts or U-shaped burrows, orientated at various angles to the bedding. There may be also complex, parallel to concentric, burrow repetitions (i.e. spreiten) with unlined walls. The structures result from burrowing by animals that were 'underground miners', i.e. were essentially deposit-feeders which, in addition, sought secure refuge. An important feature is that the radial or U-shaped burrows do not touch since the animals avoid previously mined sediment. Borings made through shells of living bivalves can be included here. These traces are made by epibenthic and endobenthic deposit feeders like polychaete worms, and are known as *Fodinichnia.* These traces are related to grazing and dwelling structures. Examples are *Chondrites, Phycodes, Rhizocorallium* and *Zoophycos* (Figs 9.39 & 40).

Q.14 *Could the structures be the dwellings of suspension feeders or carnivores?* Features which might suggest this are simple, bifurcated or U-shaped burrows perpendicular or inclined to the bedding, or branching burrows having vertical and horizontal components. Some traces lack spreiten and have mucus-cemented sand–silt or clay-lined walls. They may be borings or burrows which were more or less permanently occupied by suspension feeders or active carnivores which strengthened the walls of their homes but did not back fill them. These forms are produced by sessile or hemisessile suspension feeders (e.g. shrimps), or by predator tube-dwelling worms or arthropod scavengers (e.g. crabs) and are known as *Domichnia.* They are closely related to feeding and resting traces. Examples are *Skolithos, Arenicolites, Diplocraterion, Thalassinoides* and *Ophiomorpha* (Figs 9.39 & 40).

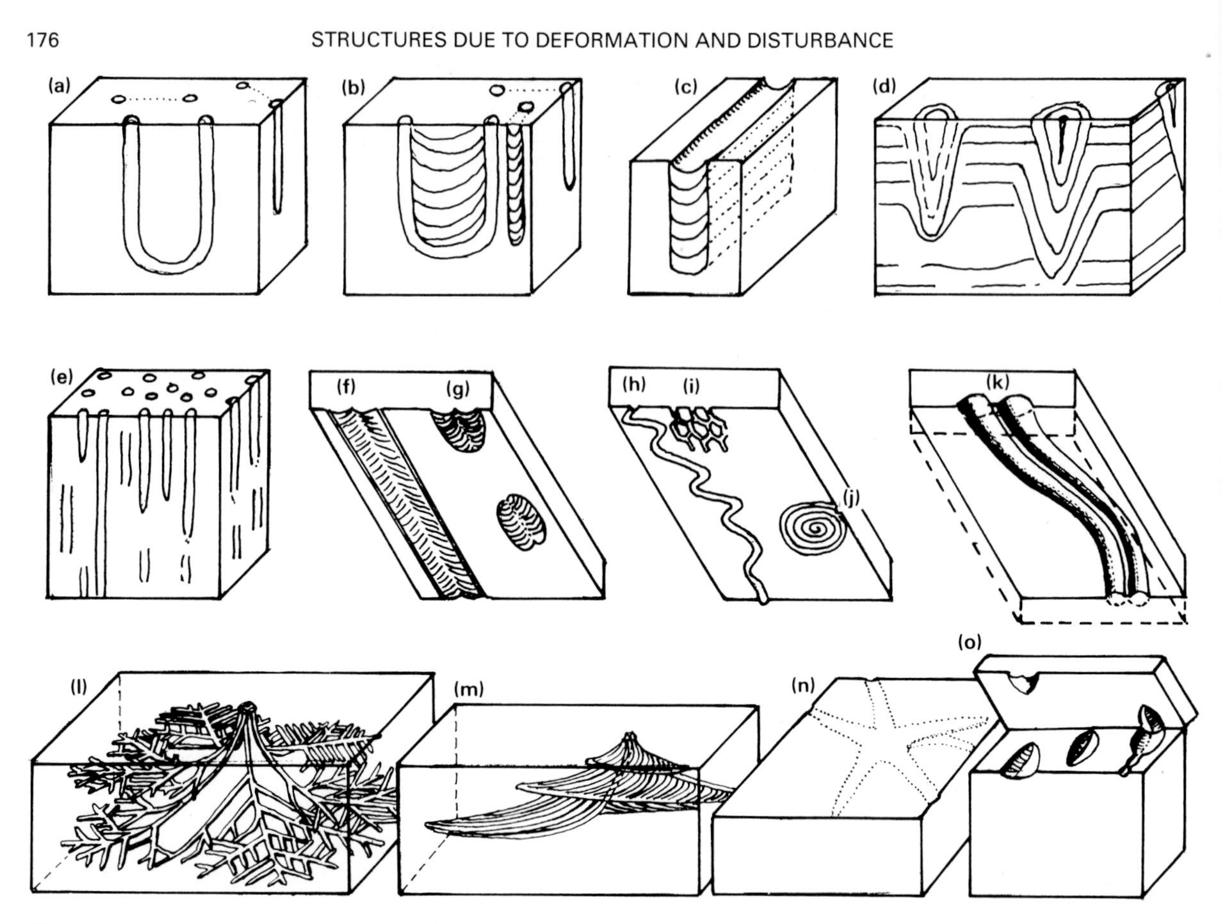

Figure 9.39 Exercise to work out the modes of behaviour of each of the trace fossils (a–o) and to classify them into one or more of the six groups described in the text and in Figure 9.38. Suggest the animal group to which each trace might belong. Indicate whether each trace fossil could be used as a way-up indicator. (a) *Arenicolites;* (b) *Corophioides;* (c) *Teichichnus;* (d) *Monocraterion;* (e) *Skolithos;* (f) *Cruziana;* (g) *Rusophycus;* (h) *Helminthopsis;* (i) *Paleodictyon;* (j) *Spirophycus;* (k) *Scolicia;* (l) *Chondrites;* (m) *Zoophycos;* (n) *Asteriacites;* (o) *Pelecypodichnus.* (Modified after diagrams in Frey and Crimes in Frey 1975 and Basan *et al.* 1978.)

Q.15 *Could the structures have resulted from the upward or downward movement of escaping organisms?* Are the traces roughly cylindrical, subvertical and lacking a lining? Are they vertically repetitive, with laminae concentric, *en echelon* or forming chevrons or nested funnels? Are there 'U-in-U' spreiten in the burrows or other types of structures which could reflect the displacement of semisessile suspension feeders (e.g. bivalves) upwards or downwards with respect to the original sediment surface as a response to erosion or sedimentation? Complete forms may be preserved in aggrading sequences. These traces are classed as *Fugichnia,* but there is complete overlap with resting and dwelling burrows, and rare transitions to both types of feeding structure because the last named are less likely to show rapid response to change of interface position. Examples are found in retrusive *Diplocraterion* and *Pelecypodichnus,* in *Monocraterion*, in *Skolithos* which possess funnels or retrusive bases, and in elongate *Ophiomorpha* (see Figs 9.37, 39 & 40).

To familiarise yourself with some of these terms, features and questions, try to sort the trace fossils depicted in Figures 9.37, 39 & 40 into one or more of these behavioural categories and, where appropriate, interpret a sequence of events. This is, however, no substitute for doing the same thing in the field or the laboratory.

Figure 9.40 A diagram illustrating common facies, environmental zones, energy–depth related trace fossil genera (b–e) and substrate related genera (a) and (f). Note that not all traces are compatible temporally, but each may have close analogues living at the present day. *Exercise*: Try to classify each trace fossil 1–62 into modes of behaviour using the six categories depicted in Figure 9.38. Referring to the same categories, suggest what general patterns of modes of behaviour characterise each of the facies associations and environmental zones (a–f). (Modified after Seilacher 1967, Crimes 1975 and Rhoads in Frey 1975 and Chamberlain in Basan 1978.)

The trace fossils are as follows (scale bars approximately 1 cm unless otherwise indicated):

(a) *Scoyenia Association*

1,2 *Scoyenia*: endichnial burrow with striated margin, and meniscus-packed infill (1) and box sections of same with spreite due to back-stuffing (2); 3, 4 *Planolites*: endichnial unbranched horizontal burrow with structureless fill in box section (3) and three dimensions (4); 5 *Cylindricum*: endichnial burrow in cross-section; 6 *Dikaka*: endichnial structure of plant in cross-section; 7, 8 *Isopodichnus*: lobate, bilobate hypichnia (*Rusophycus* type) (7) and curved double-ribbon hypichnia (*Cruziana* type) (8); 9 *Acripes*: hypichnial or epichnial track (cf. *Diplichnites*); 10 *Beaconites*: large endichnial burrow with spreiten; 11 *Anthichnium*: epichnial or hypichnial trackway; 12 *Chirotherium*: epichnial, or more usually hypichnial trackway; note variations of individual traces.

(b) *Skolithos and Glossifungites Association*

13, 14 *Skolithos*: frequent vertical, epichnial, endichnial burrows with structureless fill; 15 *Cylindrichnus*: vertical cone-in-cone shaped infilled burrow single epichnial and endichnial burrows developed at several horizons; 16, 17 *Diplocraterion:* epichnial endichnial U-burrows in box section (16) and showing protrusive and retrusive spreiten (17: left and right respectively); 18, 19 *Thalassinoides*: endichnial box section showing lined asymmetric burrow fill (18) and Y-shaped forking, bulbous terminations and position of last burrow channel (19); 20–23 *Ophiomorpha*: view from above of endichnial hexagonal burrow system (20), vertical and horizontal burrows (21), detail of lined and mamillated burrow margin (22) and its expression in box sections (23); 24 *Arenicolites*: epichnial, endichnial mud-lined, subvertical U-burrows with structureless fill.

(c) *Cruziana Association*

25 *Rusophycus*: hypichnial bilobate, ovoid cast; 26–8 *Cruziana*: hypichnial groove cast with v-shaped marks in plan (26) showing ridge in semi-relief (28) and sometimes related to *Rusophycus* in the manner shown; 27 *Diplichnites*: hypichnial groove with lines of separated marks sometimes related to *Cruziana* in the manner shown; 29 *Phycodes*: sub-horizontal grouped curved hypichnial burrows with structured infill, i.e. spreiten; 30, 31 *Rhizocorallium*: epichnial, endichnial, hypichnial horizontal or inclined burrows with spreiten in box section (30) or three dimensions (31); 32–4 *Teichichnus*: horizontal to subhorizontal endichnial burrow with vertically retrusive spreiten seen in box sections (32) or in three dimensions (33 & 34).

(d) *Zoophycos Association*

35–8 *Zoophycos*: epichnial endichnial vertical to sub-horizontal burrows with spreiten in box sections (35), in three dimensions showing high and low spiral burrows (36, 37), and in splitting/bedding plane expression (38).

(e) *Nereites Association*

39 *Helminthoida*: box sections and epichnial meandering burrow; 40 *Nereites sensu stricto*: epichnial meandering burrow with marginal lobes.

(f) *Zoophycos and Nereites Association–undifferentiated*

41 *Oldhamia*: radiating burrows; 42 *Urohelminthoida*: meandering hypichnial burrows with angular turning points; 43 *Cosmorhaphe*: unbranched burrow with two orders of meanders; 44 *Spirophyton*: interlaminar view of spiral burrow with spreiten; 45 *Spirophycus*: regularly single coiled burrow; 46 *Spirorhaphe*: spirally double coiled meandering multifloored burrow; 47 *Paleodictyon*: hypichnial hexagonal or polygonal irregular burrow network with structureless fill; 48 *Dictyodora*: regularly meandering burrow; 49 *Paleomeandron*: hypichnial burrows with two orders of lobose meandering; 50–52 *Scalarituba*: epichnial meandering meniscate burrows in different preservational expressions called *Nereites* (i.e. in epirelief), sometimes called *Phyllodocites* (50); hypichnial meandering line of bumps sometimes called *Neonereites* (52); 53, 54 *Lophoctenium*: single and multiple fan-shaped areas covered with irregular spreiten: lowly inclined burrow with spreiten.

(g) *Trypanites Association*

55, 59, 60, 62 *Trypanites*: Various tubular burrows in hard rock (or hardground); 56, 57, 58, 61 *Gastrocoenites*: flask-shaped burrows (probably of boring bivalves) with shapes of cross-sections of burrows indicated.

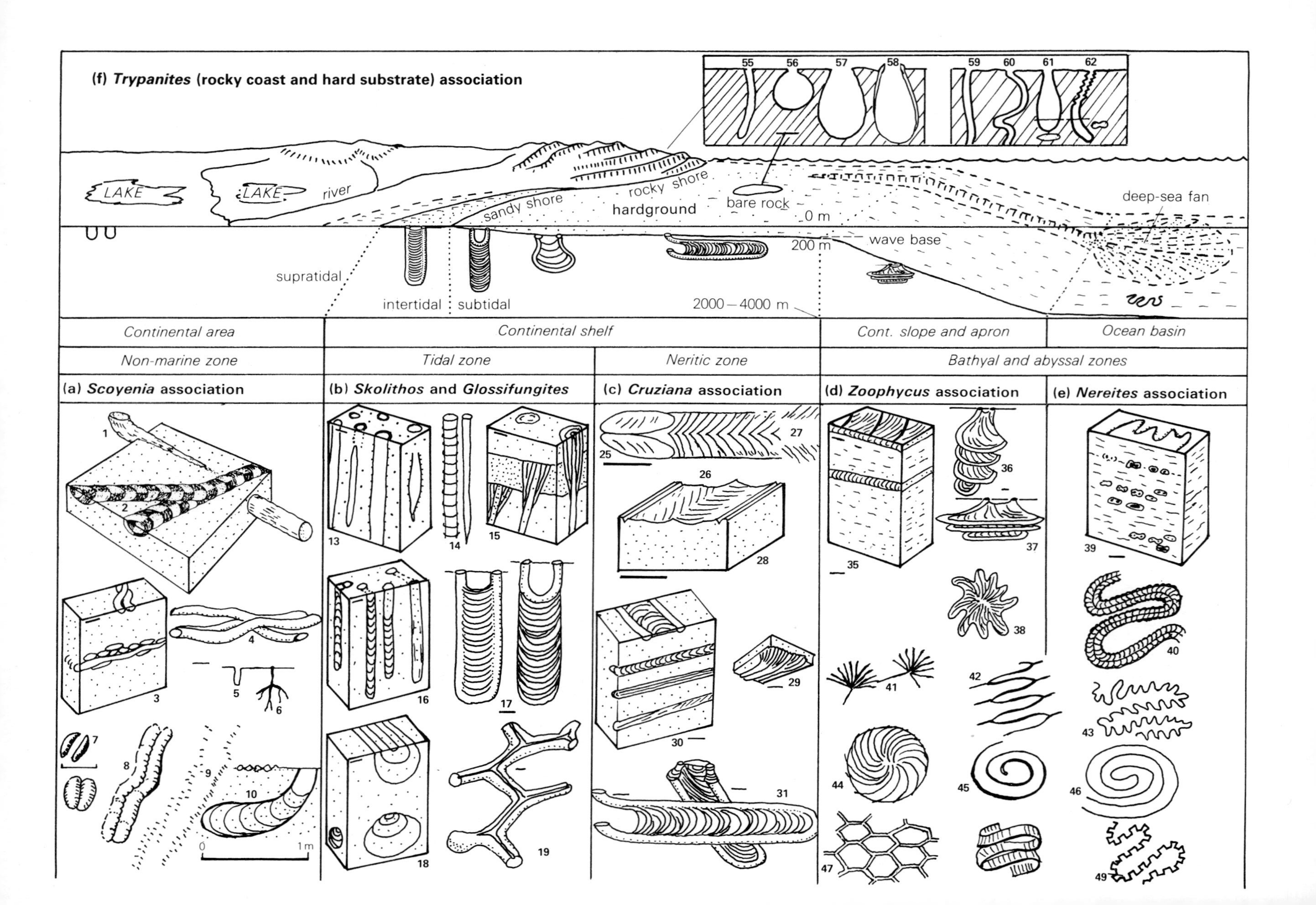

(f) Trypanites (rocky coast and hard substrate) association
55
56
57
58
59
60
61
62
LAKE
LAKE
river
sandy shore
rocky shore
hardground
bare rock
0 m
200 m
wave base
deep-sea fan
supratidal
intertidal
subtidal
2000 – 4000 m
Continental area
Continental shelf
Cont. slope and apron
Ocean basin
Non-marine zone
Tidal zone
Neritic zone
Bathyal and abyssal zones
(a) Scoyenia association
(b) Skolithos and Glossifungites
(c) Cruziana association
(d) Zoophycus association
(e) Nereites association
1
2
3
4
5
6
7
8
9
10
0
1 m
13
14
15
16
17
18
19
25
26
27
28
29
30
31
35
36
37
38
41
42
44
45
47
39
40
43
46
49

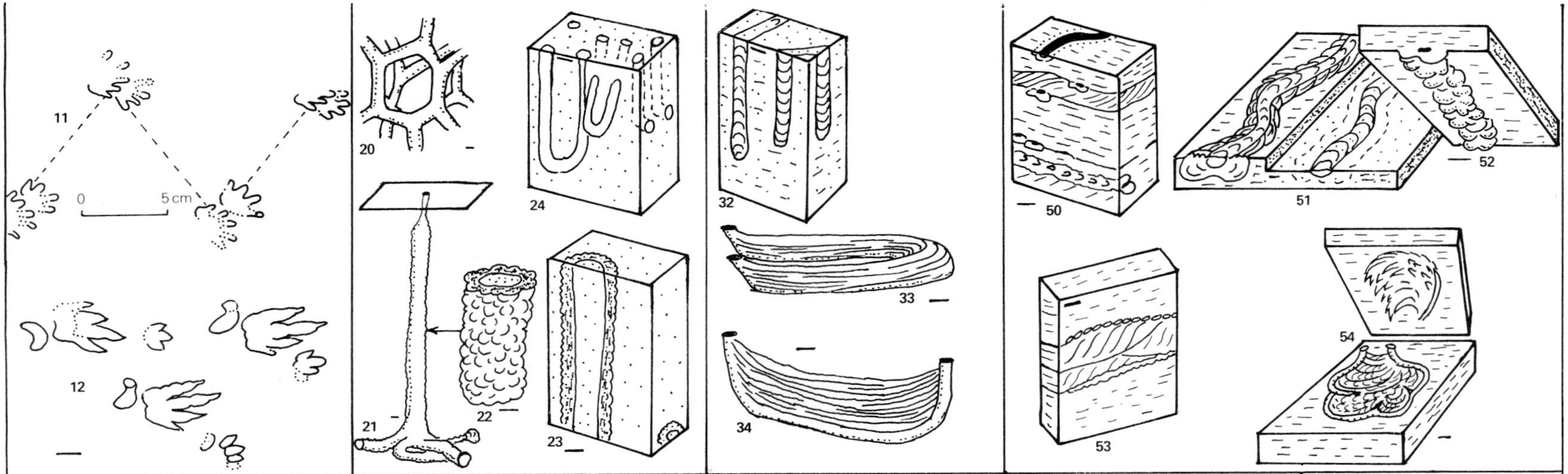

	11, 12	20, 21, 22	23, 24	32, 33, 34	50, 53	51, 52, 54
Energy (waves, wind)	variable; waves, river, air currents	higher energy, constant wave mixing	lower energy, frequent wave mixing	lower energy, frequent wave mixing and wave action	lower energy, infrequent wave mixing, some density flows	no wave mixing, some density flows, ocean current flows
Eh	oxidising, except at lake bottoms	oxidising	oxidising	oxidising	O_2 reduced, high organic content	little O_2, e.g. from density flows, even anoxic
Salinity	low, rarely high	mostly normal	normal	normal	normal	normal
Temperature	extreme variety	daily changes	daily changes	seasonal changes	$< 10°C$, no changes	down to 2°C, no changes
Light	daily changes	daily changes	daily changes	daily changes in upper part	none	none
Substrate sediment type and firmness	aqueous and marginal; sand, silt, mud	sand>mud, less firm, reworked highly mobile	mud = sand, firmer reworked, eroded	mud = sand, stable, rarer reworking and ripples	mud, stable except in failure and density flow	pelagic mud dominant, most stable but for density and ocean current flows
Diversity	lower than in marine realm	low	low	high	low	low, but higher than in (d), higher through time
Abundance	high in places	dense	dense	high	low	low, but seems higher due to slow deposition
Dominant organisms, traces	arthropods (insects) molluscs, vertebrates, small shallow burrows, trails, tracks, herbivores, carnivores	arthropods, molluscs, echinoderms, corals, 'worms' unbranched vertical burrows mostly suspension feeders	vertical and inclined ear-shaped burrows	arthropods, molluscs, echinoderms, corals horizontal crawling – grazing traces > inclined burrows mostly sediment feeders	arthropods, polychaete worms, hemichordates (e.g. acorn worms), echinoderms complex horizontal grazing and shallow feeding traces, spreiten inclined in sheets, ribbons, spirals – bioturbation sediment churners, feeders	casts, crawling, grazing traces, spreiten planar and on surface sediment grazers, farmers

Figure 9.40 – caption on p. 177.

9.4.4 The experimental approach to the understanding of the behavioural aspects of trace fossils

This approach involves the study in the field or the laboratory of the varied factors that influence the behaviour of organisms. In practice it is concerned mostly with invertebrates rather than vertebrates or plants. Most studies are with burrowers, often bivalves, and the way in which they destroy primary sedimentary structures and form biogenic structures. Studies may vary from the simple observation of the marks made by moving objects on a substrate, or the burrowing of given organisms placed upon a carefully prepared succession of particular composition and consistency (see Farrow's film), to one which stresses the functional morphology of the animal in relation to its burrow. Other studies approximate natural conditions and describe both the burrowing behaviour and its effect on the substrate, and simulate such processes as erosion and sedimentation (see for example, Schaefer, Frey and Manton in *Study list*).

9.4.5 Trace-fossil associations and their environmental implications

Trace fossils contribute increasingly to the delineation of facies in many present-day environments because they reflect interactions between organisms and particular sets of physical and chemical conditions. Traces reflect many of the processes that are the basis of the uniformitarian and actualistic models for distinct ecological and sedimentological settings. In marine environments, factors controlling distributions of organisms include temperature, food supply and the intensity of wave and current activity, and these tend to reflect relative depth. Along the shore and in non-marine environments, however, many of these factors tend to change laterally, reflecting primarily variations in energy distribution in relation to distance from input and degree of exposure.

Unlike roughly 95% of body fossils, which are drifted to varying degrees, trace fossils are found *in situ* where life activity occurred. Trace fossils are particularly appropriate, therefore, to the definition of facies and the erection of facies models. The fact that some workers distinguish 'trace-fossil assemblages' and 'associations' without reference to the sediment in which they are found is legitimate but unfortunate. Similarly, sedimentologists who erect facies and facies associations without searching for trace fossils which would render the distinctions between rocks more valid, do a similar disservice. Some workers, notably Seilacher, have included sedimentary and trace-fossil evidence in defining **ichnofacies** and **ichnofacies associations**. Using uniformitarian principles, associations of ichnofacies seen in marine sequences have been interpreted in terms of relative water depth, energy level, substrate type and firmness. Each assemblage has been named after a representative ichnogenus which does not have to be present in every example of that facies (Fig. 9.40). The aim of these studies is to put limits on the physical, chemical and biological conditions prevailing when the traces were made. A fundamental prediction of an actualistic approach is that these or similar traces would have formed whenever the particular set of conditions recurred. This prediction appears to be justified by the occurrence of trace fossils throughout the rock record.

Once processes relating to the formation of traces have been recognised, we can consider the kind of environment and deposit in which the traces were formed. Substrate consistency and energy levels, in particular, can be read quite well from traces of behaviour in the sediments, but other, more biological, factors in the environment are less readily discerned.

9.4.6 The uses of trace fossils

In **sedimentology,** trace fossils can show whether sedimentation was continuous (at relatively slow or high rates) or discontinuous (at variable rates with or without erosion). They may give an indication of substrate consistency and aeration at the time of activity, the directions of palaeocurrents and the degree of reworking of sediment.

In the recognition of **sedimentary environments**, trace fossils provide records of life *in situ* and the traces may be closely related to the depositional environment of the host sediments. Where organic activity has destroyed or masked depositional structures, trace fossils may provide the only clues about the nature of an environment. Traces are formed in the whole environmental spectrum from continent to abyss, but the narrow environmental range of many forms reflects the preference of trace makers for particular sets of ecological conditions and substrates, e.g. carbonates and muddy-sand, sandy-mud settings. Their abundance in clastic rocks, which often lack body fossils due to diagenetic processes, means that the palaeoenvironments of the rocks can be interpreted in the light of organic activity which would not otherwise be recognised. The long time range of some traces throughout the Phanerozoic permits palaeoecological comparisons of rocks of different ages.

In **palaeontology and palaeobiology**, trace fossils record the behaviour patterns of extinct organisms, e.g. the feeding, locomotive and protective activities of trilobites. Furthermore, they record the activities of organisms that had only soft parts. They increase the diversity of fossil groups known from the geological record, so that this provides a wider sample of former life forms and encourages the understanding of the evolution of fossil behaviour. They help to elucidate problems in late Precambrian rocks where body fossils are generally absent and where trace fossils record important events such as the appearance of Metazoa. The presence of traces in coarse-grained sediments, laid down where abrasion and weathering (especially oxidisation) were active in destroying hard and soft parts of organisms, helps to bridge several palaeontological gaps. Vertebrates such as the producer of *Chirotherium* (the 'hand beast' – probably an early thecodont archosaur of the Lower Trias) are known only from their footprints and trackways, which offer important evidence of the evolutionary radiation and habits of the early reptiles. Indeed the evidence of many reptiles is known only from their traces in aeolian sediments.

In **structural geology** the fact that the organisms react to gravity and light, grow, or move downwards and upwards, and are asymmetric about a horizontal plane make them excellent small-scale, 'way-up' indicators. Quantitative estimates of compaction and deformation depend upon the recognition of objects of known shape prior to deformation (Fig. 9.41). Burrows can be such objects, though great care is needed in their use. Some sand-filled burrows become flattened films on bedding planes when newly deposited mud has lost a great deal of water. U-shaped burrows are particularly useful because they allow the reconstruction of strain ellipses and aid quantitative estimates of strain due to pre-cleavage compaction, compression, rotation and cleavage distortion.

In **geotectonics**, associations of traces may help to define faunal provinces. Certain distinctive traces, for example, are attributable to trilobites of an Atlantic Province, others to a Pacific Province in the late Cambrian–early Ordovician. This evidence has contributed to current conjectures concerning the former separation of these provinces by an Iapetus Ocean.

In **stratigraphy**, the long time range of many trace fossils restricts their use for biostratigraphic or chronostratigraphic purposes, but short-ranging forms such as forms of *Cruziana* can be used to date poorly fossiliferous successions in the lower Palaeozoic. Vertebrate footprints provide a worldwide basis for a stratigraphy of the Trias (see *References*). These two

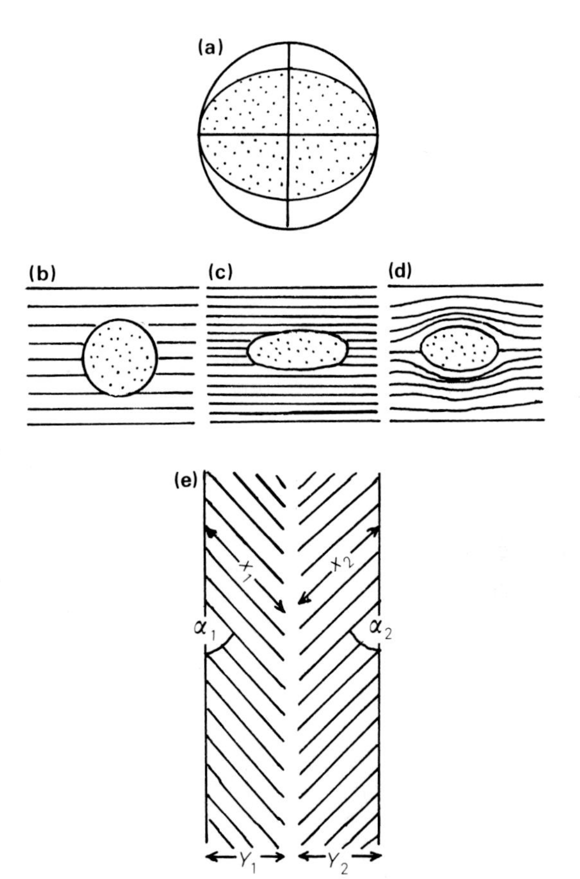

Figure 9.41 The use of trace fossils in deformational studies (after Crimes in Frey 1975): (a) the elliptical outline of burrow (stippled) may be used to determine the percentage compaction by the construction of a circle of equal area to represent the outline of the undeformed burrow, and then the comparison of the axial lengths of the ellipse and the circle may be made; (b) no compaction after burrowing, the burrow remains circular in outline and adjacent laminae undeformed if burrow infill and matrix compact equally after burrowing; (c) measurement of burrow compaction gives correct value for matrix; (d) matrix compacts more, the laminae being deflected around the burrow; compaction values for the burrow, therefore, are less than for the matrix – as in the commonest cases where the burrow infills are coarse-grained; (e) effects of tectonic deformation on *Cruziana*. In the undeformed state, normally $x_1 = x_2$; $\alpha_1 = \alpha_2$; $Y_1 = Y_2$. Any increment of compression or shear disturbs the equality and can be used to quantify deformation two-dimensionally in the plane of the trace. Note the initial assumption that the burrows are cylindrical. *Planolites* and *Chondrites* are the most frequently and best-used for these studies, though work on *Beaconites* has recently been very effective (see *References*).

examples testify to rapid evolution and hence the correlations may have considerable reliability. Trace fossils also help to generate and control palaeogeographic reconstructions for individual stratigraphic time periods.

Lastly, trace fossils are of considerable use in **applied geology**, particularly in the hydrocarbon industry. The understanding of the concepts and techniques outlined in Section 9.4 may be useful, and indeed some vital, if worthwhile predictions are to be made about depositional environments, tectonic setting, economic basement, and the extents of source, reservoir and cap rocks. Fortunately trace fossils show up well in cores wherein appreciation of their three-dimensional nature is often easier than at outcrop. Ability to recognise bioturbation as the cause of massive, somewhat argillaceous, sandstones with low porosity and permeability may be important to the success of secondary and tertiary recovery programmes. Trace-fossil studies have potential in relation to determining the mass properties of sediments required by engineers tackling problems of shifting coastal morphology (see Reineck in *Further reading*).

9.4.7 *Confusion of traces with inorganic sedimentary structures*

Traces may easily be confused with a wide range of primary and secondary structures of inorganic origin. A mark made by an active trilobite is a trace fossil; a mark made by impact of a moulted or deceased adult carapace of a trilobite is an inorganic structure, i.e. a tool mark (skip cast). Confusion is particularly common where burrows are concerned, for they have been related to gas pits (feeding trace of polychaete worms), air escape holes (burrows of amphipod crustacean sand-hoppers), conical fracture patterns (feeding burrows such as *Phycodes*), sand volcanoes and fulgurites. In other cases, current crescents have been described as the trace fossil *Blastophycus,* rill marks as *Dendrophycus,* sinuous hierarchial mudcracks as *Manchuriophycus,* interference ripples as tadpole nests, and convolute or contorted lamination as bioturbation, etc. The possibilities are endless and a student should recognise this by learning to generate and test a wide range of working hypotheses when an apparently novel trace fossil is found.

Reference and study list

References marked with an asterisk * are suitable for 16–19 year old advanced level students in schools and colleges in the UK as well as for undergraduates.

Field experience

Structures due to inorganic or organic disturbance may be found in almost any environment but, when they are found, the processes that gave rise to them have often ceased to operate. Sandy beaches and fluvial and estuarine sand bars are good places for geologists to fluidise water-laden sand by stamping, thus changing the grain packing and the pore-water pressure so that the sand becomes quick. With practice sand volcanoes may be produced. Desiccation and synaeresis cracks may be observed in dried-out and still-water ponds respectively (e.g. in supratidal or fluvial areas). Large-scale sub-aerial landslides are easy to study and may be usefully compared with slumps. Soil profiles in which concretions are forming may be seen in many temperate and tropical areas. The text below suggests how one may equip and organise oneself to study trace fossils in tidal flats of estuaries, deltas and beaches, or on river plains.

Farrow, G. E. 1975. Techniques for the study of fossil and recent traces. In *Trace fossils*, R. W. Frey (ed.), 537–54. New York: Springer-Verlag.

Perkins, B. F. (ed.) 1971. *Trace fossils, a field guide*. Louisiana State University School of Geoscience, Misc. Publ. 71–1.

Laboratory experience

Laboratory tanks may be easily used for experiments on the dewatering of sand and the formation of load and flame structures in interbedded saturated mud and sand. Indeed, these features and sand volcanoes may be produced by disturbing suitably organised sedimentary layers in containers as small as coffee jars. Resin-fixed soil profiles may be bought commercially for study in the laboratory. Laboratory work on trace fossils is suggested in the text.

*Ager, D. G. 1976. Filmstrip: *Palaeoecology*. Milton Keynes: Association of Teachers of Geology, UK. [A set of colour slides suitable for teaching a few basic principles to beginners.]

*Anon 1980. Film: *Reef at Ras Muhammad*. London: Partridge Films. [A most beautiful film concerning the ecology and sedimentology of coral reefs in the Red Sea.]

Elders, C. A. 1975. Experimental approaches in neoichnology. In *The study of trace fossils*, R. W. Frey (ed.), 513–36. Berlin: Springer-Verlag.

Farrow, G. E. 1975. Techniques for the study of fossil and recent traces. In *Trace fossils*, R. W. Frey (ed.), 537–54. New York: Springer-Verlag.

*Farrow, G. 1976. *Living sediments: prelude to palaeoecology*. Colour, sound; 20 min. Glasgow: Audiovisual Aids Unit, University of Strathclyde. [An excellent film with extensive, well written accompanying notes on the sediments and traces visible in the intertidal estuarine environment of the Solway Firth, which lies between England and Scotland. This film demonstrates basic skills and techniques in the field and laboratory, and worthwhile values and attitudes relating to philosophy and methodology. Good shots of laboratory experiments, e.g. on the escape structure of burrowing bivalves, and the making of expoxy resin peels from box cores.]

Farrow, G. E. 1976. *Study notes for the undergraduate teaching film 'Living sediments: prelude to palaeoecology'*. Glasgow: University of Glasgow, Geology Department.

Manton, S. M. 1977. *The arthropoda; habits, functional morphology and evolution*. Oxford: Oxford University Press. [Many experimental approaches and results are described.]

Marszalek, D. S. and W. W. Hay 1968. Marine aquaria for palaeontology. *J. Geol Education* **16**, 159–63.

McKee, E. D. 1969. Experiments on formation of contorted structures in mud. *Bull. Geol Soc. Am.* **80**, 231–44.

McKee, E. D., M. A. Reynolds and C. H. Baker, Jr 1962a. *Laboratory studies on deformation in unconsolidated sediment*. Prof. Pap. US Geol Surv. 450–D, D151–5.

McKee, E. D., M. A. Reynolds and C. H. Baker Jr 1962b. *Experiments on intraformational recumbent folds in cross-bedded sand*. Prof. Pap. US Geol Surv. 450-D, D155-9.

Murray, J. 1970. Film: *Abu Dhabi 1969 – a modern limestone environment*. Colour, sound; *c*. 20 min. Bristol: University of Bristol, Geology Department.

Sarjeant, W. A. S. 1971. *The Beasley collection of photographs and drawings of fossil footprints and bones and of fossil and recent sedimentary structures*. Nottingham: University of Nottingham, Geology Department.

Light reading

Heezen, B. C. and C. D. Hollister 1971. *The face of the deep*. Oxford: Oxford University Press.

Seilacher, A. 1967. Fossil behaviour. *Scient. Am.* **217** (2), 72–84.

Physical and chemical disturbance

Essential reading

*Allen, J. R. L. 1970. *Physical processes of sedimentation*. London: Allen & Unwin. [Simple explanation of load casts and sandstone balls (pp. 75–6) and slumps (pp. 196–9).]

Allen, J. R. L. 1982. *Sedimentary structures: their character and physical basis*. Amsterdam: Elsevier. [Ch. 9 of Vol. 2 deals with soft-sediment deformation.]

Allen, J. R. L. 1985. *Principles of physical sedimentology*. London: Allen & Unwin. [Ch. 10 deals with the mechanisms of certain styles of deformation.]

Allen, J. R. L. and N. L. Banks 1972. An interpretation and analysis of recumbent-folded deformed cross-bedding. *Sedimentology* **19**, 257–83.

Broadhurst, F. M. and I. M. Simpson 1967. Sedimentary infillings of fossils and cavities at Treak Cliff, Derbyshire. *Geol. Mag.* **104**, 443–8.

Chowns, T. M. and J. E. Elkins 1974. The origin of quartz geodes and cauliflower cherts through the silicification of anhydrite modules. *J. Sed. Petrol.* **44**, 885–903.

Donovan, R. M. and R. J. Foster 1972. Subaqueous shrinkage cracks from the Caithness flagstone series (Middle Devonian) of northeast Scotland. *J. Sed. Petrol.* **42**, 309–17.

Franks, P. C. 1969. Nature, origin and significance of cone-in-cone structures in the Kiowa Formation (early Cretaceous), north central Kansas. *J. Sed. Petrol.* **39**, 1438–54.

Goudie, A. 1973. *Duricrusts in tropical and subtropical landscapes*. Oxford: Clarendon Press.

Hallam, A. 1967. Siderite- and calcite-bearing concretionary nodules in the Lias of Yorkshire. *Geol. Mag.* **104**, 222–7.

King, C. A. M. (ed.) 1976. *Periglacial processes*. Stroudsburg, Pa: Dowden, Hutchinson & Ross. [Several papers on periglacial deformation.]

Lajoie, J. 1972. Slump fold axis orientations: an indication of palaeoslope? *J. Sed. Petrol.* **42**, 584–6.

Lowe, D. R. 1975. Water escape structures in coarse grained sediments. *Sedimentology* **22**, 157–204.

Lowe, D. R. and L. D. Lopiccolo 1974. The characteristics and origins of dish and pillar structure. *J. Sed. Petrol.* **44**, 484–501.

Morgenstern, N. R. 1967. Submarine slides, slumps and slope stability. In *Marine geotechnique*. A. F. Richards (ed.), 189–220. Urbana, Ill.: University of Illinois Press.

Paru, W. C. and E. H. Schot 1968. Stylolites: their nature and origin. *J. Sed. Petrol.* **38**, 175–91.

Walker, R. G. (ed.) 1984. *Facies models*, 2nd edn. Geol. Assoc. Canada. [Ch 3 (glacial), 16 (carbonate slopes) and 17 (evaporites) include accounts of deformation structures, cracking and nodule and concretion developments.]

Woodcock, N. H. 1979. The use of slump structures as palaeoslope orientation estimators. *Sedimentology* **26**, 83–99.

Further reading

Allen, J. R. L. 1977. The possible mechanics of convolute lamination in graded sand beds. *J. Geol Soc. Lond.* **134**, 19–31.

Corbett, K. D. 1973. Open-cast slump sheets and their relationship to sandstone beds in the Upper Cambrian flysch sequence, Tasmania. *J. Sed. Petrol.* **43**, 147–59.

Crans, W., G. Mandle and J. Haremboure 1980. On the theory of growth faulting. *J. Petrolm Geol.* **2**, 265–307. [A highly theoretical approach to the problem of growth fault form and formation.]

Donovan, R. N. and R. Archer 1975. Some sedimentological consequences of a fall in the level of Haweswater, Cumbria. *Proc. Yorks. Geol. Soc.* **40**, 547–62.

Edwards, M. B. 1976. Growth faults in Upper Triassic deltaic sediments, Svalbard. *Bull. Am. Assoc. Petrolm Geol.* **60**, 341–55.

Eyles, N. and B. M. Clark 1985. Gravity-induced soft-sediment deformation in glaciomarine sequences in the Upper Proterozoic Port Askaig Formation, Scotland. *Sedimentology* **32**, 789–814.

Hendry, H. E. and M. R. Stauffer 1977. Penecontemporaneous folds in cross-bedding: inversion of facing criteria and mimicry of tectonic folds. *Bull. Geol Soc. Am.* **88**, 809–12.

Hesse, R. and H. G. Reading 1978. Subaqueous clastic fissure eruptions and other examples of sedimentary transposition in the lacustrine Horton Buff Formation (Mississippian), Nova Scotia, Canada. In *Modern and ancient lake sediments*, A. Matter and M. E. Tucker (eds), 241–57. Oxford: Sp. Publ. Int. Assoc. Sediment. 2.

Horowitz, D. H. 1982. Geometry and origin of large-scale deformation structures in some ancient wind-blown sand deposits. *Sedimentology* **29**, 155–80.

Johnson, H. D. 1977. Sedimentation and water escape structures in some late Precambrian shallow marine sandstones from Finnmark, North Norway. *Sedimentology* **24**, 389–411.

Jones M. E. and R. M. F. Preston (eds) 1987. *Deformation of sediments and sedimentary rocks*. Sp. Publ. Geol Soc. Lond. 29. Oxford: Blackwell Scientific.

Kennedy, W. J. and P. Juignet 1974. Carbonate banks and slump beds in the Upper Cretaceous (Upper Turonian – Santonian) of Haute Normandie, France. *Sedimentology* **21**, 1–42.

Laird, M. G. 1970. Vertical sheet structure – a new indicator of sedimentary fabric. *J. Sed. Petrol.* **40**, 428–38.

McKee, E. D. and J. J. Bigarella 1972. Deformation structures in Brazilian coastal dunes. *J. Sed. Petrol.* **42**, 670–81.

Postma, G. 1983. Water escape structures in the context of a depositional model of a mass flow dominated conglomeratic fan delta (Abrioja Formation, Pliocene, Almeria Basin, S. E. Spain). *Sedimentology* **30**, 91–103.

Price, R. J. 1973. *Glacial and fluvioglacial landforms*. London: Longman. [Sections on periglacial disturbance structures.]

Rider, M. H. 1978. Growth faults in Carboniferous of western Ireland. *Bull. Am. Assoc. Petrolm Geol.* **62**, 2191–213.

Vandenberghe, J. and P. van den Breoek 1982. Weichselian convolution phenomena and processes in fine sediments. *Boreas* **11**, 299–315. [Beautiful illustrations of periglacial deformation associated with the decay of permafrost conditions.]

Washburn, A. L. 1973. *Periglacial processes and environments*. London: Edward Arnold. [Periglacial deformation structures well discussed.]

West, I. M. 1975. Evaporite and associated sediments of the basal Purbeck formation (Upper Jurassic) of Dorset. *Proc. Geol Assoc.* **86**, 205–25.

Woodcock, N. H. 1976. Structural style in slump sheets: Ludlow Series, Powys, Wales. *J. Geol. Soc. London* **132**, 399–415.

Wright, V. P. (ed.) 1986. *Palaeosols: their recognition and interpretation*. Oxford: Blackwell Scientific.

Biological disturbance (trace fossils)

Essential reading

*Ager, D. V. 1963. *Principles of palaeoecology*. New York: McGraw-Hill.

Basan, P. B. (ed.) 1978. *Trace fossil concepts*. Short Course Notes no. 5. Tulsa: SEPM. [The articles by Chamberlain on the recognition of trace fossils in cores, by Frey on behaviour and ecological implications, by Howard on sedimentology and trace fossils, by Seilacher on using trace fossils to determine depositional environment and by Warme & McHuron on marine boring organisms are highlighted.]

*Clarkson, E. N. K. 1979. *Invertebrate palaeontology and evolution*. London: Allen & Unwin. [A simple introduction to trace fossils (pp. 16–19).]

Crimes, T. P. and J. C. Harper (eds) 1970. *Trace fossils: Geol. J. Special issue no. 6*. Liverpool: Seel House Press. [Papers by Crimes and Martinsson are especially recommended.]

Crimes, T. P. and J. C. Harper (eds) 1977. *Trace fossils 2: Geol J. Special Issue no. 9*. Liverpool: Seel House Press.

Curran, H. A. (ed.) 1985. *Biogenic structures: their use in interpreting depositional environments*. Tulsa: SEPM Sp. Publ. 35. [A varied collection of interesting papers.]

Ekdale, A. A., R. G. Bromley and S. G. Pemberton 1984. *Ichnology: the use of trace fossils in sedimentology and stratigraphy*. Short Course Notes. Tulsa: SEPM.

Frey, R. W. (ed.) 1975. *The study of trace fossils*. Berlin: Springer-Verlag. [The most wide-ranging collection of papers on trace fossils in a single volume. Papers by Bishop, Boyd, Bromley, Chamberlain, Crimes, Dorges & Hertweck, Frey, Howard, Kennedy, Osgood, Rhoads and Simpson are all valuable.]

Hantzchel, W. 1975. Trace fossils and problematica. *Part W. Treatise on invertebrate paleontology*. C. Teichert (ed.), 2nd edn. Misc. Suppl. 1. Boulder, Co. & Lawrence, Ks: Geol Soc. America and Univ. Kansas Press.

Howard, J. D. 1972. Trace fossils as criteria for recognising shorelines in the stratigraphic record. In *Recognition of ancient sedimentary environments*, J. K. Rigby and W. K. Hamblin (eds), 215–25. Tulsa: SEPM Sp. Publ. 16.

Miller, M. F., A. A. Ekdale and M. D. Pilard (eds) 1984. Trace fossils and paleoenvironments: Marine carbonate, marginal marine terrigenous and continental terrigenous settings. *J. Paleont.* **58**, 283–598. [A very wide-ranging set of papers.]

Reineck, H. E. and I. B. Singh 1980. *Depositional sedimentary environments*, 2nd edn. Berlin: Springer-Verlag. [Often introduces the traces in the diagrams and photographs.]

Seilacher, A. 1967. Bathymetry of trace fossils. *Mar. Geol.* **5**, 413–28.

Further reading

Bromley, R. G. and A. A. Ekdale 1986. Composite ichnofabrics and tiering of burrows. *Geol. Mag.* **123**, 59–65.

Haubold, D. 1984. *Saurierfahrten*. Wittenberg Lutherstadt: A. Ziersen Verlag.

Hitchcock, E. 1859. *Ichnology of New England: a report on the sandstone of the Connecticut Valley, especially its fossil footmarks*. Boston: William White.

Howard, J. D., R. W. Frey and H. E. Reineck 1975. Estuaries of the Georgia Coast, USA. Sedimentology and biology. *Senckenberg. Marit.* **7**, 1–305.

Purdy, E. G. 1964. Sediments as substrates. In *Approaches to paleoecology*, J. Imbrie and N. D. Newell (eds), 238–71. New York: Wiley.

Sarjeant, W. A. S. and W. J. Kennedy 1973. Proposal of a code for the nomenclature of trace fossils. *Can. J. Earth Sci.* **10**, 460–75. [A discussion which sets out the problems of constructing an international code comparable with that relating to zoological, botanical and body-fossil nomenclature. It shows that a trace fossil code is bound to be very different from the other codes.]

Schaefer, W. 1972. *Ecology and palaeoecology of marine environments*. Edinburgh and Chicago: Oliver & Boyd and University of Chicago Press.

Seilacher, A. 1964. Sedimentological classification and nomenclature of trace fossils. *Sedimentology* **3**, 253–6.

Sullivan, M. J., M. A. Cooper, A. J. MacCarthy and W. H. Forbes 1986. The palaeoenvironment and deformation of *Beaconites*-like burrows in the Old Red Sandstone at Gortnabinna, SW Ireland. *J. Geol. Soc.* London. **142**, 129–36.

Extra references relating to figure captions

Goldring, R. 1964. Trace fossils and the sedimentary surface in shallow water marine sediments. In *Deltaic and shallow marine deposits*, L. M. J. Van Straaten (ed.), 136–43. Amsterdam: Elsevier.

Gruhn, R. and A. L. Bryan 1969. Fossil ice wedge polygons in southeast Essex, England. In *The periglacial environment*, T. L. Péwé (ed.), 351–63. Montreal: Arctic Institute of North America.

Krumbein, W. C. and R. M. Garrels 1952. Origin and classification of chemical sediments in terms of pH and oxidation–reduction potentials. *J. Geol.* **60**, 1–33.

Osgood, R. G. Jr 1970. Trace fossils of the Cincinnati area. *Palaeontographica Am.* **6** (41), 281–444.

Seilacher, A. 1953. Studien zur Palichnologie. I. Uber die methoden der Palichnologie. *Neues Jahrb. Geol. Palaont. Abh.* **98**, 87–124.

Webby, B. D. 1969. Trace fossils (Pascichnia) from the Silurian of New South Wales, Australia. *Palaont. Z.* **43**, 81–94.

10 The study of assemblages of structures

10.1 Introduction

In earlier chapters we have shown how sedimentary structures relate to erosional, depositional and post-depositional processes. Although the ability to interpret sediments in these terms is a vital first step, it is often more important to be able to set the processes in the context of an environment. In reading earlier chapters you may have noticed how little mention was made of environments. This omission was deliberate in order to show or imply that many structures and processes are common to a variety of environments.

In order to interpret sediments in terms of their environment of deposition, we must establish the spatial relationships of the structures and try to relate these to the spatial and temporal distribution of processes acting in particular present-day environments. We must also know something of the directional properties of sedimentary structures in order to test and refine our hypothesis of environment and to orientate our preferred environment and thereby give it a palaeogeographical significance.

We need, therefore, to record and present our observations of sedimentary structures, their directional information and their positions in measured sections in ways that will aid us in the task of environmental interpretation.

10.2 Collection and presentation of data

10.2.1 Directional data

From earlier chapters, it should be clear how one derives directional data from particular structures. For modern sediments, or for ancient ones that have undergone little or no tectonic displacement, the data can be collected and used directly. Do not become too obsessed with precision; with most directions an error of ± 10° is acceptable. When the rocks have undergone considerable tectonic tilting, it is necessary to reorientate the data by restoring the original bedding to horizontal. A simple procedure using a stereogram is used and this is outlined in Appendix A.

Once the directional measurements have been restored to their original orientation, it is usually useful to organise them for graphical presentation. This can be done in several ways. The method chosen usually depends upon the quantity of data and the variety of structures from which they were collected. Compilation inevitably leads to some loss of information; in particular the distribution of various directions, both laterally and vertically, within the sequence. Compiling directional data is a useful way of summarising the current pattern, but it is no substitute, in environmental interpretation, for relating directions to specific structures in a measured section.

Where data are few and have been collected from only a few types of structure, it is often convenient to plot each measurement as a radiating line of fixed length on a circular diagram (Fig. 10.1). Where both dip direction and magnitude have been recorded from

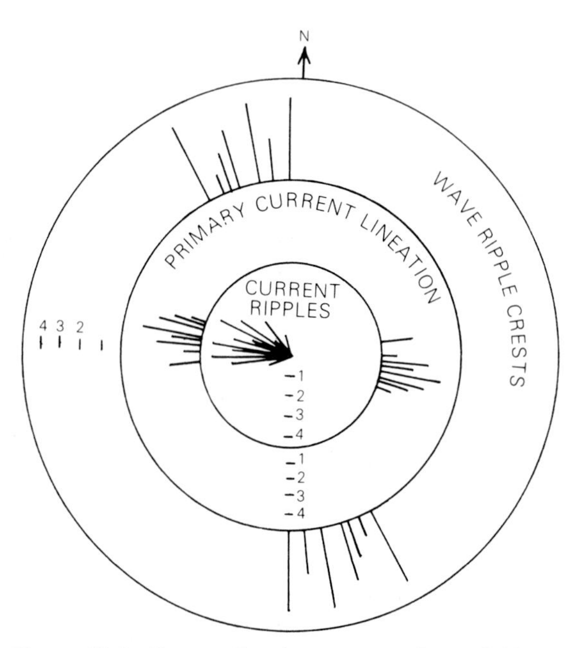

Figure 10.1 Current directions presented as radial line or 'spoke' diagrams. Note how different types of structure are separated and how those structures which give only direction and not sense of movement are plotted as double-ended lines.

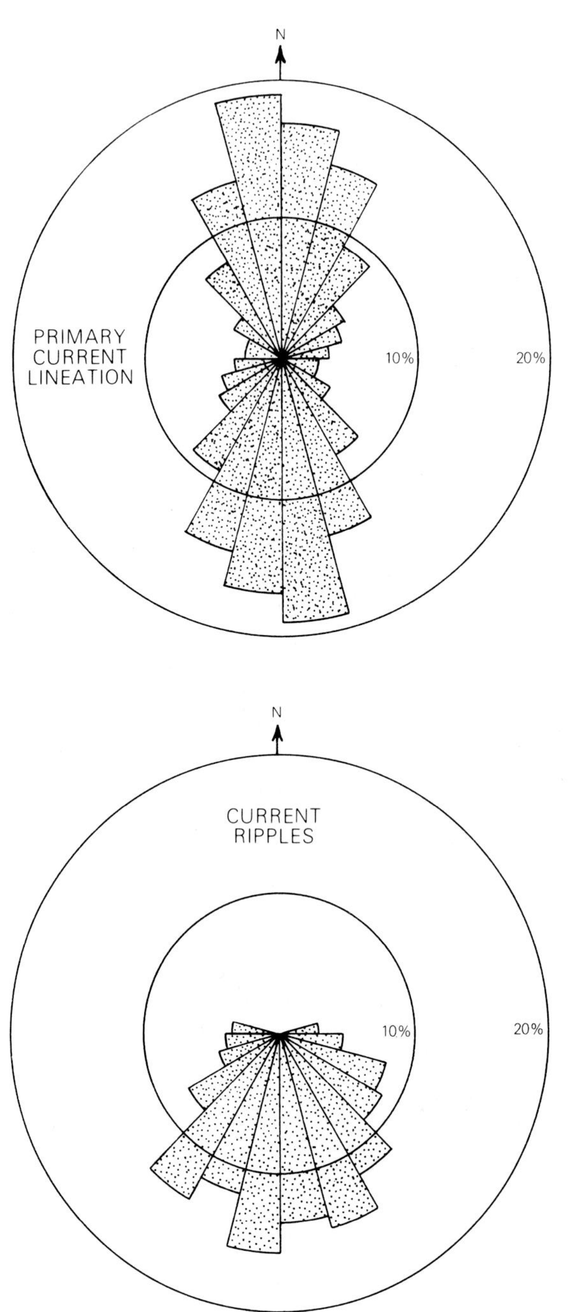

Figure 10.2 Current directions presented as 'rose' diagrams with 15° class intervals. Data from different classes of structure are separated and structures which give only direction and not sense of movement are plotted as double-ended sectors. The radius of arc of each class sector is proportional to the number of readings in the class.

cross-bed foresets, the plotting of poles to the foresets on a stereogram may be preferred to the plotting of foreset azimuths.

When data are more numerous, line or spoke diagrams become very cluttered and data are then better grouped in a circular histogram or 'rose diagram' (Fig. 10.2). The class interval for such diagrams can be set at any size, although 10°, 15°, 20° and 30° are the most commonly used, depending on the volume of data. With abundant data, small class intervals show the detailed structure of the population, but with few data, small class intervals may give a false idea of complexity. In addition, the fact that rose diagrams are usually plotted with a linear relationship between abundance and radius can lead to exaggeration of the apparent importance of more abundant classes. Modal classes may seem more abundant and well defined than they really are. What solutions can you offer for the problems of plotting data that fall on class boundaries?

Using the data in Table 10.1, and with reference to Appendix A, reorientate the cross-bedding measurements to their position before tectonic tilting. Plot the data as stereograms and as radial line and rose diagrams for dip azimuth. Assess the relative merits of the different types of plot.

Table 10.1 Dip and dip direction of cross-bedded foresets in beds with a tectonic dip. Use the procedure outlined in Appendix A to reorientate the cross bedding to its pre-tectonic orientation.

Tectonic dip	*Cross-strata*
042/28	359/39
076/52	048/34
068/32	000/05
253/30	252/50
090/38	057/24
244/49	258/69
105/42	120/22
048/47	023/59
246/38	280/34
068/40	071/17
090/46	117/39
057/37	009/36
029/36	345/35
264/48	270/64
072/61	067/45
090/20	010/06
079/64	098/62
080/42	048/37
090/12	190/12
250/16	293/38

10.2.2 Collection of data in present-day sedimentary environments

A whole range of types of observation and methods for recording data can be applied to modern sediment surfaces. Try to visit a tidal flat, a beach or an exposed river bed and carry out a systematic survey of the structures in a selected area. In some cases a single traverse across an area will be adequate, whereas in others more detailed mapping might be called for. Generally the features to be recorded and mapped are predetermined and the main problems can be of navigation and location, particularly on wide, low and featureless areas such as tidal flats.

For straight-line traverses two sighting posts placed some distance apart at one end of the traverse, and in line with it, are a great help. By keeping them in line it is possible to steer an accurate straight-line course on foot or by boat. Establishing one's position along the traverse is more difficult. If one has a topographic map at an appropriate scale, compass bearings on nearby features of known position off the line of section provide good fixes on long traverses. Measurement by tape or chain may be used over shorter distances or for more detailed work.

Mapping an area presents more complex problems. On a small scale it may be possible to mark out a measured grid; on a larger scale a series of cross-cutting traverse lines can be established by marker posts around the edge of the mapped area, like those set up for single traverses. Often, however, it will be necessary to take bearings or other angular measurements on surrounding fixed points. A sextant is an accurate and efficient way of doing this. Two angles measured between any three fixed points establish position quite accurately.

When surveying on foot, remember that surface features on loose sediment are easily ruined by footprints and so photographs should be taken at an early stage. For the same reason try to use a few strategic pathways. When working from boats, problems of disturbance are less acute, but observing the sediment surface can present problems. In shallow and reasonably clear water, a glass-bottomed box is of great use and polarising sunglasses can help by cutting down reflection. In deeper or turbid water, indirect methods of observation such as echo-sounding become essential.

Descriptions of sediment surfaces can be made at various levels of detail from the qualitative description of the type of bedform to detailed measurement of dimensions, orientations and distribution densities of particular structures. Systematic recording is often helped by a data sheet which can be completed at each locality. An example is shown in Figure 10.3. Presentation of directional data has been discussed in Section 10.2.1. These and other features can be shown on maps or profiles in a variety of ways. Rose diagrams can be superimposed on maps and maps can be contoured for quantifiable features such as bedform height and spacing, or distribution density of burrows.

Examination of internal structures of modern sediments can be achieved by digging trenches or by taking shallow cores (Fig. 10.4). Allow carefully cleaned sides of trenches to dry and structures will often show up in much more detail than on a freshly cut surface. When time is short or the water table is too high for trenching, cores of considerable length can be obtained by pushing boxes or tubes into the sediment.

When taking cores be careful to record their orientation where possible. In the laboratory, cores can be impregnated with resin to preserve them permanently and to show structures more clearly. Procedures for collecting and impregnating shallow cores are given in Appendix B.

10.2.3 The measurement of sections in rock sequences

Worthwhile environmental interpretations of sedimentary rock sequences rely heavily on measured sections. In Chapter 2 we outlined the importance and some of the problems of section measurement. One or two points mentioned in Chapter 2 are important enough to warrant repetition and emphasis here. In logging a sedimentary section you should first decide upon subdivision into lithological units. Then describe and record the thickness and internal features of each unit and determine the nature of its contact or boundary with adjacent units. Once you have decided which units to record, measurement of thickness is usually fairly straightforward. If beds conspicuously thicken and thin laterally within the extent of the exposure, record this either by noting it on the single measured section or by measuring more than one laterally equivalent section. When drawing up the section, remember to adjust the thickness of units so that the total thickness of the sequence is accurately recorded. For example, always showing the maximum thickness of units in a section through lenticular beds would seriously exaggerate the total thickness.

The features you record will vary with the nature of the sequence and also with the detail of interpretation required. We cannot stress too highly that there is no absolute standard of description. Each investigation has its own aims and timetable, and these determine the criteria for description and subdivision into units.

Locality: ______ **Sample study locality:** ______ **Map reference:** ______
Sub-environment number and type: ______
Survey map no. ______ Plane table map no. ______ Sketch map no. ______
Scaled drawing nos. ______ Film nos. ______ Photograph nos. ______
Sample bag nos. ______
Water sample nos. ______

Feature	Observations at sample locality	Observations in adjacent areas
Sediment types and conditions (1) Grain size (mm or ϕ) (2) Mineral composition (types %) (3) Colour–Munsell or Goddard scale (4) Fabric–texture (5) pH (6) Eh – depth to reducing layer (7) Salinity (8) Temperature – air : substrate	mean mode(s) orientation water sample	
Primary structures - bedforms (1) Surface { spacing (e.g. wavelength), relief (e.g. height), index } { symmetry (lee face), crest/trough (stoss face), index } orientation -- crest inclination – dip { crestline shape, degree of bifurcation } (2) Interior foresets dip, azimuth drapes (3) Secondary and other structures	1 2 3 4 5 Photo/drawing no.	1 2 3 4 5 Photo/drawing no.
Life forms (1) Permanent forms–life assemblage (a) flora (b) epifauna (c) infauna macrofauna, meiofauna (0.5 – 0.005 mm), microfauna (2) Temporary forms – life assemblage (a) marine (b) terrestrial obligate (dependent on intertidal zone) facultative (not dependent on intertidal zone) (3) Drifted forms–death assemblage	Types: depth: frequency m^{-2}: orientation; planktonic, nektonic, benthonic (vagrant, vagile, sessile, gregarious, solitary, epibiotic); attached, free, burrowing, boring, herbivore, carnivore, omnivore; *deposit feeder,* collector, swallower; *suspension feeder,* filterer, awaiter; *grazer; scavenger.* Photo no.	Photo no.
(4) Ecological relationships predator – prey, symbiosis, food chain, food web		
(5) Ichnological observations morphological preservational behavioural	Track, trail, radial trace, tunnel, spreite, pouch; relief; epichnia, endichnia, hypichnia, exichnia; resting, dwelling, feeding, grazing, crawling, locomotion, escape; orientation.	

Figure 10.3 Sample data sheet for the systematic collection of observations on a tidal flat.

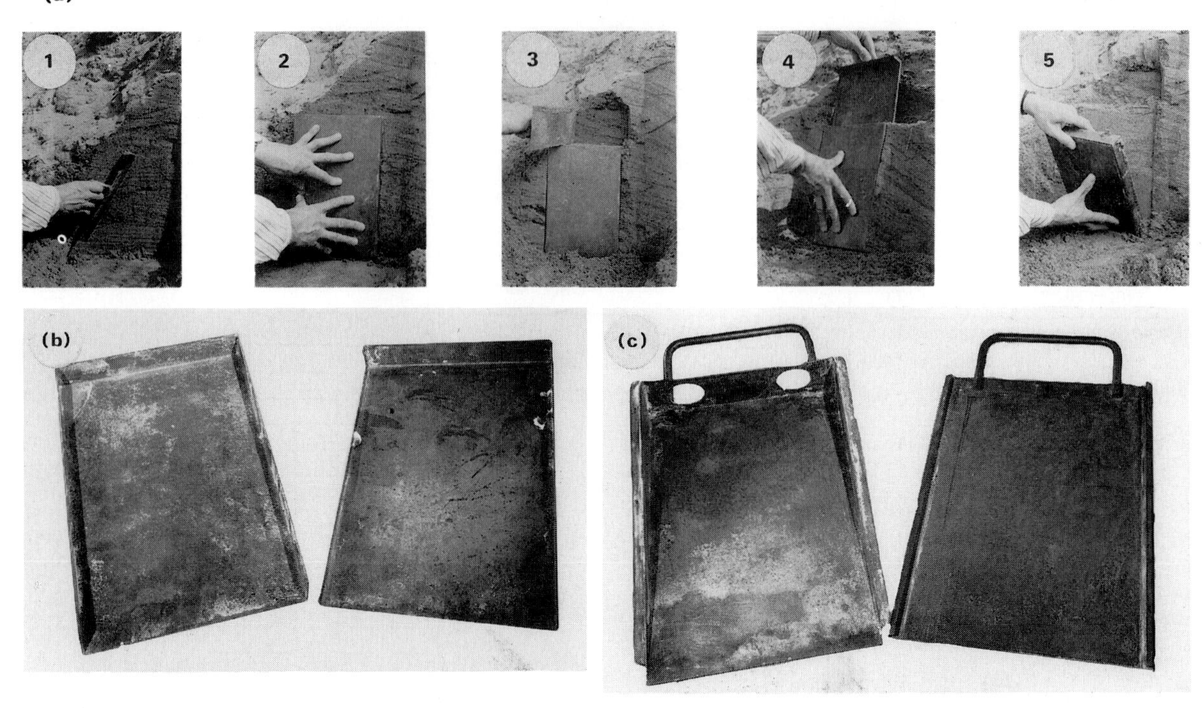

Figure 10.4 Methods of recovering box core samples from present-day sediment surfaces: (a) taking a box core from a trench wall; (b) straight-sided ('Senckenberg') box core used from the sediment surface; (c) tapering ('Reineck') box core for use in waterlogged or mobile sediment.

The feature of measured sections most commonly ignored is the nature of contacts between units. Some contacts are gradational and it may sometimes be difficult to decide exactly where a boundary should be placed. Other contacts are sharp and some may be clearly erosive, with conspicuous relief truncating underlying bedding or with erosional structures superimposed upon the surface. In Section 4.4.3 we suggest clues that may indicate an erosional contact even when these features are missing. Always consider the possibility of erosion whenever you see a sharp contact, although, of course, not all sharp contacts are necessarily erosive.

Recording of information from vertical sections demands a disciplined method of working. Some geologists prefer to draw a graphic log while in the field, either in their notebooks or on specially prepared sheets (Fig. 10.5). Others use an essentially verbal log, supplemented with drawings where appropriate and leave the drawing of a graphic log for the laboratory. Remember that the more time you spend drawing up an elaborate log in the field, the less time you spend observing the rocks.

If you use a verbal log in the field, organise it so that information can be easily abstracted later and specimens and photographs can be easily related to it. A system of columns, each devoted to separate features such as thickness, lithology and the contact with the overlying unit, is convenient. Other columns can be added for specimen and photograph numbers and for palaeocurrent measurements.

Presentation of measured sections as graphic logs is a matter of personal judgement. A glance through any sedimentological journal will show that there are almost as many styles of graphic log as there are authors. There is nothing wrong with this and it reflects to some extent the different aims and emphases of different pieces of work. Some styles of graphic logs are, however, more easily understood than others. Examples of what seem to be fairly easily understood schemes are shown in Figure 10.6. In two of these, grain size is indicated schematically by column width and the symbols for lithology and sedimentary structures are relatively straightforward and self evident. The nature of the contacts between units is also well shown. Where mixed sandstone–mudstone lithologies occur, it

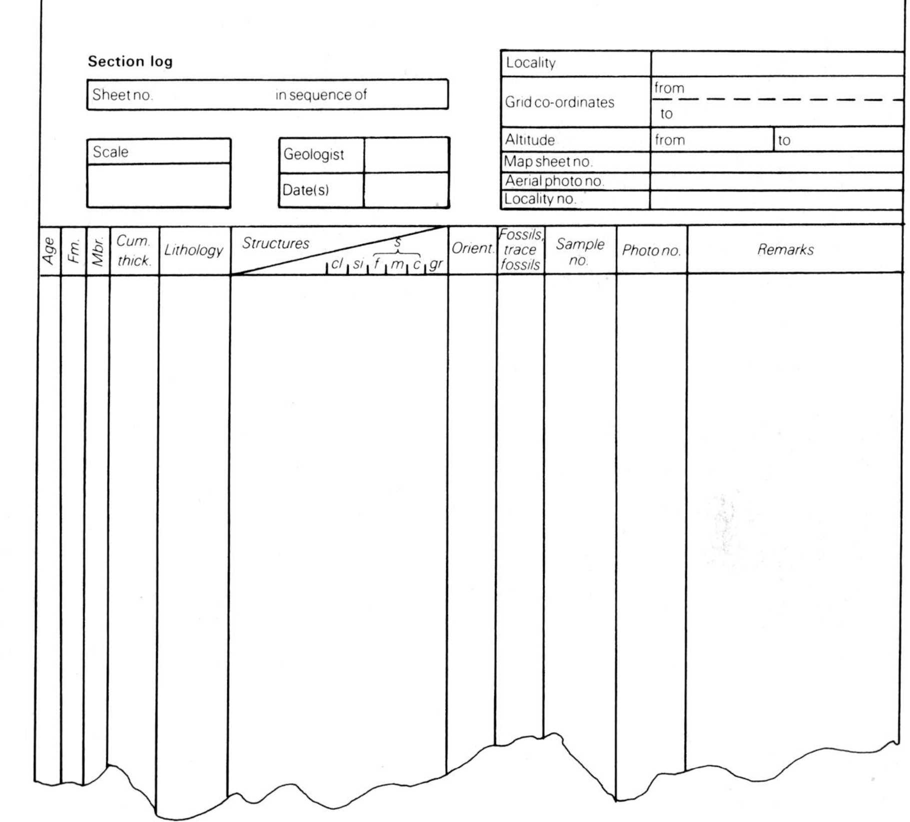

Figure 10.5 Example of a logging sheet for the recording of vertical sequences in sedimentary rock successions. (After scheme prepared for Greenland Geological Survey by L. B. Clemmensen and F. Surlyk.)

is a good idea to indicate their relative proportions in a separate 'lithology' column (e.g. Fig. 10.6c).

Palaeocurrent measurements are best recorded opposite the units from which they were taken. A well drawn graphic log incorporating all these features serves as a sound base for your own environmental interpretation and also enables others to use your data to suggest alternative or more refined interpretations.

Where a sequence shows conspicuous lateral variation, more than one vertical section may be needed. Location and spacing of the sections will depend on the complexity of the variation and on the aims and timetable of the study. With laterally continuous exposure, the full two-dimensional form of the lithological units should be recorded. Panoramic photographs of two-dimensional exposures may often help in the construction of suitable diagrams. Where erosion surfaces or bounding surfaces are apparent, be sure to establish if a hierarchy exists and take care to assign a particular surface to its appropriate level.

With discontinuous exposure, as, for example, with a series of separated quarries, stream sections or boreholes, it is normally only possible to link the sections by correlation lines on the most confidently identified bedding surfaces. Be careful when correlating sandstones or coarser units, particularly if there is any evidence that they may be lenticular. Correlation of similar sandstones at similar positions in a sequence may give a misleading impression of lateral continuity. Beds could have died out laterally between observed sections.

10.3 Analysis, interpretation and evaluation of data

10.3.1 Directional data

In looking at the pattern of current directions shown by a rose or radial line diagram you should have four questions in mind:

(a) Is the pattern unimodal, polymodal or without any obvious preferred direction?
(b) What are the dominant or mean direction or directions?

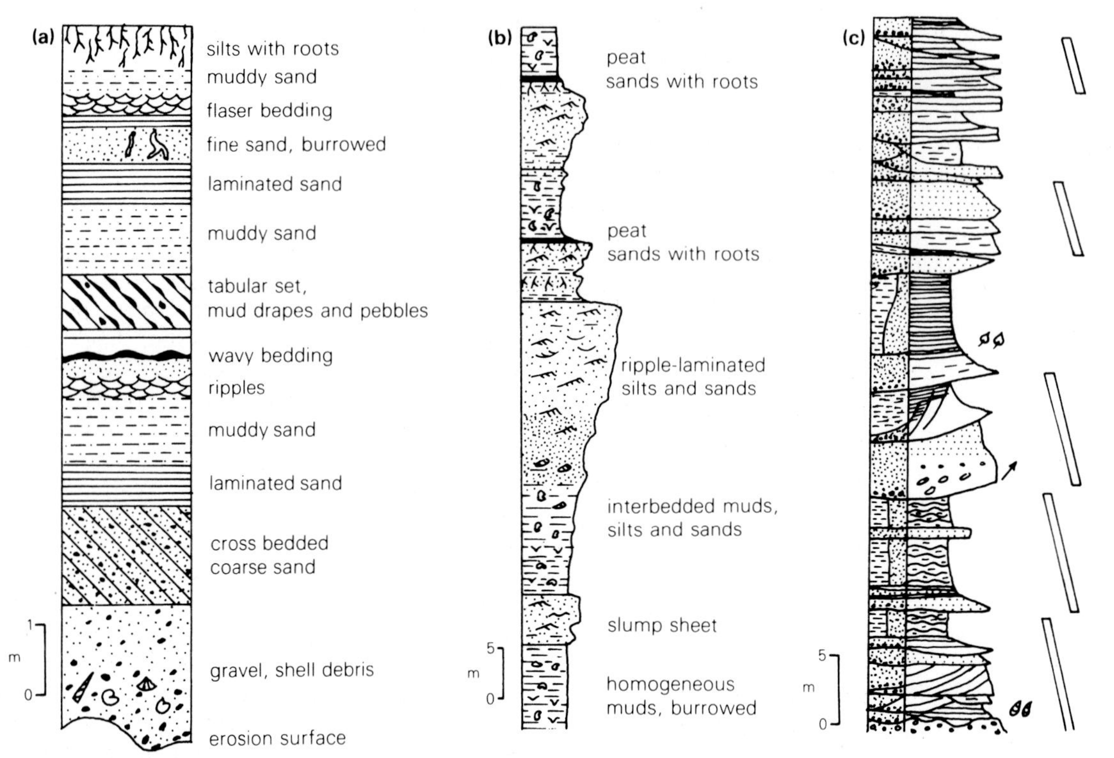

Figure 10.6 Examples of graphic logs of measured vertical sequences through sediments. Note how in (c), the proportion of sand and mud in the 'mixed lithologies' is indicated in a separate column and how major fining-upwards units are indicated. (Based (a) on Greer 1975; (b) on Coleman & Wright 1975; (c) on Surlyk 1978.)

(c) How widely scattered are the directions about the mean values (i.e. what is the spread of the directions)?

(d) Are there any systematic differences in the means and dispersions of directions derived from different types of structure?

(a) The most important feature of a set of directional data is its pattern of preferred directions. In rock sequences, this can often give information on the nature of the environment of deposition and the orientation of the regional palaeoslope. Although unimodal patterns are more common, certain environments generate bimodal or even polymodal patterns.

(b) Where the pattern of directions is clearly unimodal, it is possible to calculate a mean direction. It would, however, be nonsense to calculate a single mean value for a bimodal population, particularly one with a bipolar pattern. The single calculated mean could well be at right angles to both obvious dominant directions and therefore be totally meaningless.

Because directional data are distributed around a circle so that 360° = 0°, calculation of mean cannot be a matter of simple averaging. For example, the mean of two bearings close to but either side of north, say 350° and 10°, by simple averaging is 180° when clearly the sensible mean is 360°. To overcome this difficulty at a simple level, use a false origin. After all, beginning the scale at north is a purely arbitrary convention. If an origin outside the range of the directions is chosen, a simple arithmetic average deviation from the false origin can be calculated which can them be restored to a true bearing. Such an approach is quite satisfactory for closely grouped data. A more widely applicable method treats the data as vectors and resolves the vector component to give a **vector mean** by applying the formula:

$$\tan\bar{\theta} = \frac{\Sigma \sin \theta}{\Sigma \cos \theta}$$

where $\bar{\theta}$ is the vector mean and $\Sigma \sin \theta$ and $\Sigma \cos \theta$ are the sums of the sines and cosines of the individual readings. In such calculations, it is important to take account of the sign of the trigonometric functions and also to have seen, by inspection, the general direction in which the mean is likely to lie.

In polymodal patterns, means can be calculated for data around each mode provided that the clustering is clear with no overlapping or ambiguous readings. Where there is overlap, calculations along these lines are probably meaningless and visual inspection of the rose diagram will be at least as good a guide to preferred directions.

(c) The clustering of directions about the mean value or values can sometimes be quite close and sometimes more widely dispersed. In many cases it will be good enough to describe dispersion qualitatively by inspection of the rose diagrams. In some circumstances a more quantitative expression of dispersion may be more appropriate. The parameter most commonly used to express this is the **vector strength** which is given by the equation:

$$S = \frac{\sqrt{(\Sigma \sin \theta)^2 + (\Sigma \cos \theta)^2}}{n}$$

where n is the number of readings. High values indicate low dispersion and low values a higher dispersion.

The most obvious example of the significance of the pattern of modes is a tidal environment where a particular type of bimodality, namely bipolarity, may be produced by the ebb and flow currents. However, not all tidal settings produce and preserve sedimentary structures with a symmetrical bipolar pattern. If either the ebb or the flood current dominates by even a small amount, an asymmetrical bipolar pattern or even a unimodal one may result.

In aeolian dune deposits, wind directions are not controlled by the topographic slope and more than one modal wind direction may be recorded from cross bedding. In certain sandy river deposits, tabular cross bedding resulting from the migration of diagonal cross-channel bars shows a bimodal pattern due to transverse components in their movement. If the mean directions of cross bedding from several fluvial sandstone units in an ancient sequence show a high dispersion, sinuous rivers are suggested; low dispersion suggests straighter channels.

(d) Differences in the pattern of directions recorded from different types of structures in the same sequence can usually be detected by visual inspection of rose diagrams. It is seldom necessary to resort to statistical tests. The interpretation of such differences will clearly vary from case to case. In the rock record it may be possible to record differences in wave and current directions or differences in movement pattern of sinuous-crested dunes (trough cross bedding) and straight-crested dunes and bars (tabular sets). Study the plots of palaeocurrents presented in Figure 10.2 and your own plots of the data from Table 10.1. What sort of interpretations of processes or, more tentatively, of environment might be made from these data?

10.3.2 Mapping of modern environments

The main aims of mapping sedimentary structures in modern environments are to learn something of the distribution of hydrodynamic or wind energy within the environment and to predict the likely patterns of lithology and sedimentary structures, should deposits of the environment be preserved. The second aim has particular relevance to the application of uniformitarian principles to the interpretation of sedimentary rocks.

The most common method of investigating the distribution of water-produced bedforms on intertidal areas or on river beds is on foot at low water. Although the mapping is quite straightforward, interpretation is more complex, as the pattern observed will probably be the product of a succession of flow states. All bedforms need time to respond to a change in flow. Large bedforms, produced under conditions of strong flow, may be stranded if the water level falls rapidly. Small bedforms, such as ripples, adjust more readily and many continue to respond to the flow almost to the point of emergence.

Prediction of the vertical sequence of sediment that will be generated by a particular environment requires answers to several questions. Which of the observed bedforms is most likely to generate preserved internal structures? What is the distribution of such bedforms on the broader topography of the environment? How is the environment as a whole changing through time? In particular, is a systematic migration of sub-environments taking place? If so, it can be predicted that structures developed in topographically low areas will occur low in a vertical sequence with structures from successively higher topographic areas coming in above *in the same vertical order as their horizontal distribution* (Fig. 10.7). This method of relating the lateral distribution of surface features or sub-environments to a vertical sequence of lithology and sedimentary structures is **Walther's principle of succession of facies** and is one of the fundamental starting points for any environmental interpretation of ancient sediments (see 1.3 & 10.3.3).

One complicating factor particularly important in intertidal settings is the behaviour of burrowing animals. Animals that live below a surface subjected to particular conditions of currents, waves or emergence may extend their burrows down into layers of sediment that were deposited under conditions quite different from those now at the surface. By the time the burrowing takes place, these different conditions may have shifted some distance from the site of burrowing. In other words, burrows can cut across the vertical sequence and the animals that produce burrows in a particular unit of sediment cannot be assumed to have lived under the conditions in which that sediment was laid down.

10.3.3 Vertical sequences in rocks

The interpretation of sequences of sediments in terms of their environment of deposition is one of the main aims of sedimentology. We have seen how most sedimentary structures allow interpretation of process of erosion, deposition or postdepositional alteration and how many of these structures and processes are common to many environments. The *sequence* of processes deduced from the rock succession is one major starting point for environmental interpretation.

Before considering in detail the vertical sequence of lithology and sedimentary structures several more general features of any sedimentary succession may allow us to develop a preliminary hypothesis about the environment of deposition. The most obvious of these is the presence or absence of fossils and, if they are present, their type. They may tell us if a sequence is marine, non-marine or marginal or, if they are absent, the possibility of a continental setting may have to be considered.

In some, cases, it may be possible to use fossils as a basis for making preliminary conjectures about water depth. Before this is attempted it is important first to establish if the fossils are *in situ* or have been transported after death. This may be judged by their state of articulation, their abrasion and the way in which they lie within the sediment. If one is reasonably confident that the fauna is *in situ* or has not been transported far, then it may be possible to infer a shallower 'shelf' setting for a more shelly fauna and a deeper setting for a more pelagic fauna. Palaeontological expertise is normally required to carry forward such arguments to a more sophisticated level. The environmental interpretation of Precambrian sediments is greatly handicapped by the lack of this basic information and it is often difficult in some examples even to decide between a continental and a shallow marine origin. Evidence of sub-aerial exposure of the sediment surface in the form of rain pits, desiccation mudcracks, seatearths and soil profiles (see Ch. 9) also helps to put constraints on the range of environmental possibilities.

Once the broad environmental context has been narrowed down by such considerations, a more detailed analysis can proceed based on, amongst other things, the nature of the vertical sequence.

The law of superposition tells us that the vertical sequence of lithologies records changes in depositional conditions *through time at that point.* Such changes occur for two fundamentally different reasons. In one case, the environment remains essentially unchanged, but within it changes in conditions take place *through time* to produce distinctive sediment units. For example, a deep basin normally receiving fine-grained sediment from suspension may have this background condition punctuated by the arrival of intermittent turbidity currents which deposit layers of coarser sediment. The vertical change in lithology does not then record a change of environment but a temporary change in the prevailing processes. The idea of temporary changes in conditions within a more or less stable environment means that we may be able to recognise the products of 'normal' and 'catastrophic' deposition (see 2.2.3).

In the second case, environmental conditions remain essentially constant through time, but there is a *spatial segregation* of processes and products within the environment. A gradual shifting of the environment through time thus leads to a vertical sequence of changing lithology and structure. This second case shows the application of Walther's principle of succession of facies introduced in the previous section. The operation of this principle is best illustrated by a simple general model (Fig. 10.7). During migration of this

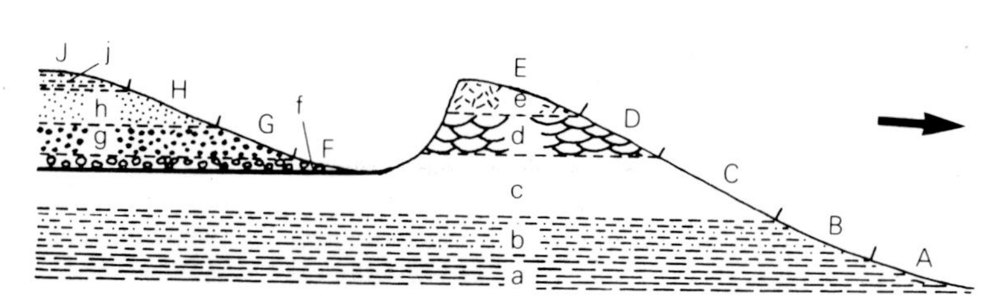

Figure 10.7 A schematic diagram to illustrate Walther's principle of succession of facies. Sub-environments A–E are on a sloping surface which is building out to the right, generating lithological units a–e. A channel cut into the top of this topography, migrates in the same direction and generates lithological units f–j by deposition in sub-environments F–J.

system to the right, sub-environments A–E generate lithological units a–e in the same order and with gradational contacts. The channel system on top also generates its own gradational sequence of lithologies f–j above an erosion surface due to the migration of channel sub-environments F–J. If one examines the sequence on the left-hand side of the diagram without taking account of the contacts between units, it would be possible to infer mistakenly that sub-environment F had been adjacent to sub-environment C. Recognition of erosion surfaces is vital and it is necessary to begin the application of Walther's principle afresh above each such surface.

10.3.4 Lateral relationships in rocks

Although the vertical sequence is vital to the interpretation of rock successions, there are many cases where lateral relationships also play a vital role in developing an interpretation. Where lateral changes have been recorded, they can commonly allow refinement of the environmental model or the resolution of uncertainties that remain from consideration of a single vertical sequence. For example, a sequence of interbedded mudstones and sharp-based, graded sandstone beds can be interpreted in terms of normal and catastrophic deposition, but such processes may take place in a variety of environmental settings. If it was found that the sequence was laterally equivalent to a channel sandstone showing abundant unidirectional cross bedding, it might be a fair inference that the overall setting was fluvial. The interbedded sequence might then be interpreted in terms of crevasse splays into an overbank flood plain or lake during floods. In contrast, if the interbedded sequence proved to be the lateral equivalent of mudstones containing only a pelagic marine fauna, the inference might be that the sandstones are turbidites, the product of catastrophic density currents into a deep-water setting.

As a second example, we can consider the lateral variability in sediments in the upper parts of a widespread, upwards-coarsening deltaic sequence. The nature of the variability may allow us to make more specific inferences about the type of delta rather than just record a general deltaic interpretation. A birds-foot delta, like that of the present-day Mississippi, has great lateral variability in its delta-top sediments because of the segregation of channels, levees, interdistributary bays, etc. In contrast, an arcuate delta, such as the Nile, has rather less variability due to redistribution of sediment along the shoreline by wave action to give a more homogeneous sand sheet.

These comments on the interpretation of vertical and lateral relationships in rock successions are beginning to carry the discussion beyond the stated scope of this book. They are intended merely as an introduction to the more complex field of facies analysis and environmental interpretation which leads, through stratigraphy, to the development of palaeogeographic reconstructions.

By way of illustration and as an appetiser for this broader field of study, Figure 10.8 shows several sequences encountered in the rock record and how, through an interpretation of processes, the environmental interpretation may be approached.

Reference and study list

References marked with an asterisk * are suitable for 16–19 year old advanced level students in schools and colleges in the UK as well as for undergraduates.

Field experience

Suggestions for appropriate activities are given in the text. Particular attention should be paid to safety when mounting investigations at the sea coast and in estuaries with high tides and strong currents, or in mountainous areas. The references relating to fieldwork in Chapter 2 may be helpful, together with the following:

King, C. J. H. 1980. A small cliff-bound estuarine environment; Sandyhaven Pill in South Wales. *Sedimentology* **27**, 93–105. [Useful results from a student project carried out over a relatively short period. The methodology is particularly clear.]
Pugh, J. C. 1975. *Surveying for field scientists*. London: Methuen.

Laboratory experience

Suggestions for the following-up of fieldwork and the analysis of data are given in the text. The logging of borehole cores by groups of students can be particularly helpful. Many departments keep such cores for teaching purposes. After all the individuals in a group have completed their logs, comparisons between their efforts and those of their tutors and discussion of differences and discrepancies can be very instructive.

The following film may be of use:

Open University 1980. *Conceptual models in stratigraphy*. Colour, sound. Milton Keynes: Open University Enterprises.

Light reading

*Ager, D. V. 1973. *The nature of the stratigraphical record*. London: Macmillan.

Essential reading

A selection from:

Bouma, A. H. 1969. *Methods for the study of sedimentary structures*. New York: Wiley.

(a) *Main facts*	*Processes*	*Environment*
Coarse, red, structureless siltstone. No body fossils. Sparse concretions of calcium carbonate. Invertebrate burrows in lower part.	Deposition of silt from suspension. Could be from air or water. Post-depositional burrowing and growth of concretions.	Continental, overbank interchannel setting accumulating sediment by vertical accretion. Concretions grow in caliche soil profile.
Fine grained red sandstone. Ripple drift cross laminated. Invertebrate burrows.	Rapid deposition of sand during bedload transport as current ripples. Post-depositional colonisation by animals.	Channel margin/levee?
Fine to medium grained red sandstone. Trough cross bedded with unidirectional palaeocurrents. Siltstone clasts on set bases. Convolute bedding in some sets. Upwards reduction in set size.	Net accumulation of sand during migration of dunes. Post-depositional liquefaction suggests rapid deposition. Erosion taking place nearby.	Laterally migrating channel with deposition on one side; probably a point bar of a meandering stream.
Sharp surface with slight relief. Scoured sole marks. Concentration of siltstone clasts above surface.	Erosion by fluid scour. Winnowing of finer sediment to leave a lag of intraformational clasts.	

(b) *Main facts*			*Environment*
Cross bedded sandstone. Sharp contact with relief and siltstone clasts.	Dune migration; strong currents. Erosion and winnowing.	prograding delta front	Fluvial distributary channel.
Siltstone with parallel lamination and graded sand beds.	Deposition from suspension. with episodic decelerating flows.		Mouth bar.
Siltstone.	Deposition from suspension.		Delta slope.
Mudstone with marine fossils.	Deposition from suspension; high salinity.		
Coarsening upwards unit.	Increase in energy; shallowing?		Minor readvance of delta?
Mudstone.	Deposition from suspension.		
Siltstone with current ripples, burrows on top.	Mixture of deposition from suspension and from bedload transport by weak currents.		Abandonment of mouth bar.
Cross laminated sandstone with silty partings.	Bedload transport as ripples with quiet interludes.	prograding delta front	Proximal mouth bar.
Silty mudstone with siltstone laminae.			
Gradationally striped silty mudstone.	Deposition from siltstone with fluctuating supply.		Delta slope.
Mudstone with marine fossils.	Deposition from suspension.		Offshore; pro-delta.

(c) Main facts	Processes	Environment
Thick, roughly parallel-sided sandstone beds. Graded and multiple graded. Intraformational and exotic clasts. Mainly massive but laminated towards top. Dish and pillar structures. Convolute lamination. Thin mudstone interbeds with deep marine fauna.	Deposition of mud from suspension on deep, quiet sea floor. Episodic large-scale currents with erosion in early stages and rapid deposition of sand later. Vigorous water escape from liquefied sand bed. (Episodic currents into a deep sea setting are probably turbidity currents.)	General upward increase in the scale of the turbidity currents may be attributed to the build-out of lobes on a submarine fan.
Thin, parallel-sided sandstone beds interbedded with mudstones. Sandstones sharp-based, slightly graded. Some are massive in lower part. Parallel and cross laminated tops. Some convolute lamination.	Deposition of mud from suspension punctuated by sudden incursions of rapidly deposited sand by currents of waning strength.	

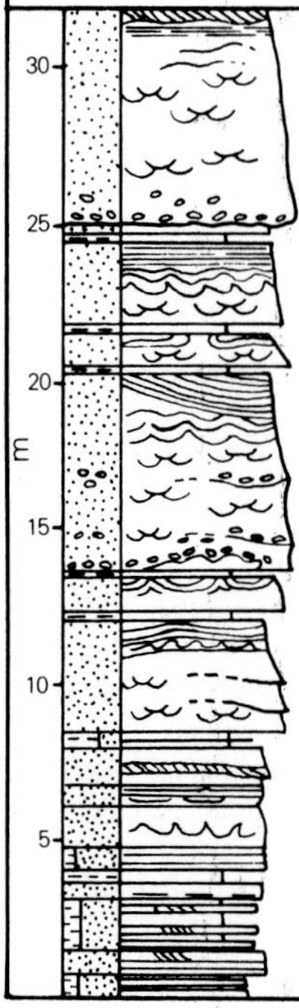

(d) Main facts	Processes		Environment
Lime sand with shells and intraclasts.	High energy erosion and redeposition.		Marine transgression.
Sand with cross bedding.	Strong currents; dune bedforms.		
Silty dolomite. Lime mud with nodular anhydrite. Enterolithic anhydrite.	Quiet deposition of lime mud. Later alteration to dolomite. Quiet deposition and displacive growth of anhydrite; high evaporation.	prograding carbonate shoreline to give emergent conditions	Supratidal flat in arid environment.
Lime mud with algal lamination, crystals and nodules of anhydrite.	Trapping of lime mud by algal mat. High evaporation and emergence. Displacive and poikilitic growth of evaporite within sediment.		Algal mat zone at about high tide level.
Lime sands. Marine fossils.	Currents of variable strength in marine setting.		Nearshore marine possibly intertidal.
Lime muds with burrows and marine fossils.	Generally quiet conditions. Occasional high energy events. Slow deposition allows burrowing.		Offshore shallow marine; calm fair weather with storms.
Lime mud.	Quiet conditions.		
Lime sand with shells and intraclasts.	High energy erosion and redeposition.		Transgressive beach.

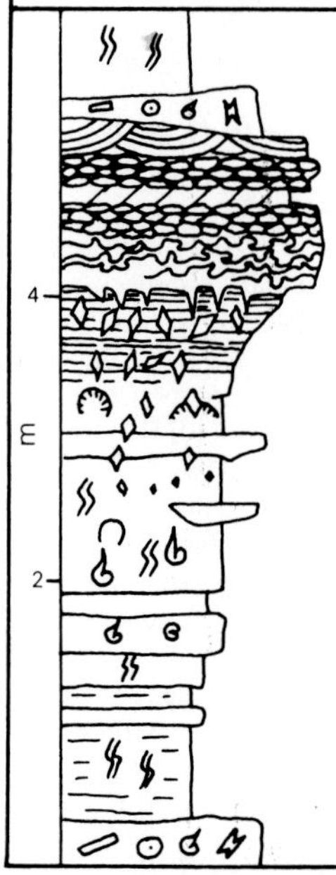

Figure 10.8 Examples of some commonly occurring vertical sequences of sediment showing their interpretation in terms of process and environment by the application of Walther's principle. (Based (a) on Allen 1964; (b) on Kelling & George 1971; (c) on Kruit *et al.* 1975; (d) on James in Walker 1979.)

Brenchley, P. J. and B. P. J. Williams (eds) 1985. *Sedimentology: recent developments and applied aspects*. Geol Soc. Lond. Sp. Publ. 18. Oxford: Blackwell Scientific. [See especially papers by Anderton (clastic facies models), McCave (shelf sediments), Stow (deep water clastics), Suthren (volcaniclastics) and Tucker (carbonates).]

Cant, D. J. and R. G. Walker 1976. Development of a braided-fluvial facies model of the Devonian Battery Point Sandstone, Quebec. *Can. J. Earth Sci.* **13**, 102–19.

Cheeney, R. F. 1983. *Statistical methods in geology*. London: Allen & Unwin.

Dunbar, C. O. and J. Rodgers 1957. *Principles of stratigraphy*. New York: Wiley. [Part IV of this book synthesises the data discussed in Parts I–III and deals with structures and facies in local and regional successions.]

Galloway, W. E. and D. K. Hobday 1983. *Terrigenous clastic depositional systems*. New York: Springer-Verlag.

Ginsburg, R. N. (ed.) 1975. *Tidal deposits: a casebook of recent examples and fossil counterparts*. New York: Springer. [A useful compilation of case histories, many using analyses of facies and current directions in combination.]

Harms, J. C., J. B. Southard, D. R. Spearing and R. G. Walker 1975. *Depositional environments as interpreted from primary sedimentary structures and stratification sequences*. Short Course Notes no. 2. Dallas: SEPM.

Krumbein, W. C. and F. A. Graybill 1965. *An introduction to statistical models in geology*. New York: McGraw-Hill.

Miall, A. D. 1982. Recent developments in facies models for siliciclastic sediments. *J. Geol. Education* **30**, 222–40.

Middleton, G. V. 1973. Johannes Walther's law of correlation of facies. *Bull. Geol Soc. Am.* **84**, 979–88.

Reading, H. G. (ed.) 1986. *Sedimentary environments and facies*, 2nd edn. Oxford: Blackwell Scientific. [All chapters give an idea of how the skills, techniques and attitudes developed in this book may be developed. Ch. 2 sets out a concise view of facies and the ways in which they may be analysed. Chs 3–13 put the ideas into practice in the context of different environments.]

Selley, R. C. 1969. Studies of sequences in sediments using a simple mathematics device. *J. Geol Soc. Lond.* **125**, 557–81. [The discussion is particularly valuable.]

Selley, R. C. 1976. *An introduction to sedimentology*. London: Academic Press. [Pages 253–313 give a review of facies and models.]

*Selley, R. C. 1985. *Ancient sedimentary environments*, 3rd edn. London: Chapman and Hall.

Scholle, P. A., D. G. Bebout and C. H. Moore (eds) 1983. *Carbonate depositional environments*. Tulsa: AAPG Mem. 33.

Scholle, P. A. and D. Spearing (eds) 1982. *Sandstone depositional environments*. Tulsa: AAPG Mem. 31.

Till, R. 1974. *Statistical methods for the earth scientist: an introduction*. London: Macmillan.

Walker, R. G. (ed.) 1984. *Facies models*, 2nd edn. Waterloo, Ont.: Geol. Assoc. Canada. [All articles (see reference lists in earlier chapters) show how the skills, techniques, attitudes developed in this book may be taken further. Walker (Ch. 1, pp.1–9), gives a clear introduction and Harper (Ch. 2, pp. 11–13) suggests a modification in the method of analysing vertical sequences.]

Further reading

Allen, J. R. L. 1964. Studies in fluviatile sedimentation: six cyclothems from the Old Red Sandstone. *Sedimentology* **3**, 163–98.

Allen, J. R. L. 1965. Fining-upwards cycles in alluvial successions. *Geol J.* **4**, 229–46.

Broussard, M. L. (ed.) 1975. *Deltas. Models for exploration*. Houston: Houston Geol Soc.

Collinson, J. D. 1969. The sedimentology of the Grindslow Shales and the Kinderscout Grit: a deltaic complex in the Namurian of northern England. *J. Sed. Petrol.* **39**, 194–221. [A description and interpretation of a major, prograding deltaic sequence. This paper complements that of Walker (1966).]

Duff, P. McL. D., A. Hallam and E. K. Walton, 1967. *Cyclic sedimentation*. Amsterdam: Elsevier.

Elliott, T. 1976. Upper Carboniferous sedimentary cycles produced by river-dominated, elongate deltas. *J. Geol. Soc. Lond.* **132**, 199–208.

Krumbein, W. C. and L. L. Sloss 1963. *Stratigraphy and sedimentation*, 2nd edn. San Francisco: W. H. Freeman.

*Laporte, L. 1968. *Ancient environments*. Englewood Cliffs, NJ: Prentice-Hall.

Matthews, R. K. 1974. *Dynamic stratigraphy*. Englewood Cliffs, NJ: Prentice-Hall.

Miall, A. D. 1973. Markov chain analysis applied to an ancient alluvial plain succession. *Sedimentology* **20**, 345–64.

Morgan, J. P. (ed.) 1970. *Deltaic sedimentation, modern and ancient*. Tulsa: SEPM. Sp. Publ. 15.

Raaf, J. F. M. De, H. G. Reading and R. G. Walker 1965. Cyclic sedimentation in the Lower Westphalian of north Devon, England. *Sedimentology* **4**, 1–52.

Read, W. A. 1969. Analysis and simulation of Namurian sediments in central Scotland using a Markov-process model. *J. Int. Assoc. Math. Geol.* **1**, 199–219.

Reading, H. G. 1971. Sedimentation sequences in the upper Carboniferous of northwest Europe. *C. R. 6é Congr. Int. Strat. Geol. Carbonif., Sheffield* 1967 **IV**, 1401–12.

Walker, R. G. 1966. Shale Grit and Grindslow Shales: transition from turbidite to shallow water sediments in the Upper Carboniferous of northern England. *J. Sed. Petrol.* **36**, 90–114. [An early example of the use of the facies concept and the development of a facies model. The story is completed by Collinson (1969) – see above.]

Walker, R. G. 1978. Deep-water sandstone facies and ancient submarine fans: models for exploration for stratigraphic traps. *Bull. Am. Assoc. Petrolm Geol.* **62**, 932–66.

Extra references relating to figure captions

Coleman, J. M. and L. D. Wright 1975. Modern river deltas: variability of processes and sand bodies. In *Deltas: models for exploration*, M. L. Broussard (ed.), 99–149. Houston: Houston Geol Soc.

Greer, S. A. 1975. Sand body geometry and sedimentary facies at the estuary–marine transition zone, Ossabaw Sound, Georgia; a stratigraphic model. *Senckenberg. Mar.* **7**, 105–35.

James, N. P. 1979. Shallowing-upward sequences in carbonates (facies models 10). In *Facies models*, R. G. Walker (ed.), 109–19. Waterloo, Ont.: Geol Assoc. Canada.

Kelling, G. and G. George 1971. Upper Carboniferous sedimentation in the Pembrokeshire Coalfield. In *Geological excursions in South Wales and the Forest of Dean*, D. A. Bassett and M. G. Bassett (eds), 240–59. Cardiff: Geologists Assoc., South Wales Group.

Kruit, C., J. Brouwer, G. Knox, W. Schollnberger and A. van Vliet 1975. *Une excursion aux cones d'alluvions en eau profonde d'age Tertaire près de San Sebastian*. Excursion Guide 23. Nice: IX Int. Sed. Congress.

Surlyk, F. 1978. *Submarine fan sedimentation along fault scarps on tilted fault blocks (Jurassic/Cretaceous boundary, East Greenland)*. Grønlands Geol. Unders,. Bull. 128.

Appendix A

Restoration of directional data from tectonically inclined beds

When sedimentary rocks are tectonically inclined, it becomes necessary to restore directions measured on sedimentary structures to their orientation prior to deformation if they are to have palaeocurrent significance. For linear structures such as flutes, primary current lineation, the alignment of ripple crests or the axes of sets of trough cross bedding, deviations induced by tectonic dips of less than 25° are sufficiently small to be ignored. More serious deviations occur when measurements are made on the foresets of cross bedded sets. Tectonic dips of greater than only 5° then necessitate reorientation.

In order to restore directions to their original attitude, it is necessary to plot and manipulate the data on a stereogram. It is necessary to know the magnitude and direction of dip both of the foresets, as they now occur, and of the overall sequence (i.e. local tectonic dip). The procedure outlined only applies if fold plunge is negligible. A more complex procedure is needed for plunging folds. Plot normals to both the foresets and the original depositional horizontal on a stereographic projection (Fig. A.1a). Rotate the points until the normal to the depositional horizontal lies on a great circle of the projection (Fig. A.1b). In order that the beds are restored to horizontal this point must be moved to the pole of the projection and, by doing that, the normal to the foreset must move a similar angular distance along the small circle upon which it lies (Fig. A.1c). The new position of this point now represents the normal to the foreset at the time of deposition and this can be converted to a direction and magnitude of dip. This direction (foreset azimuth) may then be used as an indicator of palaeocurrent direction.

For linear data in steeply dipping beds, plot on the stereogram the attitude of the lineation in space and rotate both the normal to bedding and the lineation as described above.

Exercise

Table 10.1 gives a series of measurements of local tectonic dips and dips of foresets measured in the dipping beds. Reorientate the foresets using the method outlined and plot the foreset azimuths as a rose diagram using 15° intervals. These data should be used for exercises outlined in Section 10.3.1.

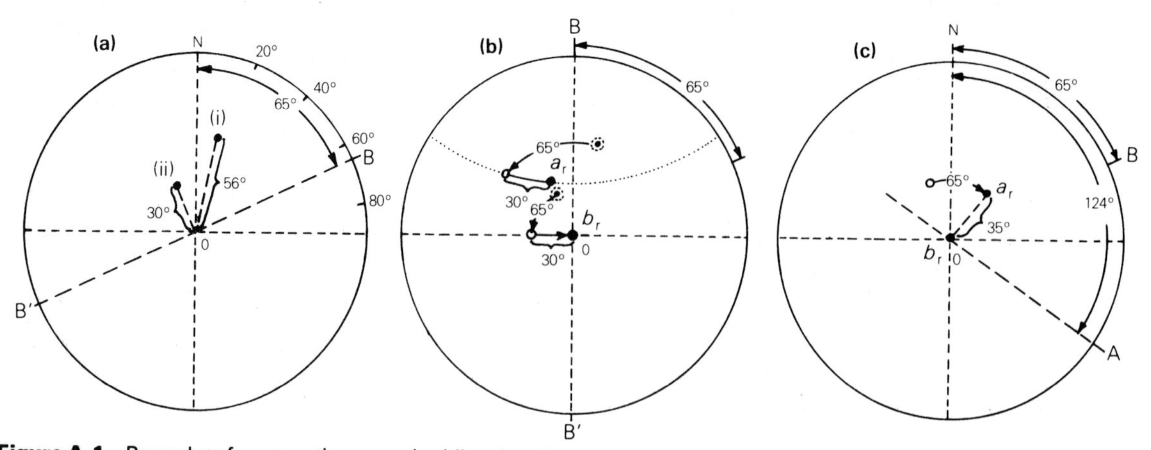

Figure A.1 Procedure for correcting cross-bedding data for tectonic tilt. Bedding is plotted as poles to bedding surfaces on lower hemisphere of stereogram. (a) Observed data; bedding; strike 65°, dip 30° SE; Cross bedding: strike 102°, dip. 56°SW: BOB'; Strike of bedding: (i) pole of cross bedding; (ii) pole of bedding. (b) Rotation of bedding and cross bedding; b pole of bedding rotated 30° to zero dip. a_r, pole of cross bedding rotated 30° around BOB' along small circle. (c) Restoration of a_r with respect to north. Result: a_r original dip of cross bedding 35°SW. OA original strike of cross bedding 124°. (After Potter & Pettijohn 1963, see Reference List, Chapter 10.)

Appendix B

Collection and preservation of structures from unconsolidated sediments as box cores and lacquer peels

The collection and preservation of sedimentary structures from unconsolidated sediment requires special techniques to consolidate artificially the sediment and to pick out the lamination and bedding. There are two main ways of doing this: by the taking of box cores and the making of lacquer peels.

Box cores

To take box cores, simple metal or plastic boxes are pushed into the sediment and then removed in such a way that they contain a relatively undisturbed sample of the sediment. This sample can then be impregnated with a glue or resin either directly in the field or later in the laboratory. If care is taken to spread the impregnating resin evenly, it will penetrate to different depths according to the slightly different porosities and permeabilities of individual layers and laminae.

The simplest corer is the so-called 'Senckenberg box' (see Fig. 10.4b) which can be used on sub-aerially exposed surfaces or on vertical faces. On present-day surfaces it is pushed vertically into the sediment and then dug out after insertion of the cover (see Fig. 10.4a). On a vertical face of a pit or trench it is pushed in horizontally in an upright position. The cover is then slid into place vertically after slight excavation of the top of the box.

A more complex, and slightly more difficult corer to use is the tapering 'Reineck box' (see Fig. 10.4c). This is valuable in shallow water or where the water table is too high to permit the use of a Senckenberg box. The corer is pushed vertically into the sediment and is followed by the cover. The flanges of the box and the grooves in the side of the cover hold the two parts of the corer together, but sediment can obstruct sliding of the flange. The box and the cover are than pulled vertically out of the sediment giving a downwards-tapering wedge-shaped core which can later be impregnated with resins. After the resin has hardened, glue a sheet of hardboard or thick cardboard to the exposed surface. When this has set, carefully free the board and its attached sediment layer and remove it from the box. Brush or blow off any loose sediment from the newly exposed surface. If permeability differences are appropriate, the internal lamination should be picked out in relief.

Suitable resins and glues are listed in the book by Bouma (see *Study list, Chapter 10*) but you should experiment with different products as they come on to the market and with different degrees of dilution. Some glues are soluble in various solvents and this can be useful if, for example, you wish to investigate grain-size distribution of particular laminae. Cutting out the laminae from the box core and dissolving the glue with a suitable solvent can give loose grains suitable for sieving and other grain size measurement.

Lacquer peels

Lacquer peels can be taken from the walls or floors of trenches. The surface should be carefully scraped flat and then sprayed, using a garden spray, with a dilute solution of an appropriate lacquer. Lacquers which use volatile organic solvents such as acetone are often used and, in such cases, the surface can be ignited after spraying, causing the sediment to dry out and the lacquer to penetrate more deeply. Several sprayings may be used. The result is to cement and harden a surface layer. However, in order that the layer can be removed, it must be strengthened by reinforcement. This is done by carefully plastering several layers of lacquer-soaked bandage or gauze on to the surface. When the lacquer is thoroughly dry, the peel can be carefully removed, often with the support of a rigid board. Loose sediment can then be removed from the exposed surface and the surface fixed by further spraying. Peels have the advantage over box cores of allowing the sampling of larger areas and being lighter to carry. This preparation, however, makes rather bigger demands on field time.

A full account of the types of lacquers and the possible variations in method can be found in the book by Bouma (see *Study list, Chapter 10*).

Index

Numbers printed in italics refer to text figures; numbers printed in bold refer to text tables.